Work Study and Ergonomics

This textbook studies the fundamentals of work study and ergonomics in a single volume. It examines the theories of human physiology and cognitive sciences and also evaluates the application of these theories to design a work environment that optimizes work potential and reduces threats of work-related disorders. It discusses the strategies to design effective work processes and congenial work environment in order to enhance human well-being and efficiency. The book also explains the ergonomic tools and techniques including biomechanics, work posture assessment tools, anthropometry and work physiology.

Using live examples from the industry, the author discusses the principles of work study including string diagram, method study, work sampling and man-machine system. He illustrates why it is important to 'fit the job to the man' rather than continuing with conventional practices that 'fit the man to the job'. Multiple choice questions, exercises and case studies are interspersed throughout the book for better understanding and assessment.

Lakhwinder Pal Singh is a faculty in the Department of Industrial and Production Engineering, Dr B R Ambedkar National Institute of Technology Jalandhar. He has more than fifteen years' experience in teaching and research, and has published nearly forty-five papers in national and international journals and conferences. His research interests include human factors engineering, production and operations management, lean manufacturing, occupational health and environment, logistics and supply chain management.

Work Study and Ergonomics

Lakhwinder Pal Singh

CAMBRIDGE
UNIVERSITY PRESS

CAMBRIDGE
UNIVERSITY PRESS

4843/24, 2nd Floor, Ansari Road, Daryaganj, Delhi 110002, India

Cambridge University Press is part of the University of Cambridge.

It furthers the University's mission by disseminating knowledge in the pursuit of education, learning and research at the highest international levels of excellence.

www.cambridge.org
Information on this title: www.cambridge.org/9781107503366

First published 2016

Printed in India by Shree Maitrey Printech Pvt. Ltd., Noida

A catalogue record for this publication is available from the British Library

Library of Congress Cataloging-in-Publication Data
Singh, Lakhwinder P., author.
 Work study and ergonomics / Lakhwinder Pal Singh.
 pages cm
 Includes index.
 ISBN 978-1-107-50336-6 (pbk.)
 1. Work design. 2. Human engineering. I. Title.
 T60.8.S56 2016
 620.8--dc23
 2015030275

ISBN 978-1-107-50336-6 Paperback

To my parents

Contents

List of Figures

List of Tables

Preface

This book has been especially written and prepared for undergraduate and post-graduate students of industrial engineering, production engineering, industrial and production engineering, mechanical and industrial engineering or mechanical engineering, pursuing a one semester course on work systems design, work safety and ergonomics, work design and ergonomics, work study and ergonomics, industrial ergonomics, occupational health and ergonomics, or industrial engineering and ergonomics. Normally the course is offered as a single semester course. But in the present text it has been the author's endeavour to cover not only what can be taught in a single semester but also provide some additional material needed by an ergonomist or an occupational safety engineer. The present text is supplemented with the case studies and live examples from the industry. Hence, human factor engineers or occupational safety engineers or work study engineers will benefit from the knowledge on practical applications of *Work Study and Ergonomics*.

Chapter 1 explains the basic concepts, definitions, scope and importance of productivity, reasons for lower productivity, and methods to improve the same followed by integration of work study and productivity. Chapter 2 describes the role of human factors in work study, and the qualities of a good work study man. Chapter 3 deals with the concepts, definitions and applications of method study, procedural steps to conduct methods' study, different tools and techniques for methods' study; followed by principles of motion economy, therbligs and work simplification. Subsequently Chapter 4 focusses on basic techniques of work measurement, work sampling, and stopwatch time study procedure, equipment and forms for time study, allowances, rating scales, calculation of basic and standard time; followed by some numerical problems. As the standard time calculated through work sampling or stopwatch time study is useful for deciding wage rates and incentives for the workers and employees, hence Chapter 5 is dedicated to various wage and incentives schemes and programmes.

By the end of Chapter 5 the students would have acquired an understanding of work study, its importance and its role in industry. Consequently they will be able to conduct motion studies, time studies, set time standards and apply them in fixing wages and incentives for the employees/workers of an organization. Given this, Chapter 6 of the book will introduce the students to the basic concept of ergonomics, to the historical evolution, scope and objectives of ergonomics, and then follow up the discourse by introducing the man machine system and its performance determinants. After an introduction to ergonomics, students need to be introduced to concepts of work physiology. Chapter 7 describes the structure and features of the human body, metabolism and measurement of physiological functions followed by work load and energy consumption. Chapter 8 focuses upon the musculoskeletal system and work related musculoskeletal disorders (MSD), followed by the biomechanics of lifting, and revised National Institute of Occupational Safety and Health (NIOSH) lifting equation. Thereafter, Chapter 9 elucidates the assessment of distal upper extremities disorders using staring index method. It is also important to enable the students to assess the work related MSD risk due to work posture; Chapter 10 is dedicated to work posture assessment tools like Rapid Upper Limb Assessment (RULA), rapid entire body assessment (REBA), the procedural steps to apply these tools and is supplemented by a case study from the small scale industry. Ergonomics interventions are very important in offices; Chapter 11 explains work place design, seat design and anthropometry followed by a case study.

Chapter 12 probes the effect of vibration, noise and temperature on performance. Occupational noise exposure assessment is taken up in Chapter 13, where the basic theory of noise, its measurement, and hearing protective devises, are introduced. The effect of occupational noise exposure mainly results in NIHL ; Chapter 14 gives a brief introduction to noise induced hearing loss (NIHL), assessment of hearing loss, procedures for audiometry, followed by a case study. After assessing occupational noise exposure and noise induced hearing loss the student needs an introduction to heat stress and respirable suspended particulate matter (RSPM). Chapters 15 and 16 are dedicated to an assessment of heat stress and RSPM levels in the industry. Chapter 17 describes occupational health and safety, present scenario of research into occupational health in India. At the end of this book, Chapter 18 presents a case study on the cardiovascular health of steel industry workers.

The author hopes that the information provided in the book will be useful for an appreciation of the importance (and application) of work study and ergonomics, in the industry and a variety of occupations.

Acknowledgments

This project would not have been possible without the support, advice and suggestions of colleagues, friends and family members. It is my moral and divine duty to acknowledge the contribution of every person whose effort has made this project possible. Before acknowledging anyone else, I bow my head in worship to my spiritual Guru and the almighty God for giving me the internal strength and self discipline for this assignment. I take this opportunity to express gratitude to the reviewers for their valuable suggestions for the book. I extend sincere thanks to Professor Debkumar Chakrabarty (IIT Guwahati), Dr D. Majumdar (DIPAS, Delhi), Professor S. Gangopadhya, Professor K.K. Deepak (AIIMS, Delhi), Professor A. Bhardwaj (NIT, Jalandhar) for their scholarly advice and encouragement. I owe sincere gratitude to Professor S. K. Das (Ex. Director NIT Jalandhar), Professor I. K. Bhatt (Director, NIT Jalandhar), Ajit Singh (Registrar, NIT Jalandhar) for providing a positive work environment. I am also grateful to my head of department, colleagues and friends for their continuous moral support. My family deserves a special mention: my parents, brothers and sister, my better half Poonam, my lovely kids Isha and Vansh, my niece and nephews, who stood behind me as a pillar of strength. I also extend my gratitude to the management of different SMEs for providing me the opportunity to conduct some case studies for supplementing the text.

I express my gratitude to the team at Cambridge University Press: Gauravjeet Singh Reen for his excellent ground work and syllabus research, that facilitated in finalizing the table of contents. I always turned to him for suggestions wherever I was stuck and he was always available to answer my queries. Members of the academic editorial team Hardip Grewal and Shikha Vats were also very cooperative and gave me advice and suggestions as and when required. Last but not the least I am thankful to all my students and teachers who have taught me and made me what I am today.

Productivity and Work Study

1.1. Introduction

A significant proportion of the Indian population is still struggling for its basic needs: food, shelter, clothes, security and health services. A nation can only raise the level of satisfaction with respect to these basic needs only if the return from resources is maximized or the productivity is improved. Then, the economy will grow and help provide a better quality of life. Productivity at an enterprise is assessed in terms of output of a production variable per unit of input. Fundamentally, it is used to measure the output of resources such as manpower, machinery, materials and money, in producing goods and services or commodities to produce income or profitability. There are ample examples of Japanese manufacturers increasing their productivity, after the Second World War when they faced the dilemma of vast shortage of material and human resources. The problem of Japanese manufacturers was entirely different from their Western counterparts. Later, in the mid-1940s, it was recognized by the president of Toyota Motor Company that American companies were outperforming them by a factor of ten. Therefore, in order to make a move towards rapid movement, Japanese leaders, such as Toyota, Shingo and Ohno devised a new process-oriented approach that is known as the 'Toyota Production System or Lean Manufacturing'. The main objective of this approach is to reduce wastes, which in Japanese terms is called 'Muda' and maximizing the activities that add value to the good or services in the customer's perspective. The term value can be attributed to anything in a product or service for which the customer is willing and ready to pay. The value addition in goods and services will increase the overall profitability of system, company or whole supply chain. According to the *lean philosophy*, at an organizational level or throughout the supply chain there are three types of activities:

1. Non-necessary and non-value adding activities
2. Necessary but non-value adding activities
3. Necessary value adding activities

The first type of activities that constitutes majority (about 60 per cent) of all the activities, includes internal handling of material at the shop floor or packing and unpacking of semi-finished components and materials. The second category of activities is necessary but does not add value to the goods and services. These contribute about 35 per cent of the total supply chain activities; therefore labelled necessary, non-value adding activities. Example of such activity is cutting of a wire piece and mild steel rod on a machine into a standard size. These cut sizes will further be processed to change into usable goods such as shirt hangers and dumbbell rods, respectively. The third category is the necessary and value-adding activities, which constitutes merely 5 per cent of the total activities in an organization. Example of such activity is bending the previously cut piece of wire and shaping it in the form of a shirt hanger; similarly, performing operations such as turning, knurling and facing on a lathe machine to make a dumbbell rod. Thus, the value adding activities contribute significantly towards increasing the productivity. Moreover optimizing the necessary but non-value adding activities and reducing to less numbers is also indicator of increased productivity.

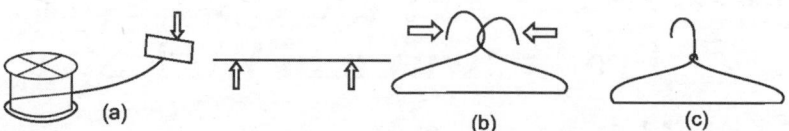

Fig. 1.1: (a) Cutting of wire in standard size, (b) Bending of wire into shape of hanger,
(c) Final shape of hanger after paint/plastic coating.

In addition to the above example, there is a case of one small scale foundry unit, which produces a family of hydraulic valves. The manager of the firm was looking after all the operations starting from moulding, pouring, trimming, chipping/chiselling, grinding, etc. The manager observed that the productivity of the firm was lower due to bottleneck at a workstation where the worker was removing the extra metal from the parting line of the casting by using chisel and hammer tools. The manager was much concerned to increase the productivity of that particular workstation, so that the overall productivity could be improved. So, he started looking for better and efficient tools (than the chisel and hammer), so that the bottleneck can be removed. This is one good example of *non-necessary and non-value adding activity*. The manager hired a consultant and asked him for designing better tools or devise new method so that efficiency of the worker can be increased. The manager took the consultant to the live work place; and after that the consultant immediately said that there was no need to design any tools; in fact, there was hardly any need of this operation even. Being surprised the manager asked for the reason. The consultant explained that the actual problem was not due to the tools or efficiency of the workers, but somewhere else, i.e., due to the loose clamping of moulding boxes (cope and drag), the molten metal leaked out at the parting line between the upper and lower moulds. Therefore, before pouring the molten metal into the mould, there was a need to assure that the side clamps were tightly fastened, and if needed extra weight be used so that when the molten metal was poured, the upper half of the mould would not be lifted due to the pressure of the molten metal. Now, the activity like removing the extra metal at the parting line is unnecessary and non-value adding; however, clamping of the moulding boxes is necessary but non-value adding activity. Therefore, the higher productivity is the outcome of value adding activities. The work study is mainly focused upon eliminating the non-value adding activities and improving the efficiency of value adding activities. Nowadays, most of the supply chains or the value chains are focused upon eliminating the non-necessary and non-value adding activities, and optimizing value adding activities.

One of the most important reasons for the decline of an enterprise or a supply chain is low productivity. Failure to achieve the targeted level of productivity leads to the higher cost per unit, hence higher prices and low competitiveness in the market. In the current scenario of a competitive world, many companies are striving to maintain the competiveness in the market. The productivity improvement at the local level cannot solely increase the productivity and profitability of a supply chain. Thus, it is very important to device a competitive strategy that enables not only an individual stage (enterprise) but also the whole supply chain to improve the productivity and profitability.

1.2. Definition of Productivity

Productivity is simply defined as 'the ratio of output to a given input', and represented as follows:

$$\text{Productivity} = \frac{\text{output}}{\text{input}}$$

This definition can be applied to an enterprise, a sector of economic activity or the economy as a whole. The term '**productivity**' is useful to assess or measure the extent to which a certain output can be extracted from a given input. While this appears simple enough in cases where both the output and the input are tangible and can be easily measured, productivity can be more difficult to estimate once intangibles are introduced. Let us elaborate further the meaning of the term productivity with an example as follows:

A handcrafter working eight hours a day produces 800 crafts a month using a traditional old method and can sell each craft for $2.0; hence, his monitory output value is $1600/month.

- Let us assume that as a result of a change in the method of work he was able to produce 1000 crafts in a month instead of 800 with the same equipment and hours of work. His productivity calculated in terms of number of crafts produced will then have increased by 25 per cent.

- Now, let us assume that due to 25 per cent surplus production, as a result he was unable to sell all 1000 crafts and had to lower his price from $2.0 per piece to $1.80 per piece. If he wants to assess his productivity gain, the crafter may be more interested in using monetary terms rather than using the number of crafts produced. He could then disagree that the value of his output used to be $800 \times 2 = \$1600$ per month and is now $1000 \times 1.80 = \$1800/$ month while his input has not changed. Hence his productivity gain is calculated as follows:

$$\text{Productivity} = \frac{\$(1800 - 1600)}{\$1600} \times 100 = 12.5\%.$$

From this example that is deliberately kept simple, one can make two observations. In the first case, productivity was used to measure increase in output expressed in numbers of crafts produced, although in monetary terms it gives different values in each case. In other words, it can be interpreted that depending upon the interest of an individual in measuring productivity, the nature of the output and input will vary accordingly. Secondly, in this example while actual production increases from 800 to 1000 crafts, productivity in monetary terms does not show the same improvement. This means that we have to differentiate between increased production and increased productivity, which in this example is measured in terms of monetary gains.

Let us continue with our example and assume that the crafter decided to replace his old manual method by an improved technique (machine). This requires an additional cost to him as an investment of $12,000 which he reckons should be amortized or repaid over ten years. In other words, the cost of this investment will be $1200 per year for ten years, or $100 a month. He also would need some supplies and maintenance, which would cost him $100 a month more than what he would have paid for the wood. Let us also assume that his production remained constant at 1000 crafts a month. Measured in monetary terms, the value of his output is:

$$\text{Production} = 1000 \times 1.80 = \$1800/\text{month}$$

From \$1800, \$100 will be deducted for capital investment and another \$100 for maintenance, which makes a total of \$200. Therefore his monetary gain is:

$$\text{Gain} = \$1800 - \$200 = \$1600.$$

In this case, his productivity expressed in monetary gain has not yet improved, while originally he was producing only 800 crafts, he sold them for \$2 each, thus arriving at the same monetary digits.

However, the crafter may wish to argue that due to the new technique quality of his crafts has improved, that he will have less rejection level and that the users' satisfaction will increase over time, so that he may be able to increase his price again. Furthermore, his own sense of satisfaction at work has improved, as it has become much easier to operate with the new machine/method. Here, the definition of the output has been enlarged to encompass quality and a relatively intangible factor, that of consumer satisfaction. Similarly, the input now encompasses another intangible factor such as ease in making crafts or satisfaction at the workplace. Thus productivity gains become more difficult to measure accurately because of these intangible factors and because of the time lag that needs to be estimated until or unless satisfaction of end users will permit an increase in prices of the crafts produced with the new technique.

This simple example helps us to understand that there are many interrelated factors affecting the productivity of an organization. Many people have been misled into thinking of productivity exclusively as the productivity of labour, mainly because labour productivity usually forms the basis for published statistics on the subject. At the same time, it is very significant to mention that, in a society or a country, improving productivity or extracting the best possible output from available resources does not mean exploitation of labour, but controlling of all the available resources to stimulate a higher rate of growth that can be used for public betterment, a higher standard of living and an improved quality of life. The detailed productivity issues are beyond the scope of this book, but emphasis is given on work study as it applies to every individual organization.

1.3. Productivity of the Individual Organization

Productivity of an organization may vary as it may be affected by a number of external as well as internal factors. The internal factors are within the control of managers of an enterprise and these basically include the number of limitations within the operations of an organization. The external factors are beyond the control of any employer. These factors comprise availability of raw materials and skilled labour, taxation and tariffs posed by the government, infrastructure in hand, capital availability and interest rates, etc.

1.3.1. The output and input factors in an organization

In a typical enterprise the output is normally described in terms of products/services provided. Therefore, products are articulated in digits and valued based upon the compliance to predetermined quality standards. Whereas in a service firm like a public or private transport company or may be a travel agency, the output is expressed in terms of the services rendered. Therefore, for a transport company, this could be expressed in terms of the number of passengers or tons of load/km carried. However, for a travel agency it could be an average value of tickets per customer. Hence, in order to achieve the best level of productivity, manufacturing and service enterprises should focus upon consumers' or users' satisfaction, such as number of complaints or rejects.

The enterprise also arranges certain resources or inputs which are used to produce the desired output. These are mainly classified as follows:

1. **Land and buildings:** Land and buildings are constructed or hired in a suitable location.
2. **Materials:** Arrangement of materials that can be converted into saleable products. These include raw materials, semi-finished components and complementary materials that are required for manufacturing and packing of finished products.
3. **Energy:** Energy is one of the vital resources, hence arranged and managed in its various forms such as electricity, solar power and fuel like gas, oil, coal, etc.
4. **Machinery and equipment:** These are required for the various operations of the enterprise. It also includes transport and handling, heating or air conditioning, office equipment, computer terminals, etc.
5. **Manpower:** This includes the arrangement of experienced and trained manpower (male and female) for planning, controlling and carrying out the various process activities, procurement of materials and selling of finished products, maintaining the accounts and other maintenance work.

Over and above, it includes funding and procurement of land, machines, tools, supplies, etc.

1.4. The Role and Responsibility of the Management of an Organization

The management of an organization is accountable for considering that the enterprise resources cited above are combined in the best possible way to achieve the highest productivity. The planning, organizing, directing and controlling these resources and balancing one resource against another are the tasks of the management. If the management fails to do so in an effective manner, the enterprise will fail in the end. In such a case, the five resources become uncoordinated just like the efforts of five horses without a driver and the enterprise a driverless coach that moves forward incoherently, gets held up for lack of material, lack of equipment, because machines or equipments are badly chosen and even more badly maintained, or because energy sources are inadequate or employees unwilling to contribute their best. Figure 1.2 illustrates this management function.

In its quest for higher productivity, an efficiency-minded management acts to influence either one or both of the two factors, the output (i.e., products and services) or the input (i.e., the five resources at its disposal). Thus the management should aim to produce a larger quantity with better quality or higher value products or services with the same input, or it may achieve a better result by changing the nature of the input such as investing in advanced technology, information systems and computers or by using an alternative source of raw material or energy. However, it is very uncommon that one manager or a small team of top managers can by themselves attend to the normal running of an enterprise and at the same time devote enough thinking and energy to the various issues raised in way of improving the productivity. More frequently, they will rely on specialists to assist them in this task, and among them is the work study practitioner. In the subsequent part of the chapter, we shall see how work study and productivity are related.

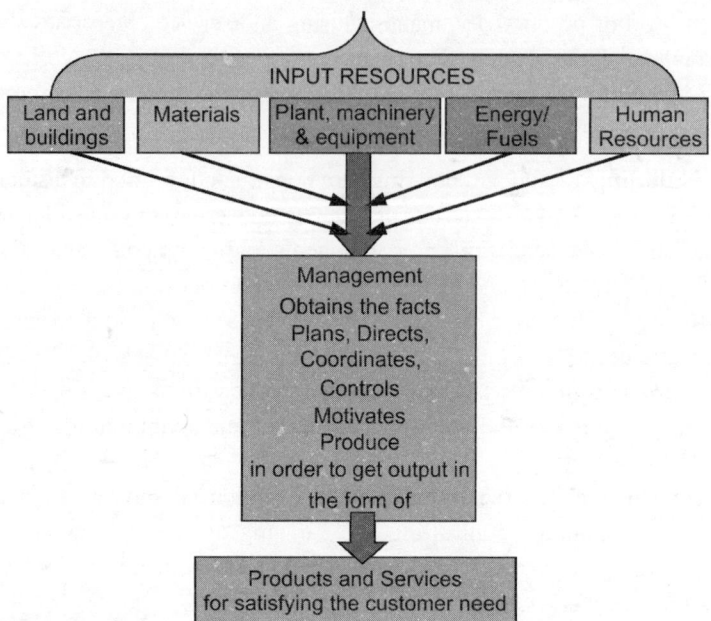

Fig. 1.2: Role and responsibility of management in coordinating the resources of an enterprise.

1.5. Work Study and Productivity

In the previous sections, we pointed out that management regularly calls specialists to help it in improving the productivity. One of the most powerful tools they can use is work study. *Work study is defined as the systematic examination of the existing methods of carrying on activities so as to improve for the effective use of resources and to set up standards of performance for the activities being carried out.*

Work study is mainly focused at investigating the way an activity is being carried out, simplifying or modifying the method of operation to reduce unnecessary non-value adding activities in terms of rework, wastage, and finally fixing the standard time for an activity. Therefore, the relationship between productivity and work study is noticeable. This fact could be understood by a simple example; that if the work–study engineers merely rearrange the sequence or simplify the method of operation with no additional expenses and lower down the time of an activity by 20 per cent, then productivity automatically increases by an equivalent value, i.e., 20 per cent. In order to understand, how work study performs to slash down the time and reduce costs of a certain activity, it is essential to analyze the different constituents of time, i.e., how the total time of a certain job is made up.

The time taken by a worker or a machine to carry out an operation or to produce a given quantity of a certain product is enhanced by the three main constituents as illustrated in Figure 1.3. Ideally, there is a basic work content of the product or operation. The basic *work content* is defined as the amount of work 'contained in' a given product or a process measured in '*work-hours*' or '*machine-hours*'.

A *work-hour* is defined as is the labour of one person for one hour, whereas a *machine-hour* is the running of a machine or part of plant for one hour.

Therefore, the basic work content is the time taken to manufacture the product or to perform the operation in ideal conditions i.e., when everything is perfect be it the design of product or service is perfect, the process or method of operation, and when there is no loss of working time from any other reason during the period of the operation for reasons apart from permissible rest pauses. In other terms, the basic work content is also defined as the irreducible minimum time theoretically required for producing one unit of output.

This is clearly an ideal situation, which is not practical, even though it might sometimes be approached, particularly in line manufacturing or process industries. Generally, actual operation time is relatively very high due to the excess work content.

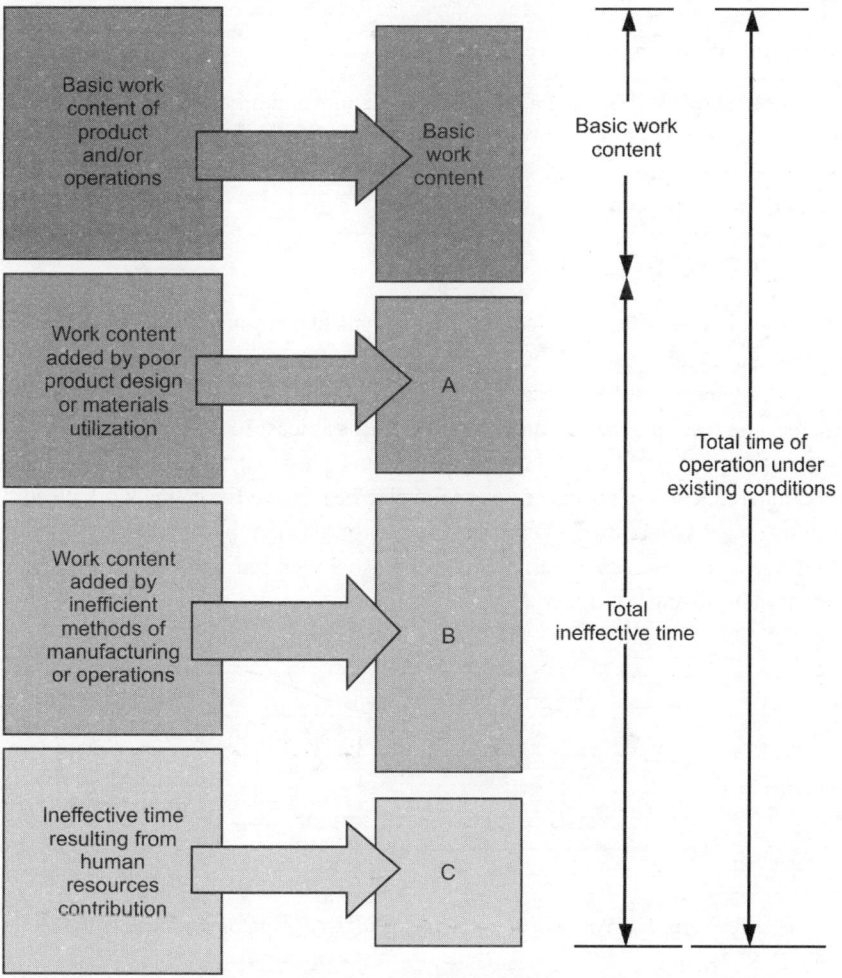

Fig. 1.3: Constituents of operation time.

1.5.1. Work content enhancement

The **work content** is enhanced on shop floor due to different causes; these can be classified broadly into three areas as follows:

A. Design specifications of products and improper utilization of materials
B. Methods of operation or way of doing the operation
C. Human resources or manpower

A. *Work content enhanced due to poor design specification and improper utilization of materials*

There are several ways in which unnecessary time and waste (resulting in higher cost of the product) can be attributed to poor design of the product or its parts, or to incorrect quality control; these are described as follows:

A-1. *Poor design or frequent changes in the design*

This is the situation when the design of a product may require a number of poor quality (non-standard) parts that lead to a long time of assembly. Too many varieties of products and lack of standardization of products/components may demand for the production in small batches, which results in loss of time, as the operator alters from one batch to another.

A-2. *Wastage of materials*

This refers to a situation when the design of the components of a product may demand material removal of too much amount to get them to their final shape. In addition to the wastage of material, this also increases the work content of the job. Hence, the cutting operation specifically need a vigilant examination to watch whether the resulting waste can be minimized or at least reused. For example, a small scale industry is manufacturing dumbbell for health gymnasiums: the bad design of the component (dumbbell rod) requires 5 mm reduction of radius i.e., extra metal to be removed from the work piece (Figure 1.4). Thus the design demands more turning cuts with an appropriate depth of cut. However, if the design is changed and the provision is made for fitting a thick washer with press fitting or joining process, there will be a reduction in wastage of material as well as time.

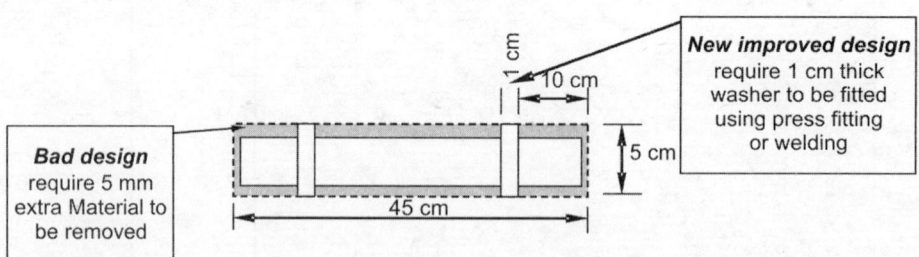

Fig. 1.4: Wastage of material due to bad design of product.

A-3. *Inaccurate quality standards*

The excessively high or low quality standards can result in increased work content. Nowadays in the competitive environment, firms like engineering and automobile industry are moving towards six-sigma standards and thus more emphasis is being given upon high quality of components with tight tolerances, which require extra machining and consequently leading to wastage of materials. On the other hand, e.g., in small scale industry, setting too loose tolerances may result in a large proportion of rejection. Therefore, deciding on the appropriate quality standard and the method of quality control is one of the significant efficiency deliberations.

B. *Work content added as a result of inefficient methods of operation (B)*

Work content may also be added due to a poor method of carrying out the operations, resulting in unnecessary movements of persons or materials, which result in ineffective time and higher cost. Similarly, such ineffective time can be due to inappropriate methods, material handling, poor maintenance of machinery or equipment resulting in frequent breakdowns, or poor inventory control causing delays because of an absence of products or parts or may be the higher costs as a result of overstocking. Work content increased due to inefficient methods of operations can be mainly attributed to the following six factors:

B-1. *Poor layout design and utilization of space*

Where there is use of space for any operation, there is an investment. Therefore it is very important to ensure proper utilization of space for realizing the reduction in operational cost. This is necessary particularly when a firm is increasing and it requires an increased working area. In addition, a proper layout design of the work place area will also reduce the waste movement, time and effort.

B-2. *Inadequate materials handling*

In a manufacturing firm throughout a production process, materials are always moved from one place to another in the form of raw material, semi-finished components and finished products. The unnecessary movement and handling of materials can be avoided by using the most appropriate handling equipment and can save time and effort.

B-3. *Quick changeover from one product to another*

The proper planning and control of production operations can ensure that one production batch or order immediately follows another so that idle time of machinery, equipment or labour is eliminated or at least minimized. Nowadays, single minute exchange of dies (SMED) is one of the most popular concepts being followed by many enterprisers like hand tool manufacturers, which enables them to reduce the change over time for shift from one component to another.

B-4. *Ineffective method of work*

In an enterprise, there may be a well planned sequence of operations but some or many of them may be done in an awkward way; resulting in unproductive time. Through an examination of operations being carried out and by developing an improved method, the unproductive time can be reduced.

B-5. *Poor planning of inventory*

In each of the operation going on in a firm, raw material is usually ordered and stocked ahead of time. However, in between the various stage of the production operation, semi-finished components and various parts are temporarily stocked as work-in-progress (WIP) material till further processing. All of these inventories correspond to a tied-up investment. This can be minimized by using proper inventory control systems like MRP (materials requirement planning) and at the same time it will ensure that operators do not run shortage of the necessary material.

B-6. *Frequent breakdown of machines and equipment*

Any enterprise running with poor maintenance policy and practices often faces the occurence of machinery and equipment breakdown and consequently due to waiting for repairs there exists an idle time. This can be corrected by establishing a preventive maintenance system. Moreover augmented maintenance campaigns can also be very fruitful to ensure the smooth running of machines and equipment.

C. *Work content resulting mainly from the contribution of human resources (C)*

Workers in an enterprise can influence the time of operations voluntarily or involuntarily as follows:

C- 1. *Absenteeism and lateness*

In case the administration of a company falls short in providing a safe and satisfactory work conditions, it is quite possible that the workers might respond by absenteeism, lateness or may intentionally work at a slow pace. For example, in a small scale casting industry, the hot humid and dusty environment is inevitable, which may cause health deterioration and hence lead to absenteeism and lateness.

C-2. *Poor workmanship*

Due to the lack of proper training of the workers, the company may experience the poor workmanship, which results in rejections and rework. Losses may also occur in the form of material wastage by such untrained workers.

C-3. *Accidents and occupational hazards*

When a firm fails to provide a safe and healthy work place, the chances of accidents or occupational illnesses increase, which finally affect the morale of and increase absenteeism among employees.

The impacts of all the above mentioned factors (A to C) are shown in Figure 1.5. If these factors can be eliminated, which is an ideal situation and can never happen in real life (as shown in Figure 1.6), there will be minimum input of time and cost for the production of a given output and maximum productivity can be achieved. It is very important to note that the work study specialist has to keep all these factors in his/her mind while examining an operation and to develop an improved method.

1.6. Interrelationship of the Various Methods Used to Reduce Ineffective Time

None of the methods discussed above can be implemented in isolation, since each factor has an effect and is affected by others. It is not possible to plan programmes of work without standards provided by time study. Simultaneously, production planning will become easier if an effective employees' policy and a well-applied incentive schemes are implemented to support and motivate the workers to perform consistently. Standardization of products or parts will reduce the efforts for inventory control as less variety of materials need to be bought and held in stock.

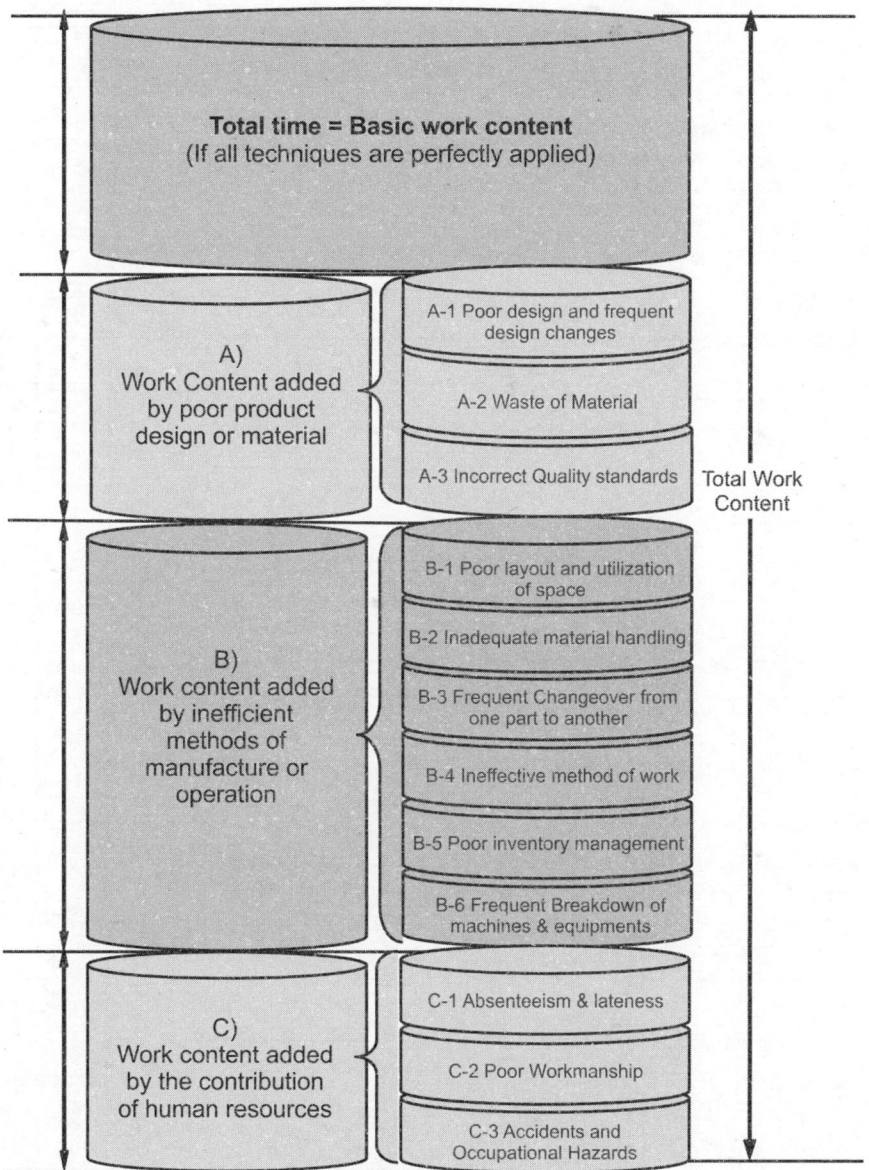

Fig. 1.5: Basic and added work content.

Some of the methods and techniques that can be used to provide information (either manual or computerized) on how productivity may be improved are discussed here briefly. Work study utilizes this kind of information to develop new methods of work and to measure workloads and duration of tasks.

Why is work study valuable?

The examination and development of new methods at the workplace is a common routine followed by good managers ever since human effort was first organized on a large scale. Managers of outstanding

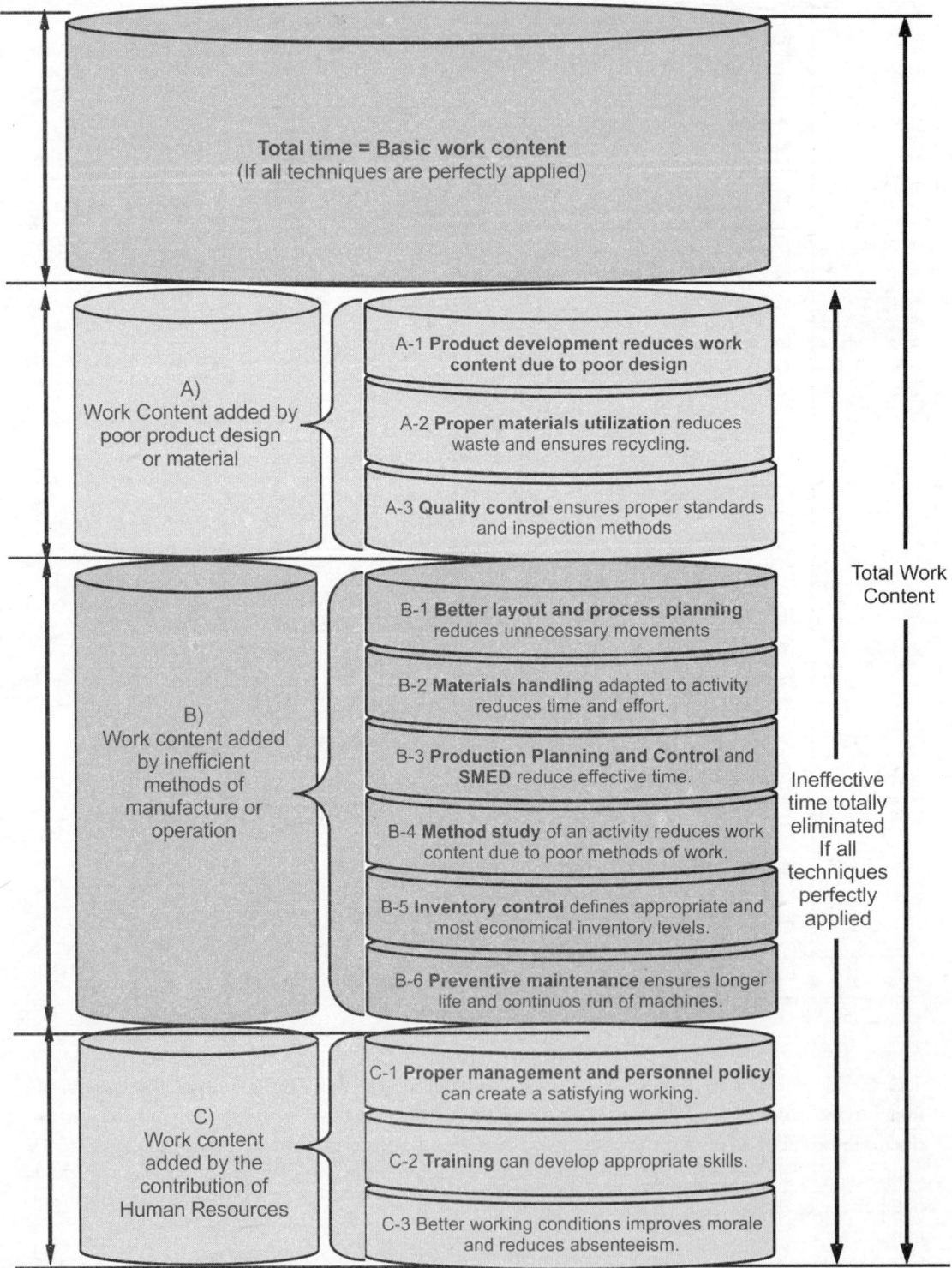

Fig. 1.6: Management techniques to reduce the added work content.

ability have always been competent to make prominent progress. Therefore, the key importance of work study is in the fact that, by carrying out its systematic procedures, a manager can achieve results as good as or better than mangers of the past.

Work study is successful since it is systematic both in the examination of the problem being considered and in the development of its solution. A systematic examination takes time. It is therefore necessary in all firms to separate the job of work study manager from the task of general management since the factory managers /supervisors are very much busy in their day-to-day work with the organization's many human and material problems, and thus are never free to devote time for work study activities. They may be capable of doing the same, but managers can rarely afford to devote a long time (without disturbance) to study a single activity on the shop floor. This means that it is more or less infeasible for them to obtain all the facts about the concerned activity. Unless or until all the details are known, it is impossible to be sure about the alterations in procedure and whether they will be fully effective. This means that work study should always be the responsibility of somebody who can carry it out full time, without direct management duties.

So far some aspects of the nature of work study and why it is such a valuable tool of management have been briefly discussed. There are other causes to be added to the above. They are as follows:

1. It is a way of increasing the productivity of a plant or operating unit by the restructuring of work, a method which normally incurs little or no expenses on facilities and equipment.
2. It is systematic and therefore no factor affecting the efficiency of an operation is overlooked, whether an analysis of original method or developing the new one, and ensures that all the facts about that operation are available.
3. It is the most precise way so far evolved, of setting standards of performance, based upon which the effective planning and control of production is accomplished.
4. It has significant contribution in the improvement of safety and working conditions at work by revealing hazardous operations and developing safer methods of performing them.
5. When work study is properly applied, it can result in significant savings which start at once but continue as long as the operation (new improved) continues.
6. It is such a universal tool that it can be applied everywhere, wherever work is done or the plant is operated, not only in manufacturing shops but also in offices, stores, laboratories and service industries such as wholesale and retail distribution and restaurants, even in hospitals and on farms. At the same time, it is relatively cheap and easy to implement.
7. It is one of the most effective tools of investigation available to management, which can directly attack on inefficient (non-value adding) operations and activities.

Work study is systematic, and since it involves investigation through direct observation of all the factors affecting the efficiency of a given operation, it will reveal any shortcomings in all the concerned activities affecting that operation. For example, observation may exhibit that the time of an operator on a production job is being wasted due to non-availability of material or due to the breakdown of the machine. This highlights a failure of material control or a failure on the part of the maintenance engineer to carry out proper maintenance procedures. Similarly, time may be wasted through short batches of work, necessitating the constant resetting of machines, on a scale that may only become apparent after prolonged study. This points towards poor production planning or a marketing policy which requires a review.

Work study acts like the knife of a surgeon, which exposes all the activities of a company and their functioning, whether good or bad. Therefore, it must be handled with skill and care. Nobody likes being exposed and unless the work study specialist handles people tactfully, one may face a situation of enmity between the management and workers, which makes it impossible to do the job properly.

Managers and supervisors have usually failed to accomplish the savings and improvements which can be effected by work study because they have been unable to devote themselves consistently to such things even they were given proper training. It is not enough for work study to be systematic. To achieve significant results it must be applied continuously, and throughout the organization. The savings achieved on individual jobs seem large at the individual level, but are generally small when compared with the activity at the company level. The complete effect can be realized in a company only when work study is applied everywhere, and when everyone adopts a positive attitude towards successful implementation of work study, i.e., intolerance of waste in any form, whether material, time, human efforts, power (electricity, fuel), etc.

Summary

This chapter mainly focuses upon the basic concept of productivity, the concept of lean philosophy and describes the various constituents of work content and their remedial measures.

Multiple Choice Questions

1. Work study is used in
   - (a) industries and hospitals
   - (b) transport
   - (c) design
   - (d) all of these
2. Work study is most useful where
   - (a) productivity activities are involved
   - (b) industrial relations are to be improved
   - (c) to prove a worker is wrong
   - (d) to improve image of a manager
3. Productivity can be increased by
   - (a) Increasing the output at the same level of input
   - (b) Reducing the input with same level of output
   - (c) Increasing output significantly with slight increase of input
   - (d) Any one of these
4. The study used to determine the most economical and effective method of performing a job and also time required by trained employee working at a normal pace to perform the job, is called
   - (a) time study
   - (b) motion study
   - (c) motion and time study
   - (d) method study
5. Some of the important elements related to the design of work places are
   - (a) proper illumination and noise
   - (b) maintaining new methods
   - (c) operation chart and flow chart
   - (d) flow diagram and string diagram
6. If a work content of 10 hours has to be produced at a rate of 400 a week, and the normal working is 40 hours, the number of operators required is
   - (a) 120
   - (b) 150
   - (c) 100
   - (d) 10

7. Time standard developed by time study can be used for
   - (a) plant layout
   - (b) budgetary control
   - (c) equipment selection
   - (d) wages and incentives
   - (e) all of these

8. Product assembly is
   - (a) the final stage of production
   - (b) intermediate stage of production
   - (c) initial stage of production
   - (d) something else

9. The advantage of productivity is reaped by
   - (a) the entrepreneur
   - (b) the labour
   - (c) the society
   - (d) all of these

10. In the early development of production management attention was concentrated on controlling
    - (a) factory labour cost
    - (b) factory material cost
    - (c) material handling cost
    - (d) factory overheads

11. Maximum productivity can be increased through control and effectiveness of
    - (a) labour management
    - (b) material management
    - (c) management of manufacture
    - (d) management of sales

12. A mathematical study of production structure of industries where the different sectors of industries are inter-dependent on each other is called
    - (a) input output analysis
    - (b) value analysis
    - (c) product analysis
    - (d) industrial analysis

Answers

1. (d) **2.** (a) **3.** (d) **4.** (c) **5.** (a) **6.** (c) **7.** (e) **8.** (b) **9.** (c) **10.** (a)
11. (b) **12.** (a)

Human Factors in Work Study

2.1. Introduction

Work study is described as a management tool to achieve higher efficiency concerned chiefly with manual work. In order to survive, the industry must use latest technology and most efficient method improvised with a consistent aim of producing best quality goods at lower prices.

One way to improve is through efficient utilization of plant, equipment and labour. Work study is specifically a study of work, describing partition of work into smaller units followed by rearrangement of these units to provide the same effective use at a minimum cost. Work study gives methodology and calculates time period required for the work involved in the process. Robert Owen, Taylor and Bedaux had their vital contributions to this particular topic. According to International Labour Organization (ILO), work study is defined as follows:

Work study is primarily concerned with finding the best ways of performing job and establishing benchmarks based on such ways. It is the technique of method study and work measurement employed to ensure the best possible use of human and material resources in carrying out a specified activity.

In other words, work study can be defined as "*the systematic investigation of the methods, conditions and effectiveness of industrial work and thus to determine the way in which human efforts can be applied most economically*".

2.2. Historical Glimpse Related to Work

Industrial revolution started around 1770s, when James Watt gave steam engine and Henry Maudslay devised the screw-cutting lathe. In factories, automated machines started replacing workers to enhance production rapidly and more accurately. The interchangeability enabled the factories mass manufacturing. Better machines, modern parts, special tools, fixtures, etc., and then unique products replace custom-fabricate common products. Henry Ford (1863–1947) started mass production through assembly lines.

The concept of scientific management came in late 1800s; Frederick W. Taylor (1856–1915) was the father of scientific management, whereas Frank (1868–1924) and Lillian Gilbreth (1878–1972), were the father and mother of motion studies. Frank Gilbreth was born in Freeport, Maine on 7 July 1868. After finishing school, Frank Gilbreth began to work for a construction company as an apprentice brick layer. At age 27, Frank Gilbreth became the chief superintendent of the company.

By the time of his promotion, Frank Gilbreth developed a number of methods, both technical and procedural, for improving efficiency at the workplace, particularly in his field of construction. In 1895, Frank Gilbreth decided to set up his own construction company in Boston and by 1900, was running a successful business with branches throughout the USA and a branch office in London.

Frank Gilbreth did implementation of motion study to the art of brick laying. He observed that the workers developed their own unusual ways of working; individuals did not always use the same motions in the course of their work. He sought the best way to perform tasks and standardized it and then he developed a method which reduced the number of motion needed to lay a brick from 18 to 4.5. They developed a more systematic and sophisticated method of time and motion study for industry, taking into account the limits of human physical and mental capacity and the importance of a good physical environment. They discovered that basic motion elements (therbligs) are used in all works and always a single best method to execute certain undertaking. *Thus **Motion Study** is the way of finding the best approach to execute an assignment and **Time Study** is a technique considered as yardstick for a job.* These standards are to be used in industry in order to calculate the labour incentives (bonus payments for higher outputs) and use of data collection, record keeping, and cost accounting. Thus the objective is to improve the (labour) productivity.

2.3. Objectives of Work Study

Primary function of work study is to increase the productivity and reduce the waste. Thus, it should be capable of the following:

1. Investigation and analysis of the situation.
2. Scrutinization of shortcoming in production processes.
3. Most efficient use of existing plant.
4. Recommending and implementing the improvements.
5. Efficient use of human efforts.
6. Measurement of work values.
7. Set standards for labour cost control.
8. Initiating and maintaining incentive bonus schemes.
9. Standardization of material and machines used.
10. Determination of the time required by an ideal operator to perform the task with efficiency.

2.4. Importance

Work study is a way of enhancing the productivity of a factory by changing the method of work, by involving a very limited or no capital disbursement. Hence, it is done in a systematic way without leaving any aspect of production ignored. It is the most skilful way of investigating standards of performance, therefore comes out in

1. Saving and adequate usage of assets.
2. Improved safety.
3. Reduced training time and.
4. The fact that it can be enforced anywhere.

2.5. Segments of Work Study

Work-study is one of the productivity improvement methods, it consist of two main segments; method study and work measurement. Work-study analyze, simplify and improve the method of working to reduce the work contents and establish the standards of performance. The process of productivity improvement through work study is illustrated in Figure 2.1.

1. Study Mechanism (Method/Motion Study)
2. Task Analysis (Work Measurement)

Study Mechanism is systematic recording, critical analysis, development and implementation of new methods to perform job to reduce costs with regard to existing or proposed jobs.

It is called *Method /Motion Study*, as it deals with the movement of goods and the activities and events in the transformational process. Method study is examination of work performance and its subsequent rearrangement to produce the same results with less effort. Work performance depends on number of interrelated operations and activities. Any combination of these activities is called as Method. The prime objective of this expert approach is to analyze the way of increasing the level of performance plus resulting in minimization of effort and cost for same or better output.

In other words, *motion study* is the careful investigation of motion of body or its parts employed in doing a job. The purpose of motion study is to eliminate or reduce ineffective movements, and facilitate and speed up the effective movements. Thus, it reduces production time and raises output, which increases productivity.

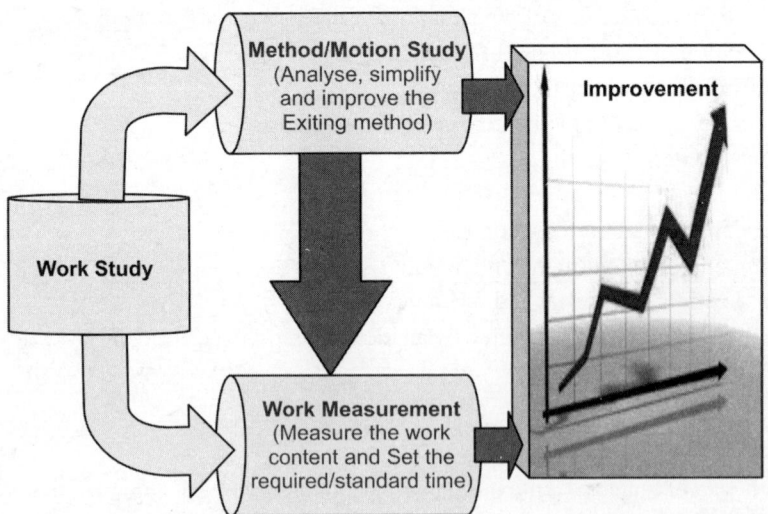

Fig. 2.1: Productivity improvements through work study.

Work Measurement is also known as *Time Study,* as it deals with the measurement of time taken to perform a task. Work measurement is identified as a scientific method of fixing standard of production or standard time for a given job. In order to manage the work effectively, first it must be measured. Work measurement means the application of a set of techniques intended to establish the amount of work to be done by an operator at a given time, under a specific condition and at defined level of performance.

2.6. Human Factors Consideration in the Application of Work Study

Success in application of any tools or techniques is generally influenced by the people who apply them and depends on those for whom they are applied. This is mainly true for work study where the main focus is on investigation of manual work, with an aim to economize the human efforts. Investigation involves different questions associated to the efficiency and effectiveness of the work under consideration, which may cause frustration to the worker or the group of workers working on the job. It is thus necessary that right at the beginning the objective of conducting such a study be made very clear to all concerned workers and supervisors by the management and the work study man. At the study stage, the work study man has to be intentional in recording all relevant data related to the work. Successful implementation mainly depends on the prevailing relationship between the workers and the management. It is always easier to implement any change in an organization where mutual trust between the worker and the management exists.

Therefore, the human factors play a very significant role in the successful conduct and implementation of work study in a company. The management who is responsible for spelling out the objectives and planning of work, the supervisor who translates these plans into day-to-day operations and continuously monitors them, the workers who carry out the operations and the work study man who conducts the study, each of them has contributed positively if the study has to succeed. The objective of this section is to look at the ways in which these different interacting groups can contribute. Accordingly, the roles of the management, supervisors and the workers are discussed in the next section. In the final section, we discuss the qualities necessary for a work study man.

2.7. The Role of Management and Supervisor in Work Study

Before the discussion on roles of each individual, it is essential to have an understanding of how the management and the supervisor form two diverse groups in an organization. As the management puts forward the objectives and plans of different activities, the supervisor's job is to translate these plans into daily operations and supervise the progress by ensuring that the workers perform up to the desired performance level, and make relevant tools available to them. Therefore, the supervisor acts as the liaison between the management and the worker. Now let us first, discuss the roles of management.

2.7.1. Role of the management

The role of management for successful application of work study can be summarized as follows:

1. The management must clearly define the organizational goals and objectives. Otherwise, the workers or supervisors may set their own goals and objectives, which may be inconsistent with those of the management. Since work study intends to identify a different method for doing work, this better method is defined with regard to the organizational goals.
2. The management should plan in such a way that certain ineffective time is minimized, which could be utilized otherwise. In case there is some ineffective time due to management, and it is identified as a part of work study, management should be open to suggestion and criticism.
3. The management must attempt to maintain good relationship with the workers and provide a good work environment for them. This helps building up a mutual trust which is useful for any study to be successful.

2.7.2. Role of the supervisor

The roles of supervisory staff in work study applications are summarized here:

1. In order to maintain the liaison between the management and workers, the supervisor must clearly communicate the organizational objectives to the workers. At the same time, he should be able to portray a clear picture to the management about practical problems of the shop floor, so as to enable the management to set realistic goals.

2. As a person who is much closer to the actual jobs than the management, he should be fully aware of the different aspects of the work and its limitations. This will help him identify potential areas of improvement and he could be of help in selecting a proper work for study.

3. Since supervisor is responsible for action on the plans, he has to be associated with the study right from the selection of the job, analysis and its implementation. This requires that he should be open to ideas. Let us not disturb the system and *status quo* is usually a preferred choice of worker as well as the supervisor. The supervisor should be aware of this and contribute to the study by sharing his expertise on the work with the work study man.

2.7.3. Role of worker

Worker plays a crucial role in successful application of any study, since he/she is an ultimate person who in reality performs the job at the shop floor. His ethics and attitude, his behavior as an individual and also in group must be considered for the purpose of the study. His role in the study is summarized below:

1. The worker should not ignore his job or waste time without any cause. As we have observed in previous chapter, ineffective time may occur, duo to negligence of the worker. Hence, he must be aware of the fact that the resulting lower productivity will affect him in the long run.

2. The worker should take interest in the work and also take the initiative with the work related factors. Some time, it is possible to select the jobs to be studied through the initiative taken by the worker. Individual, formal or informal groups or sometimes unions can be used as a platform to initiate study.

2.7.4. The work study man

In the previous sections we have observed how the management, the supervisors and workers contribute in the perspective of work study. This concluding section is intended to observe the qualities necessary in the work study man. While the management, supervisors and workers are insiders, so far as the information is concerned, the work study man is the only outsider. His job thus becomes the most complicated in terms of coordinating among all three groups and affecting change in the system. Before describing the role of a work study man, following are some of the hints that should be understood by him to act rationally;

(a) Enhancing productivity should be dealt in an equitable way without much emphasis being placed on productivity of labour. In most of the enterprises in developing countries and even in industrialized countries, vast increases in productivity can normally be affected through the application of work study to improve plant utilization and operation to make more effective use of space and to secure greater economy of materials before the question of increasing the productivity of the labour force need be raised. The importance of studying the productivity of all the resources of the enterprise, and not confining the application of work study to productivity of labour alone, cannot be over emphasized. It is very natural that workers would dislike the efforts being made to improve their efficiency while they can see glaring inefficiency in the part of the management. For example, there will be no use of reducing the time taken by the worker to do a certain job or of imposing a production output on him by well applied work study, if he is held back by lack of materials or by frequent machine breakdowns resulting from bad planning by his superiors?

(b) The person must be fully willing and must give his best for the purpose of study. He should be very honest, frank and transparent; he should not attempt to hide what is being done. He should frankly answer the questions and provide information obtained from studies. *Work study, honestly applied, has nothing to hide.*

(c) Seniors must be given comprehensive details regarding the study with suitable reasons in order to adopt the improved and more suitable method or technique. Thus, by asking workers the right questions and by inviting them to come forward with explanations or proposals, several work study specialists have been rewarded by dues or ideas that had never occurred to them. We know a worker has affectionate understanding regarding his job that can escape a work study man. One person along with foreman must be engaged with professionals carrying out work study and then forming a committee to analyze the implementations and conclusions. While interacting with the worker, the work study man should ensure that he in no way undermines or challenge the authority of the supervisor in the eyes of the worker.

(d) Discussing ideas and views among workers makes feel them as an integrated part of the organization. In many instances, a foreman, a worker or a staff specialist contributes useful ideas that assist the work study man to develop an improved method of work. This should be acknowledged readily and the work study man should resist the temptation of accumulating all the glory for himself.

(e) It is important that person doing work study must keep in his mind regarding improving job satisfaction not enhancing production only and that he should devote enough attention to this issue by looking for ways to minimize fatigue and make the job more satisfying. In recent years, several enterprises have developed new concepts and ideas to organize work to this end and to attempt to meet the workers' need for fulfilment.

The above guidelines on the job requirements of a work study man can be translated into certain qualities and qualification that one should look for in him. These can be summarized as follows:

(a) Education: The person taking charge of work study must at least have good secondary education or the equivalent school-leaving examination. It is unlikely that anyone who has not had such as education will be able to grasp things during work study course, despite a few omissions. However, if a work study man is also to be involved in studying other production management problems, a university degree in engineering or management or the equivalent becomes an important asset.

(b) Experience: It is mandatory for applying candidates to have practical experience in their respective industries including time periods in various processes of the industry. This will facilitate them to figure out various challenges while dealing with ordinary persons. Practical experience will also fetch appreciation from foreman and peasants and an engineer's background enables a man to adapt himself to most other industries.

(c) Personal qualities: Whosoever is going to undertake amendments in techniques must possess an inventive turn of mind, be able to come up with elementary system and tools in respect of saving time and efforts and also should be boosting the co-operation among engineers and technicians in advancing these tools. The type of man who is good at this is not always so good at human relations, and in some large companies the methods department is separated from the work measurements department, although both are under the same boss. The following are the essential qualities:

(i) Sincerity and integrity: The work study man must possess qualities like sincerity and honesty in order to win confidence and respect from people with whom he has to deal.

(ii) Enthusiasm and motivation: He must be fully interested in his job, giving importance to what he is doing, make other people feel his positivity and dedication.

(iii) Sympathy with people: He must be able to get along with people at all levels to ensure that he understands their point of view and be motivation behind them.

(iv) Diplomacy and negotiation: Efficient skills of dealing with people come by understanding people, respecting their feelings by generous and thoughtful words. Without tact no work study man can proceed far. Hence, the work study man must possess a negotiation skill and diplomacy to get success.

(v) Good looking: He must look handsome, well-kept, attractive, precise, enthusiastic and capable of handling any situation. This makes him confident enough among the people with whom he has to work.

(vi) Self-confidence: This can only be achieved with proper training and incorporating work study efficiently and successfully in order to win respect from his seniors and top management.

The qualities, for instance, interacting or dealing with people, can be further improved with right ways of training. More often this aspect of the training of work study man is neglected, the assumption being that if the right man is selected in the first place, that is all that needs to be done. In most work study courses, more time should be given to the human side of applying work study. From above, it can be concluded that these requisites must be enforced with professional techniques. Requirements to make a person good manager are same as that of good work study person. Work study is an accomplished way for advancing as higher authorities. Generally, it is not easy to find out people with these qualities but accurate selection of people for work study can provide excellent results not only in terms of enhanced productivity but also in terms of better relations among workers in the organization.

Summary

This chapter is mainly focused upon the definition of work study, historical glimpse about the concept of work study, objectives of work study and finally human factors consideration in implementation of work study in an organization.

Multiple Choice Questions

1. Primary function of work study is to increase the productivity and reduce the waste and it should be capable of
   (a) recommendation and implementation of the improvements
   (b) scrutinization of shortcoming in production processes
   (c) setting standards for labour cost control
   (d) all of these
2. Work study mainly includes
   (a) method study   (b) stop watch time study
   (c) work measurement   (d) all of the above
3. Method study is used to
   (a) provide a basis for setting piece rates or incentive wages
   (b) determine standard cost
   (c) determine the number of machines a person may run
   (d) compare alternative methods
4. The role of management for successful application of work study is to
   (a) clearly define the organizational goals and objectives
   (b) maintain the liaison between the management and workers
   (c) be fully aware of the different aspects of the work and its limitations
   (d) none of the above
5. Who is known as the father (mother) of motion studies?
   (a) Frederick W. Taylor   (b) Frank and Lillian Gilbreth
   (c) Henry Ford   (d) None of the above
6. Which of the following is/are the most important trait/s of a work study man?
   (a) sincerity and integrity and self-confidence
   (b) sympathy with people
   (c) diplomacy and negotiation
   (d) All of the above
7. Time standard developed by time study can be used for
   (a) production planning and scheduling   (b) capacity planning
   (c) wages and incentives   (d) all of these
8. The advantage of work study occurs to
   (a) the management only   (b) the labour only
   (c) the industry   (d) all of these
9. The human factors consideration in work study will ensure
   (a) active participation of workers   (b) reduced fatigue to workers
   (c) increased performance of operators   (d) all of the above

Answers

1. (d)   **2.** (d)   **3.** (d)   **4.** (a)   **5.** (b)   **6.** (d)   **7.** (d)   **8.** (d)   **9.** (d)

Method Study (Motion Study)

3.1. Introduction and Background

Frank (1868–1924) and Lillian Gilbreth (1878–1972) are known as the father and the mother of method/motion studies. Frank Gilbreth was born in Freeport, Maine on 7 July 1868. After finishing school Frank Gilbreth began to work for a construction company as an apprentice brick layer. At an age of 27 years, Frank Gilbreth became the chief superintendent of the company. By the time of his promotion, he developed a number of methods, both technical and procedural, for improving efficiency at the workplace, particularly in his field of construction.

In 1895, Frank Gilbreth started his own construction company in Boston and by 1900, was running a successful business with branches throughout the USA and also a branch office in London. Frank Gilbreth implemented motion study to the task of brick laying (Figure 3.1).

Fig. 3.1: Method suggested by Frank Gilbreth.

He observed that the workers developed their own unusual ways of working and they did not use constantly the same motions during their work. He sought one best way to perform tasks and standardized it and then he developed a method which reduced the number of motion needed to lay a bricks for construction of a wall. Later on they developed a more systematic and sophisticated method of time and motion study for industry, taking into account the limits of human physical and mental capacity and the importance of a good physical environment. They discovered that all works are composed of basic motion elements (therbligs).

3.2. Definition of Method Study

It may also be called *Motion Study*, as it deals with the movement of goods and activities and events in the transformational process. Method study is examination of work performance and its subsequent rearrangement to produce the same results with less effort. Work performance depends on a number of interrelated operations and activities. Any combination of these activities is called method.

Method study can be defined as the systematic recording, critical analysis, development and implementation of new methods to perform job to reduce costs with regard to existing or proposed jobs.

The function of *method analyst* is to study the way in which the level of performance would increase resulting in minimization of effort and cost for same or better output.

In other words, *motion/method study* is the careful analysis of body/limbs motion employed in doing a job. The purpose of motion study is to eliminate or reduce ineffective movements, and facilitate and speed up effective movements. Thus it reduces production time and raises output, which increases productivity.

3.2.1. Indicators of method study

Given below are some common indicators, which show that there is a necessity of method study in an organization:

1. Bottlenecks in system resulting in long delivery time and unbalanced work flow.
2. Idle time of plant or labour.
3. Higher absenteeism and poor moral of workers.
4. Inconsistency in sales and earnings.

3.3. Procedure of Method Study

The following are six procedural steps of method study, that also are illustrated in Figure 3.2.

1. Selection of job to be studied.
2. Collection, recording and presentation of necessary information.
3. Analysis of existing methods.
4. Develop a new method.
5. Install the new method.
6. Maintain the new method, if any modification is required repeat the steps 2–5.

3.3.1. Select and define

The selection of job to be investigated under method study is a very crucial task for the management. It should be worth studying and the objectives to be achieved must be defined. Before selecting a job, the following aspects are to be considered:

(a) Economic or cost aspects
(b) Technical aspects
(c) Human aspects/reaction

(a) Cost is the prime aspect to be considered for a job to be selected for method study. Indication of potential cost reductions and a comprehensive definition of the problem can be obtained by carrying out preliminary studies. This also helps to determine the efficiency levels of working groups and associated plant, equipment and materials. Otherwise, if the economic importance of the job is likely to stop in near future, it will be useless to carry out the study.

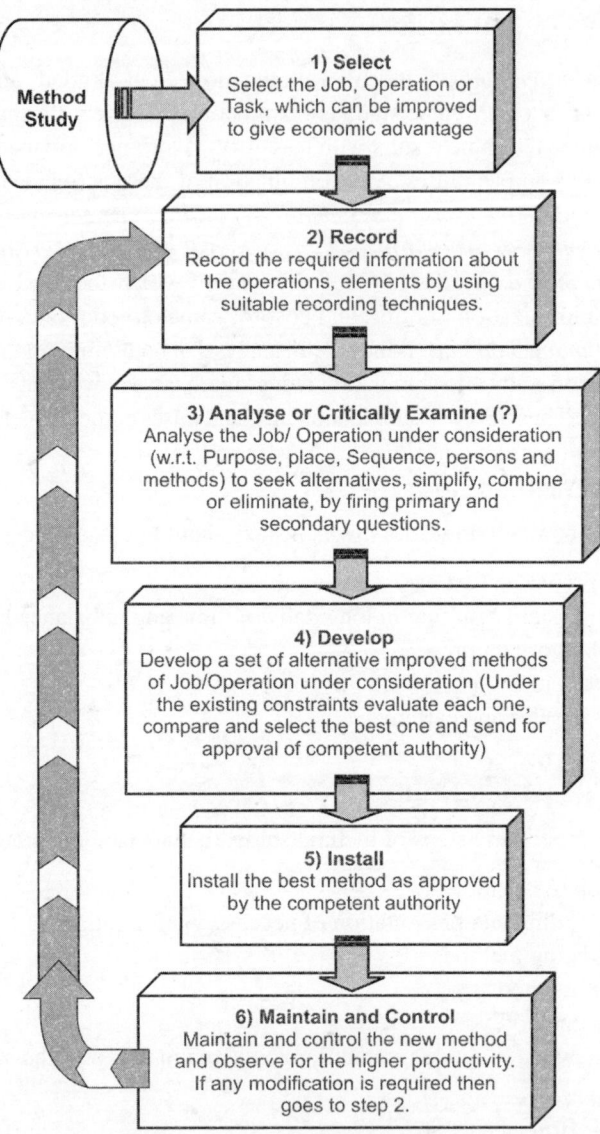

Fig. 3.2: Flow diagram of method study.

(b) About *technical aspects*, one should make sure that adequate technical knowledge is available to carry out the study. The technical knowledge provides a clear vision and the feasibility of the improvements in the existing method.

(c) As discussed in the previous chapter that although employee attitudes (*human reactions*) towards the investigations and changes are the most difficult to predict, but these are very essential to be anticipated. The supervisor will feel that his authority has been challenged because of the intrusion of outside control in his department. Workers will have natural resistances to change with the fears of laying-off and transfers; therefore, all reasonable assurances should be given. Method study may perhaps be more gladly accepted, if properly explained to all concerned and healthy industrial relations exist.

Method study will be more eagerly accepted, if it is started on jobs unpleasant to workers, i.e., hazardous jobs; or those with lifting of heavy weights; dusty environment; or respiratory hazards like chrome plating; hot and humid environment; high noise level, etc. If method study successfully removes the unpleasant features of these jobs and reduces effort and fatigue of workers, they will naturally become cooperative instead of being reluctant to method study personnel.

The jobs with the following objectives should be placed at priority to be selected for method study:

(i) To reduce manufacturing cost.

(ii) To avoid bottlenecks that hold up other production operations.

(iii) To reduce fatigue incurred by workers in order to increase their efficiency.

(iv) To avoid poor use of materials, labour or machine capacity that results in high scrap and reprocessing costs.

(v) To improve upon operations involving repetitive work using a great deal of labour and liable to run for a long time, etc.

3.3.2. Collection, recording and presentation of necessary information

After a particular job has been selected for study, the next step is to record all the relevant information and procedure pertaining to the present or existing method in detail and in the form of a chart to obtain a clearer picture about the same. The success of an investigation depends upon the accuracy with which the facts are presented because they will provide the basis of both the critical examination and the development of the improved method. It is, therefore, essential that the record be clear and concise. All the facts must be related to the situation under investigation and should be obtained and arranged for appraisal and comparison. There are number of recording techniques to collect the necessary information about the job/operation under consideration. Therefore some common techniques are described in the subsequent section.

3.3.2.1. *Commonly used recording techniques*

Various recording techniques have been evolved in order to clarify working situation. They comprise

(a) Charts:

(i) Outline process chart

(ii) Flow process chart: man type, material type and equipment type

(iii) Two handed process chart

(iv) Multiple activity chart

(v) Travel chart

(b) Diagrams:

(i) Flow diagram

(ii) String diagram

(iii) Cycle graph

(iv) Chrono Cyclegraph

(c) Motion and film analysis (Micro-motion and Memo-motion studies):

(i) Simo chart

(ii) P.M.P.S.

(d) Layout models, two and three dimensional:

These recording techniques have been discussed in detail in Section 3.7.10. of this chapter.

3.3.2.2. *Analysis of existing method (Examine)*

Once the required information about the existing method of operation is obtained, the next step is to critically analyze the same. Therefore, the questioning technique is the means by which the critical examination is conducted, each activity being subjected in turn to a systematic and progressive series of primary and secondary questions. The process of critically examining the existing method is exhibited in Figure 3.3 given below.

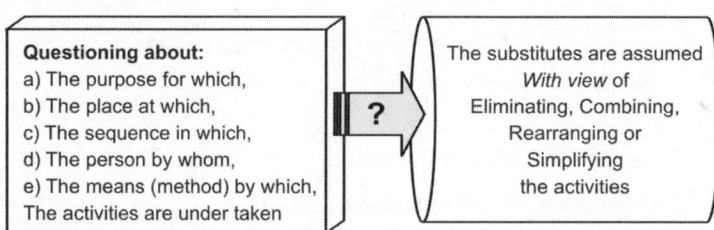

Fig. 3.3: Analysis of existing method using questioning technique.

The *primary* (P) and *secondary* (S) *questioning* technique is exhibited in Table 3.1. The primary questions indicate the facts and the reasons underlying them. The secondary questions cover the second stage of the questioning technique, during which the answers to the primary questions are subjected to further query to determine whether possible alternatives of place, sequence, persons and means are practicable and preferred as a means of improvement upon the existing method.

Table 3.1: Primary and secondary questions for analysing the job.

Variable of Investigation	Primary Question	Secondary Question
Purpose	What is done? Why is it done?	What else might be done? What should be done?
Place	Where is it done? Why is it done there?	Where else might it be done? Where should it be done?
Sequence	When is it done? Why is it done then?	When might it be done? When should it be done?
Person	Who does it? Why does that person do it?	Who else might do it? Who should do it?
Means	How is it done? Why is it done that way?	How else might it be done? How should it be done?

These questions, in the above sequence, are asked systematically every time a method study is undertaken. Let us take one example of an operation to move bars (rectangular) to blank cutting machine in a hand tool manufacturing unit; the details are shown in Table 3.2.

Table 3.2: Primary and secondary questions for handling raw material for blank cutting in a hand tool manufacturing unit.

Primary Questions		Secondary Questions	
Established Facts	*Challenge – Why?*	*Alternatives*	*Proposal – Improvement*
What is done? Bars moved to blank cutting machine.	*Why?* To cut bars into blanks.	*What else might be done?* Buy the blanks from outside.	No change.
Where is it done? Between store and machine.	*Why there?* Normal place for storing and cutting bars.	*Where else might it be done?* Move machine to store or store bars by machine.	No change.
When is it done? After requisition is issued to worker.	*Why then?* First task-material is not issued without requisition.	*When might it be done?* Issue requisitions direct to stores before issuing job ticket to worker.	Issue several jobs together-say minimum of one day's work.
Who does it? An operator of blank cutting machine.	*Why that person?* Custom and practice.	*Who else might do it?* Store helper, shop helper.	Shop helpers only.
How is it done? 'Carried Manually'.	*Why is it done that way?* Normal practice.	*How else might it be done?* Move with the help of hand truck, fork lift truck and roller bed.	Hand truck.

3.3.2.3. *Brainstorming session*

The specialized technique of *creative thinking* session that came into importance in the pre-war days in America has been used to ensure that at the stage of method study concerned with seeking alternative methods, a greater flow of new ideas is generated. To conduct a *brainstorming session, a* group of people is gathered together, presented with a specific problem and each individual is asked to come out with the first idea. Ideas suggested by one member of the group are built on and improved by the other members. The session results in a good accumulation of ideas; however most of them may completely be impractical, but few worthy of literal consideration. These few ideas are more likely to generate a solution of the problem than would have been achieved by entirely conventional methods.

In order to conduct the session effectively the following points should be considered:

(i) Careful definition of the problem: For a general type of problem, better results are likely to be realized if it can be divided into parts.

(ii) Careful selection of the team: The group should have same managerial or intellectual level along with background and knowledge of problem under discussion.

(iii) Conducive environment: The careful selection of the team will certainly help and better results may perhaps be achieved if they have had an opportunity to make interaction amongst each other in a comfortable environment. The team should be aware of outcome of the session and, where possible, should be present when the ideas are critically examined.

3.4. Develop the Best Method

Once the optimal solution is identified or found, the next step is to record the proposed method through the critical examination. At the same time, all levels of management and workers should be consulted and asked for their suggestions and objections. Afterwards, modified proposed method should be tested and thus can be claimed for its advantages over the original method. A report on proposed method to the management for its final approval and subsequent implementation must be prepared. Once a new method has been proposed after critical examination, prepare a record of the same on a flow process chart so that it can be compared with the original or existing method.

In order to show that improvements are realistic, comparison of the new proposed methods with the existing (old) is essential, and it is therefore very essential, that all activities are portrayed in the same scale and at the same level. The proposed method must be realizable w.r.t. to the following aspects:

(i) Feasibility: economical and technical
(ii) Safety and effectiveness
(iii) Acceptable to design, production, quality control and sales departments

The proposed new method, that appears to be sound, requires further study and may require experimental work to develop it. In addition, it requires considering all aspects of manufacture to get the best possible return from the ideas. The development of the new method is simplified to a certain extent by extracting and working with the key activities that are suggested by the conclusion of the critical examination. In fact many ideas, developed after critical questioning, require minute expenditure, which is one of the major desire abilities of work study. It may be necessary to modify a procedure, get someone else to do a job, etc. Still, it requires cooperation of others. The work study engineer must consult all levels of management and even workers. He should note down and clarify their objections and make them feel concerned in his ideas throughout the time of development. Resistance to change will be much less, if people are asked to participate in the development of better methods.

3.5. Install the Best Proposed Method

The proposed method is installed in two stages i.e., preparation and implementation (or actual installation).

1. *Preparation* involves three stages: plan, arrange and practise. For the planning aspect, one person should be made responsible for the task, definite schedule should be fixed for each stage to complete and the same should be followed as strictly as possible. Arrange stage involves checking that all the necessary tools and equipment are available and the services are made available. It also includes checking the availability of all material, supplies and services. The selection of workers for the proposed method and imparting the necessary training to them is also another important aspect of arrangement. Then, finally informing the every concerned individual, about the plans and schedule for the installation. Practice involves a trial run of the proposed method using supervisors, work study officers and the appropriate workers.

2. *Implementation* or actual installation involves the introduction of developed method as standard practice. During the initial period after installation, a close watch should be kept so that the problems associated with the proposed method can be carefully studied and removed. Installation phase is complete as soon as the newly installed method starts working smoothly and satisfactorily and gives encouraging results.

3.6. Maintain the Installed Method

Once the new method has been installed, it should be maintained in its specified type and workers should not be allowed to slip back into the old method or unjustified deviations from the specified procedure. The proper functioning of the installed method should be ensured by periodic checks and reviews. This enables the management to confirm whether the method being practised is the same or it has deviated from the approved one. Reasons for the deviation should be discovered and the necessary adjustment should be made in the procedure to revert to the approved one. In addition, the views of the workers, supervisors and other persons related with the installed method can greatly help in discovering further improvements. In a number of cases, the condition may change from time to time and this may mean that some of the assumptions upon which the improved method was built up, are no longer valid. Therefore, the method should be reviewed at regular intervals to make allowances for any change. A change may arise as a result of suggestion schemes.

3.7. Process Analysis

Process analysis means the study of the overall process in a factory (plant). It analyses each step of the manufacturing process and aims at improving industrial operations. Process analysis aids in finding better method of doing a job and this is achieved by eliminating unproductive and unnecessary elements of the process or through modified layout of facilities. The process is analysed with the help of process charts and flow diagrams.

Various steps involved in process analyses are:

1. Select the process for analysis.
2. Breakdown the process into operations and sub-operations.
3. Construct a process chart and flow diagram.
4. Analyse the process chart and flow diagram by subjecting each and every step to questioning procedure as discussed in Section 3.3.2.2.
5. Reconstruct the process chart and flow diagram for the modified (proposed) procedure.
6. Test the proposed method for all the advantages claimed for the same.
7. Explain the new method to workers and put it into operation.

Gilbert devised several systems of analyzing work into different district elements. One was the familiar process chart, breaking work down into five basic elements of operation, inspection, transportation, storage and delay. The analysis of work was carried further in the development of the basic elements of hand work which were termed therbligs.

3.7.1. Process chart

A process chart may be a diagram, a picture or a graph which gives an overall view of a process. It helps in visualizing various possibilities of alteration or improvement. A process chart records graphically or diagrammatically the operations (in sequence) associated with a process. The chart portrays the process with the help of a set of symbols and helps in better understanding and examining the process with a purpose to improve the same.

3.7.2. Process chart symbols

Charts are generally represented by symbols. Symbols produce a better picture and quick understanding of the facts. There are mainly five basic symbols that are used to record different types of events in a process chart, these are given in Table 3.3 given below.

Table 3.3: Description of five basic symbols used in process chart.

	Event	*Symbol*	*Description*
1.	Operation	◯	Operation represents an action or step in the procedure. An operation involves a change in the location or condition of a product. Examples: cutting a bar on a power hacksaw or bending a piece wire into a shape of hanger i.e. a value adding activity.
2.	Storage	▽	Storage represents a stage when a finished product or raw material waits for an action or when an item has been retained for quite some time for reference purposes. Storage shows an authorized control over an item. Examples: milling cutter lying in tool store or finished product/components in a stock room.
3.	Delay or Temporary Storage	D	Delay occurs when something stops the process and a product waits for the next event. It is a temporary halt in the process. Examples: power failure, waiting for the lift or a traffic jam.
4.	Transport	⇨	Transport indicates the movement of an item from one location to another. The item may be the material, equipment, an operator or his hands only. Examples: beverage flowing through a pipe line, components travel from one work station to another, e.g., mild steel rods being sent from stores to machine shop, etc.
5.	Inspection	☐	Inspection is an act of checking for correctness of the quantity or quality of the items. Inspection is not normally expected to change the shape or other characteristics of an item. Examples: gauging a piston pin or checking the hardness of a carburized mild steel piece. It is a non-value adding activity.

In addition to the above mentioned basic symbols, there are symbols for combined activities also. The important events have the outer symbol as shown in Table 3.4.

Table 3.4: Description of process chart symbols being used for combined activities.

	Event	*Symbol*	*Description*
6.	Operation-cum-transportation	⊜	Example: Articles are being painted as they are transported by the chain conveyor.
7.	Inspection-cum-operation	⊡	Examples: Powder milk tin is being weighted (inspection) as it is filled. Both the events occur simultaneously and are controlled automatically.

3.7.3. Outline (Operation) process chart

An outline process chart is intended to do broad analysis. It records an overall picture of the process and states only main events or steps sequence-wise, there are following functions of an outline process chart and the same is shown in Figure 3.4.

(a) It helps in visualizing and comprehending the full process so that necessary improvements may be made if required.

(b) It shows relationship between the different activities.

(c) It considers mainly operations and inspections, i.e. it makes use of only two symbols.

(d) Each operation and inspection is numbered from the beginning to the end of the chart.

(e) Description of operations and inspections are written on the right-hand side of the symbols.

(f) Actually an outline process chart is the first step for the beginning of a detailed analysis.

Example 3.1: Flow process chart for task like repairing and changing a punctured tyre of a car

Task	:	**Repairing and changing tyre of a car**
Chart begins	:	Unscrew nuts
Chart ends	:	Screw the nuts
Charted by	:	_____
Date	:	_____

1. Unscrew nuts of the punctured wheel	1	Unscrew nuts of the punctured wheel
2. Fix the screw jack under the car	2	Fix the screw jack under the car
3. Remove the punctured wheel	3	Remove the punctured wheel
4. Remove the tube from tyre	4	Remove the tube from tyre
5. Inspect the tube for puncture	5	Inspect the tube for puncture
6. Repair/replace the tube	6	Repair/replace the tube
7. Check for any nail inside the tyre and fix the tube	7	Check for any nail inside the tyre
8. Insert tube in the tyre wheel	8	Insert tube in the tyre wheel
9. Inflate the tube and check pressure	9	Inflate tube and check pressure
10. Fix tyre with hub and tighten nuts	10	Fix the tyre with the hub and tighten nuts

Fig. 3.4: Outline process chart.

Now consider that a new technology has been introduced in the market with tubeless tyres, let us make outline flow process chart for repairing a puncture of tube less tyre. The summary of symbols used reveals that the total number of symbols has been reduced from 10 to 7, i.e., there is 30 per cent reduction in the total number of activities to be performed for repairing a punctured wheel as shown in Figure 3.5 and summary is given in Table 3.5.

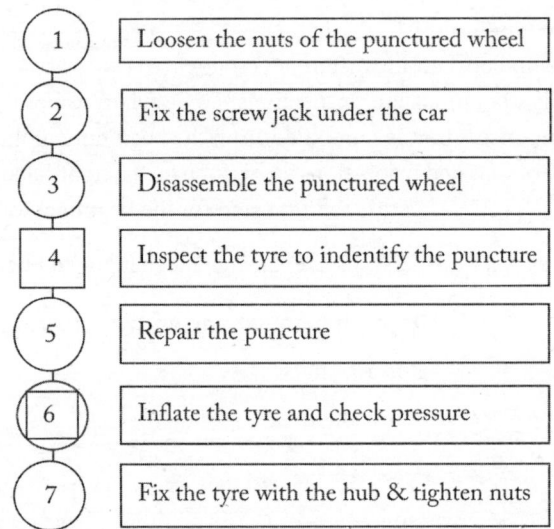

Fig. 3.5: Outline process chart of repairing a punctured wheel (tubeless) for a car.

Table 3.5: Summary of symbols used in outline flow process charts for repairing a punctured wheel of a car.

Sr No.	Description of the Symbol Used	Frequency of Symbols Used for (with Tube)	Frequency of Symbols Used for (Tubeless)
1.	Operation	7	5
2.	Inspection	2	1
3.	Operation cum inspection	1	1
4.	Total	10	07

3.7.4. Flow process charts

A *Flow Process Chart* is a detailed version of outline process chart and it records all the events. It sets out a sequence of flow (of a procedure or product), records all the events in sequence using process chart symbols, marks distances travelled and time taken for completing an activity, and mentions other important (or key) points, if any. There are three types of flow process charts, namely:

(a) Man type flow process chart.
(b) Machine/equipment type flow process chart and
(c) Material type flow process chart.

Man type flow process chart records the activities of an operator, i.e., what an operator does. *Equipment type process chart* records the manner in which the equipment is used, and *Material type process chart* records what happens to the material, i.e., the changes the material undergoes in location or condition. A portion of the material type flow process chart for cutting of mild steel rod is shown in the following Figure 3.6.

Activity	Symbols					Distance moved (meters)	Time (minutes)	Remarks, if any
	○	□	▽	D	⇨			
Steel rod (45mm dia) lying in store.			●			-	-	-
Moved to hacksaw cutting machine					●	10	03	By trolley
Wait, cutting machine being set				●		-	02	-
A piece of 14" is cut	●					-	02	-
Wait for trolley				●		-	05	-
Moved to machine shop					●	20	03	By trolley
Inspected before machining		●				-	01	-
Machining is done to maintain the size	●						05	

Fig. 3.6: Material type flow process chart for cutting of mild steel rod.

Example 3.2: A hand tool manufacturing firm is supplying hand tool kit boxes to the customer. The material type flow process chart for a spanner of size 17 mm starts with blank cutting operation on rectangular (flat) bar, followed by hot forging (at 800° C), trimming, punching, grinding and heat treatment process, etc. as shown in Figure 3.7. The martial flows in a lot of 500 pieces. If we carefully observe this chart, there are total 27 activities, out of which only 11 activities are necessary value-adding activities, and rest 16 activities are non-value-adding activities such as inspection, process delay 360 minutes due to necessary time for cooling, so that cold trimming can be performed, and repeated transportation due to bad layout design. It is also observed form the chart that a single lot of 500 spanners of size 17 mm take 1499 minutes of total time to complete, which means almost 24 hours. Throughout the production, lot is moved a total distance of 164 meters from one section to another, which takes around 104 minutes. In addition to these the flow of lot is delayed for 360 minutes (six hours) so that after hot forging the temperature of work pieces becomes at par with the ambiance level. Also the 480 minutes are consumed for stone barreling. It can be observed from Figure 3.7 that out off 1499 minutes only operations like hot forging, trimming, punching, grinding, heat treatment and electroplating are necessary value adding, and rest are non value adding activities. The summary of various symbols and time taken is mentioned in Table 3.6, which reveals the percentage of time taken by each activity and distance moved.

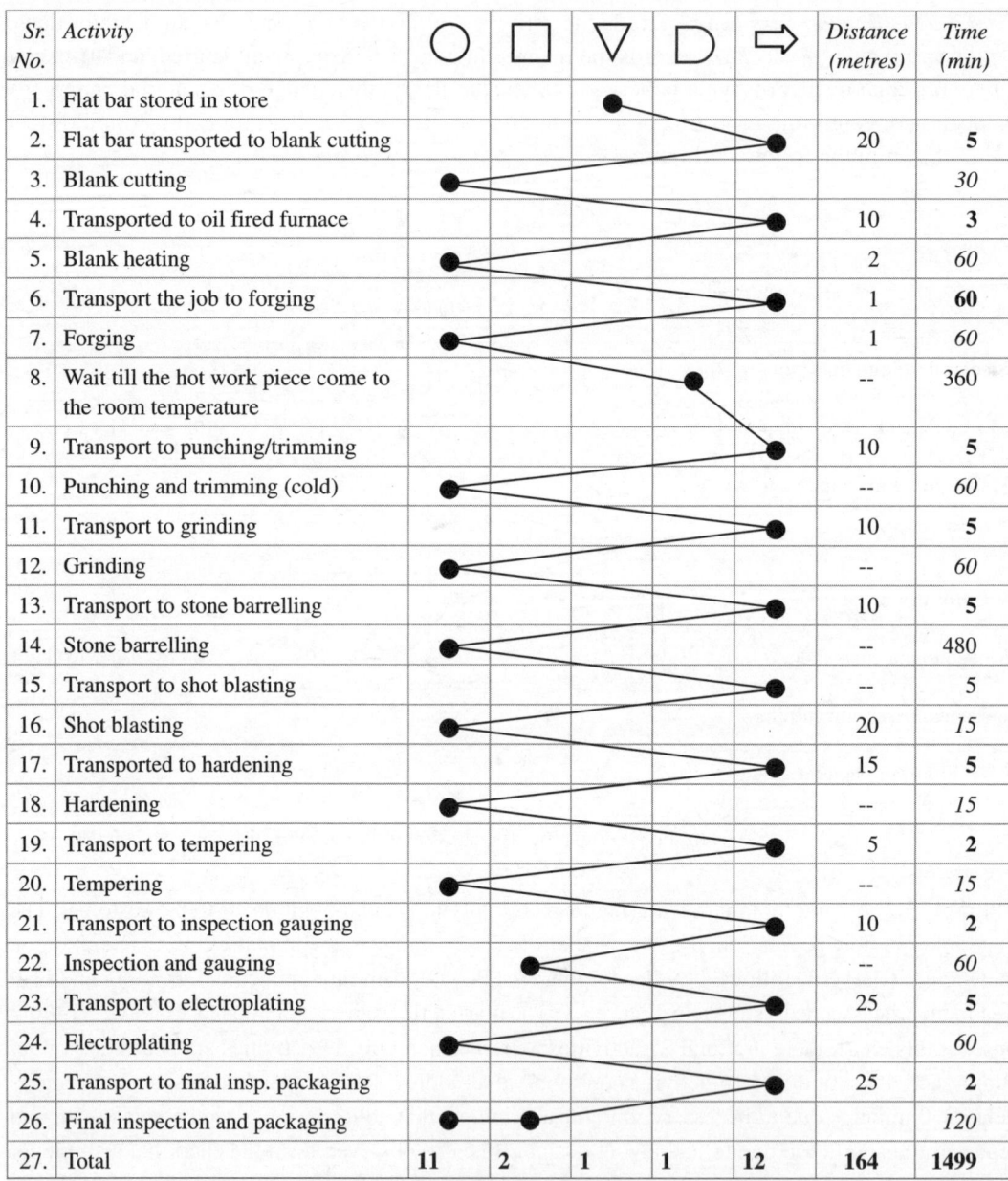

Sr. No.	Activity	○	□	▽	D	⇨	Distance (metres)	Time (min)
1.	Flat bar stored in store			•				
2.	Flat bar transported to blank cutting					•	20	**5**
3.	Blank cutting	•						*30*
4.	Transported to oil fired furnace					•	10	**3**
5.	Blank heating	•					2	*60*
6.	Transport the job to forging					•	1	**60**
7.	Forging	•					1	*60*
8.	Wait till the hot work piece come to the room temperature				•		--	*360*
9.	Transport to punching/trimming					•	10	**5**
10.	Punching and trimming (cold)	•					--	*60*
11.	Transport to grinding					•	10	**5**
12.	Grinding	•					--	*60*
13.	Transport to stone barrelling					•	10	**5**
14.	Stone barrelling	•					--	*480*
15.	Transport to shot blasting					•	--	*5*
16.	Shot blasting	•					20	*15*
17.	Transported to hardening					•	15	*5*
18.	Hardening	•					--	*15*
19.	Transport to tempering					•	5	**2**
20.	Tempering	•					--	*15*
21.	Transport to inspection gauging					•	10	**2**
22.	Inspection and gauging		•				--	*60*
23.	Transport to electroplating					•	25	**5**
24.	Electroplating	•						*60*
25.	Transport to final insp. packaging					•	25	**2**
26.	Final inspection and packaging	•	•					*120*
27.	Total	**11**	**2**	**1**	**1**	**12**	**164**	**1499**

Fig. 3.7: Material flow process chart for manufacturing of spanner.

Table 3.6: Summary of symbols used in material flow process chart.

Parameter	○	□	▽	D	⇨
Number of symbols	11	2	1	1	12
Time taken (min)	917	120	0	360	104
% Time taken	61	8	0	24	7
Distance moved (m)					164

3.7.5. Two-handed process chart

(a) A two-handed process chart records the activities of both left and right hands (of an operator) as related to each other, while the operator is doing the job.

(b) The activities of the two hands can be synchronized by providing a time scale on the chart.

(c) Once the activities have been recorded, they are examined with a view to eliminate unnecessary non-value-adding, uneconomical and non-essential motions. The remaining motions are simplified and re-sequenced to achieve better performance and operator comfort.

(d) This chart is used for analyzing repetitive jobs of short duration and assembly work.

A two-handed process chart can be employed to give a bird's eye view of the complete process (by considering operations only) as well as the detailed description (by using all the five process chart symbols) as shown in Figures 3.5 and 3.6.

Activity: Assembling nut and bolt				
Left Hand	*Symbols*			*Right Hand*
	LH	*RH*		
Pick Up Bolt	◯	◗		Idle
Hold	▽	◯		Pick Up Nut
Hold	▽	⇨		Transport To Left Hand
Hold	▽	◯		Assemble (Screw Up)

Fig. 3.8: Two-handed process chart.

L.H.	Numbers	R.H.	Numbers
◯	1	◯	2
▽	3	▽	--
⇨	--	⇨	1
◗	--	◗	1

Fig. 3.9: Summary of two-handed process chart.

3.7.6. Multiple activity chart

Where a number of workers work in a group or an individual operator handles two or more machines, their activities have to be coordinated for achieving proper results. A multiple activity chart records simultaneously the activities of all the workers and machines on a common time scale and thus shows inter-relations between them. Multiple activity chart are of the following types:

(i) *Man–machine chart*: one man handling one job or one machine.

(ii) *Man–multi-machine chart*: one man handling a number of machines.

(iii) *Multi-man–machine chart:* a group or gang doing collectively one job as in riveting.

The purpose of multiple activity charts is to:

(i) detect idle times being enforced on machines and workers.
(ii) optimize work distribution between workers and machines.
(iii) decide number of workers in a group.
(iv) balance the work teams.
(v) examine the interdependency of activities.
(vi) ultimately develop an improved method of accomplishing a task to have an effective labour cost control.

3.7.6.1. *Construction of multiple activity charts*

(i) A separate vertical bar or column to represent each subject (which may be a machine or operator) is required.
(ii) A common time scale is provided for all the subjects.
(iii) Activities of each subject in relation to those of the others are marked in the respective column.
(iv) Previously conducted time studies provide the time values for each activity.
(v) A brief description of each activity is marked on the chart.
(vi) Working and idle times are marked differently on the chart.

Analysis

A typical man machine type multiple activity chart is shown in Figure 3.10. The summary of the same chart is given in Table 3.7, which reveals the total idle time. After constructing the chart, it is tried to rearrange the work cycle to minimize men or machine idle times, simplify the operations, combine or eliminate some of the elements, etc. A multiple activity chart finds its applications mainly in plant repair and maintenance, construction jobs and planning team work.

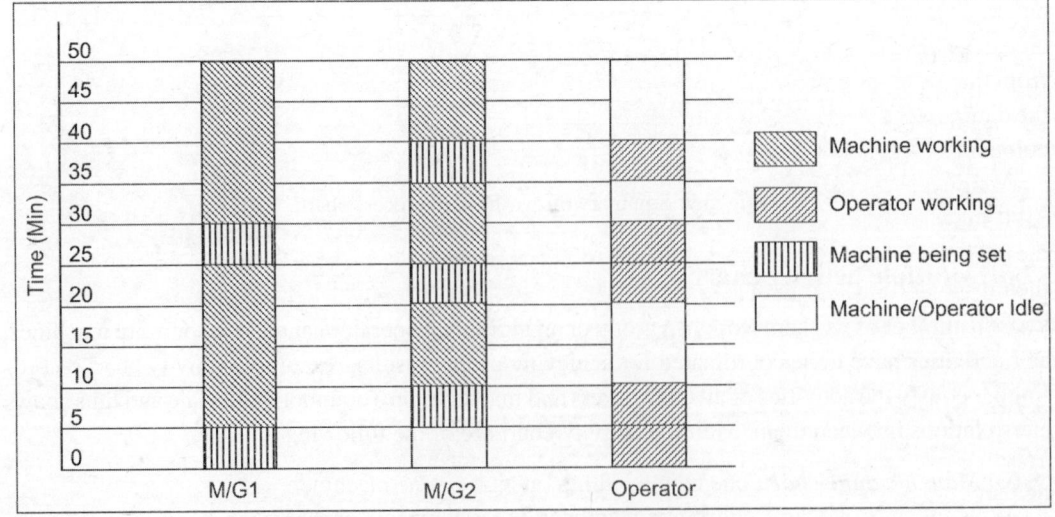

Fig. 3.10: Man-machine type multiple activity chart.

Table 3.7: Summary of multiple activity chart.

Parameter	Working Time (Minutes)	Idle Time	Setting Time
Machine -1	40	Nil	10
Machine-2	30	5	10
Operator	25	25	

3.7.6.2. *Gang process chart*

Gang process chart is another type of multiple activity chart that portrays the relationship of the activities carried out by different members of a group (or gang) with respect to one another while doing a job such as riveting. The aim (of the chart) is to reduce idle and ineffective time and to improve the efficiency of the gang operations. A gang process chart looks like a man – machine chart with the difference that instead of one operator, there are number of operators or workers.

3.7.7. Flow diagram

In a manufacturing shop, overhauling or a repair shop or in any other department there are movement of men and materials from one location to another. Though, process chart indicates the sequence of events, they do not illustrate the movements of men, material, etc. while the work is being accomplished. The number of movements if minimized results in a lot of saving both in cost as well as efforts required to do a job. The path of movement (i.e., movement between two locations and the number of times a movement is repeated) can be visualized by drawing a diagram: it may be a string diagram or flow diagram. A string diagram is preferred over a flow diagram, if path of movements are very much occupied (congested) and difficult to trace on a flow diagram. A flow diagram is more suitable for simple cases.

A *Flow Diagram* is a drawing or a diagram which is drawn to scale. It shows the relative position of production machinery, jigs, fixtures, gangway, etc., and marks the path followed by men (workers) and materials. The steps taken to draw a flow diagram include a) draw to scale the plan of the work area, b) mark the relative positions of machine tools, benches, store, racks, inspection booths, etc. c) from the different observations, draw the actual (path) movements of the materials or the worker on the diagram and indicate the direction of movement. Different movements can be marked in different colours (for the better understanding). Process symbols may also be added on the diagram.

A simple flow diagram of manufacturing a dumbell rod in a machine shop is shown in Figure 3.11, in which raw material from the store moves to station A where cutting operation is performed; the cut pieces are sent to work place B where turning, facing and knurling operations are carried out; then it moves to place C for inspection; it is further sent to bench D where it halts for a shortwhile and ultimately goes out of the shop.

3.7.8. String diagram

When there are many repetitive paths, a flow diagram becomes very congested and the same is neither easy to trace nor to understand. Under such circumstances, a string diagram is preferred over the flow diagram. Therefore, a *string diagram* is a model or scaled plan of the shop, in which every

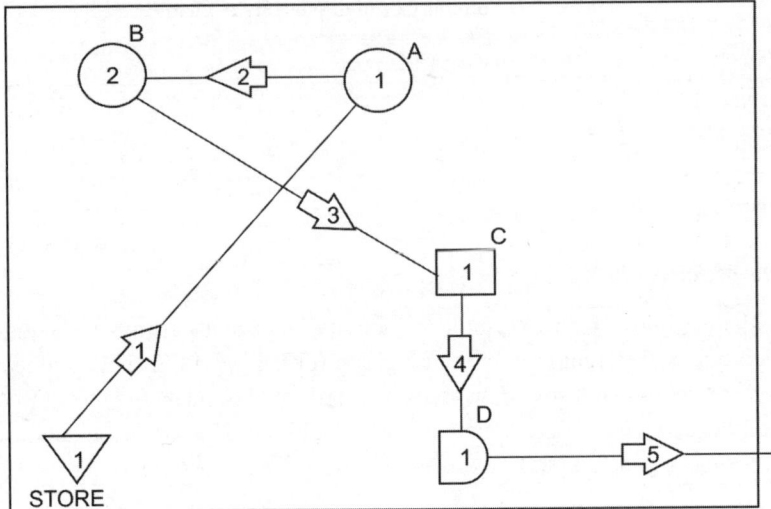

Fig. 3.11: Flow diagram.

machine or equipment is marked and a peg or pin is struck by or in the area representing a facility. A continuous coloured thread or string traces the path taken up by the material or workers while performing a particular operation.

Construction of a flow diagram

(i)   Draw the scale-layout of the shop (working) area and mark various features, such as machinery, work benches, store, etc.

(ii) Mount this scaled drawing on a soft board and strike pins or pegs at all the places which form the path of the workers and materials. More pegs may be struck in between the facilities to trace more or less, the actual path of men and materials.

(iii) A continuous coloured un-stretchable string, taken from the first to last peg, is wound to mark the path followed by workers or materials.

As many as 15 times, a thread can be taken round each peg easily and yet it will not be difficult to comprehend the various movements. The thread when measured gives approximately the total distance travelled by a worker or the material. Figure 3.12 shows a string diagram. A string diagram pictures the movements. It is very useful in dealing with complex movements and plant layout and design problem. It indicates clearly back tracking, congestion, bottlenecks, and over or under-utilized paths on the shop floor. It also measures the distance involved and points out whether a work station is suitably located. In addition, it traces existing path of movements for necessary modifications, if any. It is preferred when movements are not regular w.r.t. their frequency and distance travelled. It shows the pattern of movements and thus helps in deciding the most economical routes to do a particular operation. It is advantageous in studying the movement of:

(a) an individual operator handling a number of machines,

(b) a group or gang moving from one machine or work bench to another and

(c) workers/material in an assembly or repair shop.

Drawbacks

In case the workers or materials move in some irregular or curvilinear path, it is not possible to trace exactly the same on the string diagram and thus no estimate can be made regarding the total distance travelled by the workers or the materials.

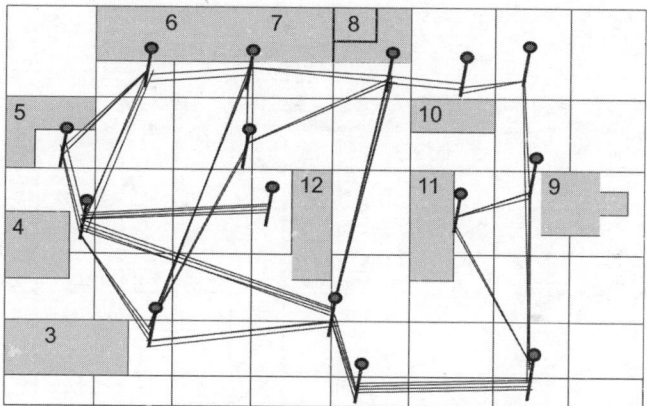

Fig. 3.12: String diagram (M-1 stands for Machine-1)

3.7.9. Travel chart

Whenever it is required to examine complex inter-departmental movements of men or material, the flow diagram becomes a jumble of lines and very difficult to understand and interpret. In fact, a flow diagram is not suitable when one has to study quantitative factors such as weight, distance and frequency of movement. The string diagram, though it proves to be more effective than a flow diagram in studying movements, takes rather a long time to construct and if lot many movements along complex paths are involved, it too becomes a web of criss-cross lines. An alternative and moreover, a quicker and more manageable recording technique to use on such circumstance is the *Travel Chart.*

Definition

A travel chart is a tabular record for presenting quantitative data (e.g., weight, distance, frequency, etc.,) about the movements of materials, workers or equipment between any numbers of places (department) over a given period of time. A travel chart is a record of the amount of distance traveled by the worker or material-in-process while going from machine to machine or from one department to another.

The amount of travel depends upon the frequency of movements between sections or departments. Travel chart is always a square, having within it smaller squares. Each small square represents a wok station; for example, a place visited by an attendant to distribute and collect letters or any correspondence, etc. A travel chart can be constructed as follows:

1. Draw a square grid as shown in Figure 3.13 and mark various departments in order to show to and from movements.
2. Assume that raw material or worker travelled from the department D to C; a '√' mark is placed at the junction of the columns joining department D to department C to record this movement.

3. Similarly '√' marks are placed at other appropriate points in the square grid for all other movements that take place during the selected period of time for study.

4. The number of √ marks in each vertical and horizontal column is added up and bar charts plotted in order to provide at-a-glance picture of the frequency of movements to and from each department.

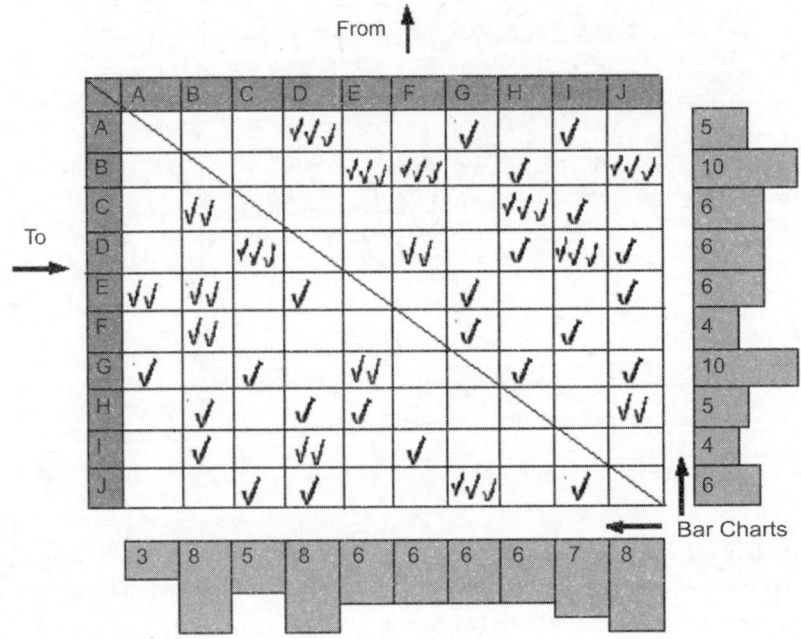

Fig. 3.13: Travel chart.

As regards construction, a travel chart differs from a flow or string diagram in the sense that while plotting, one does not actually trace the path of movement, rather just places a √ mark in the appropriate square to record a move. Knowing the existing plant layout and the frequency of movements from one department to another, a travel chart helps improving the plant layout so that the total distance travelled can be minimized.

3.7.10. Templates and models

Instead of the scales plans of the shop facilities (as in a string diagram), their templates (i.e., two-dimensional models) and three-dimensional models present a more realistic picture of the working area (i.e., shop). These enables us to effectively visualize and record congestions, bottlenecks and back tracking, if any, so that we can plan more economical routes for men and materials. Templates and models are very helpful for plant layout and design.

(a) Templates: They are used to develop plant layout. They are two dimensional or block templates made up of cardboard, coloured, paper and celluloid. They are made to scale and are placed on the scaled outline plan of the building. Templates or cut-outs show the plans of the various facilities like machinery, fittings and work benches and thus known as graphic technique. These templates can be placed and attached with a tape. Either a on a board or on a crossed hatched surface or on a graph paper: hence the name graphic technique. These templates have flexibility in use and can be moved on the graph paper from place to place in order to evaluate various feasible positions for different machine. It is better to photograph a layout before shifting the templates to try another layout of facilities. The templates save a lot of time and labour which otherwise would be spent in making drawings for each alternative plant layout arrangement. They visually present various characteristics, advantages and limitation of a layout. Coloured templates have still better vision effects. Figure 3.14 shows a block templates and a two dimensional template.

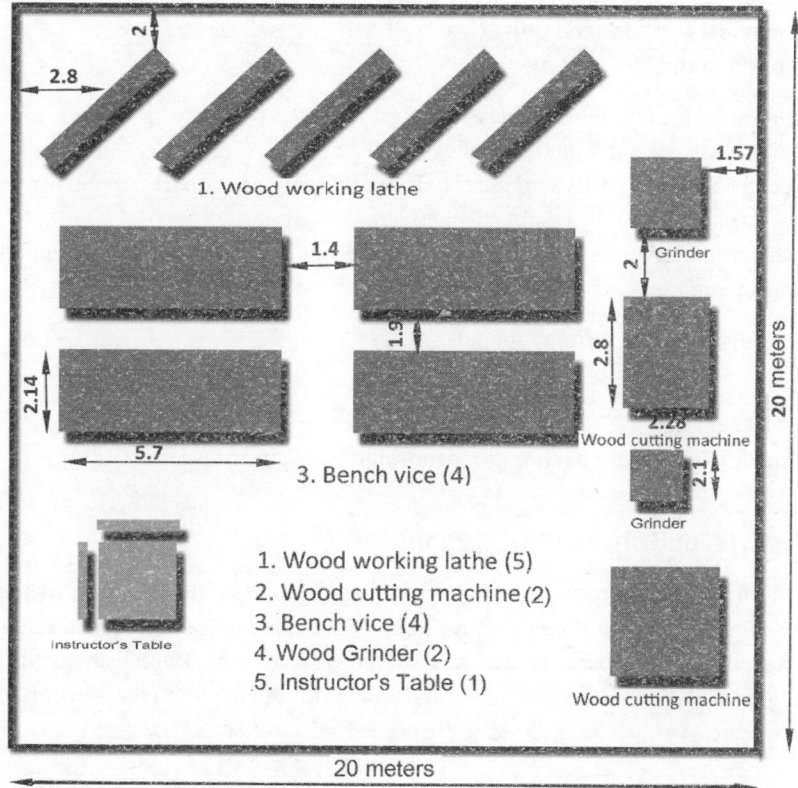

Fig. 3.14: Block templates for layout.

Two-dimensional templates give machine outline and its detail whereas a block template shows the boundary of the maximum projected area of the machine. Templates, though simple and inexpensive, do not give real situation effects which is obtained through the use of three-dimensional models or block models.

Advantages and disadvantages of two-dimensional templates

The main advantages of two-dimensional templates are that they are least expensive, and they can be readily interpreted and followed by technical hands. Moreover, it is very easy to make their duplicate copies. At the same time, two dimensional templates have their own disadvantages; e.g., non-technical persons find it difficult to grasp the clear picture. It is not possible to visualize the overhead facilities.

(b) Three-dimensional models: They are scale models of a facility and more near to the real situation, as besides length and width they show the height of a facility also. Models are easily understood by persons who are not familiar with plant layout practice; models are generally made up of wood or die cast plastic. They show minor detailed and can be mounted on a thick plastic sheets acting as the floor notes and elevations. Models can be made for production machines, workers materials handling equipment or any other facility. Models are much more effective and fast as compared to drawing or templates especially when multi-storey plant layout is to be designed. Multi-storey models can be made of plastic or acrylics. However, these are expensive, but have resulted in substantial saving in laying out for chemical factories and refineries.

Advantages of three-dimensional models:

(i)   Layout is easier for laymen to understand.
(ii) Layout can easily be explained to management.
(iii) Models can be shifted easily and quickly to study operations management.
(iv) Overhead structures can be easily checked.
(v)   Lighting, ventilation and safety features can be easily incorporated and imagined.
(vi) They convey more or less a real solution.

Disadvantages of three-dimensional models:

(i)   They require more storage area.
(ii) They are expensive.
(iii) It is difficult to take them to shop floor for references purposes.

3.7.11. Cyclegraph and chrono-cyclegraph

Cyclegraph and chrono-cyclegraph were introduced by Gilbreth. They belong to the family of techniques employed for the recording the movements. These are photographic methods and are very accurate and detailed. They are used to trace especially those movements which are too fast for human eyes to pursue, a typical path traced by cyclegraph and chrono-cyclegraph is shown in Figure 3.15.

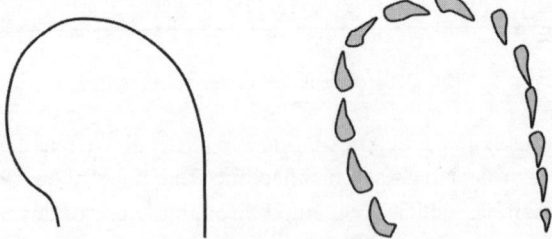

Fig. 3.15: Cyclegraph and chrono-cyclegraph.

Cyclegraph: A cyclegraph is systems that can record the paths of movements of limbs say the hands or any other part of a worker while he/she is performing a specific task. In cyclegraph system, little pea light bulbs are attached to the hands or fingers of the worker. When the worker moves his hands to perform a task, these bulbs trace continuous lines of movements on the photographic plate exposed to the view. A cyclegraph is able to show all the three aspects, i.e., workers, workplace layout and the path of movements; however, it fails to indicate about the direction and speed of movement.

Chrono-cyclegraph: A cyclegraph indicates neither the direction nor the speed of the movements. Therefore, in order to overcome this difficulty *chrono-cyclegraph* was introduced which has a system of interrupting the light source. This system, instead of recording the paths of movements as continuous lines, records the same in the form of pear shaped spot on the photographic plate. The pointed end of the pear shaped spot shows the direction of the movements and the elongated shape and spacing of these spots indicates whether the speed of working is increasing or decreasing. More elongated spots with longer gaps between them indicate higher speed of movement while working. The speed of movement can be simply determined by knowing pace at which the light source gets interrupted.

Advantages of cyclegraph and chrono-cyclegraph:

(i) They can record all sorts of complex and unrestricted patterns of movements which are not possible to trace otherwise.
(ii) They can capture most of the patterns of the movements.
(iii) They are helpful in training and evaluating the workers.
(iv) They can trace very fast motions.
(v) They are not very costly for the information they provide.

Limitations:

(i) Light source tied to the hands of the workers may not make him feel convenient while working.
(ii) It needs sufficient photographic practice to achieve good results.
(iii) Because of their specialized nature these techniques are not very common.

3.8. Principles of Motion Economy

In order to develop a better method, a set of rules based upon research was designed and formulated by Frank Gilbreth and finally were reorganized and improved by Barnes, Lowry, Maynard and others. A better method of doing a job is that which consumes minimum of time and energy for limbs motion in order to complete the task, i.e., by economizing the use of motion and efforts. It is noticeable that in industries, mostly in mass production, a large number of various motions are repeated by the workers. Hence, it is very essential to study the basic pattern of these motions with a view to develop certain principles that could be adopted to make these motions harmonious, easier, economical and effective while performing the tasks. There are six basic principles of motion economy which are as follows:

1. Principle of minimum movement: According to this principle, the objects, tools, control levers and knobs, etc., should be arranged and located in such a manner that the movement on the part of the worker while using them is minimized. It also indicates that the limbs movements should be as less as possible.
2. Principles of simultaneous and symmetrical movements: This states that there should be proper balance on the movements of both the arms and hands during the performance of work, i.e., the movements of both the hands should be simultaneous and symmetrical. Refer to the case of assembly of a nut and bolt as shown in Figures 3.19 and 3.20 in which proper balancing of motions in both the hands are achieved.
3. Principles of rhythmic movements: This principle states that when the job is of repetitive type, the movements of the operator should follow same pattern each time. This will enable him to achieve the efficiency and preciseness.

4. Principle of natural movement: The movement of limbs is more natural if they move in symmetrical and opposite directions. The movements of both the hands should be such that they should begin and end at the same time.

5. Principle of habitual movement: The positioning of the objects and tools in the workplace lay out should be such that, the sequence of movement follow a natural, habit forming pattern. There should be full scope for an operator to gain swiftness of working and developing a tempo in it.

6. Principle of smooth movement: There should not be any sudden and sharp changes in the movements, i.e., movements should be curved and continuous. Smooth habit forming movements would reduce faulty and slipping motions.

According to Barrens, the various rules under principle of motion economy are as follows:

A. *Rules concerning human body*
B. *Rules concerning workplace layout and material handling*
C. *Rules concerning tools and equipment design*
D. *Rules concerning time conservation*

A. *Rules concerning human body*
   (a) Both the hands should be engaged in productive work.
   (b) Both the hands should begin and terminate their motion at the same time.
   (c) Motion of hands or arms should be symmetrical, simultaneous and in opposite direction.
   (d) Attempt should always be made for smooth, continuous and curved motions. Frequent and sharp directional changes should be avoided. If possible work movement should be rhythmical and automatic.
   (e) Motion should be simple and involve minimum number of limbs with the purpose to perform the work in the shortest duration of time with minimum of fatigue.
   (f) Wherever desirable, a worker should apply momentum to assist himself, as the same (momentum) is not possible to be achieved by his own muscular efforts. A worker may use mechanical aids to assist him to overcome muscular efforts.
   (g) Wherever feasible, ballistic movements should be preferred over control ones because they are easy, fast and more precise, for example; striking hammer to drive a nail into a wall.

B. *Rules concerning workplace layout and material handling*
   (a) There should be definite, fixed and easily accessible locations for material and tools.
   (b) As far as possible materials, tools and other mechanical device should be kept close to the workplace.
   (c) Gravity should preferably be employed for delivering materials at the workplace.
   (d) An assembled or final product should preferably be dropped on a conveyer (or chute) near the work place so that gravity not the operator delivers the jobs at the required place; hands should normally not be employed for non-productive work.
   (e) Tools and materials should preferably be located in the order or sequence in which they will be required for use. It reduces mental strain (on the operator) and the process becomes less mechanical.
   (f) Good illumination is necessary for proper seeing, fast operating and reducing accidents.
   (g) In order to impart rest to some of his limbs, an operator may sometimes sit or stand while working. This requires a relationship between his chair height and height of the table or work place.

(h) In order to reduce fatigue, the sitting arrangement of the worker should be comfortable and adjustable.

(i) All heavy parts should be lifted by mechanical devices.

C. *Rules concerning tools and equipment design*

(a) Jigs fixtures and foot operated devices should be employed to reduce the work load on the hands.

(b) Whenever possible, those tools should be used which can perform more than one operation. This saves a lot of time otherwise wasted in searching and picking the number of tools one after another. A tool shown in Figure 3.16 can carry out more than one operation.

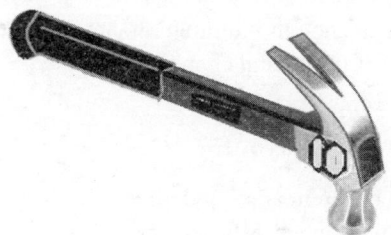

Fig. 3.16: A tool (claw hammer) can carry out two operation namely
(i) hammering, (ii) drawing outs nails.

(c) Preferably, tools and materials should be preplaced and located near the workplace. It saves the time that is otherwise wasted in searching and bringing the tools for doing a job.

(d) There should be maximum surface contact between the tool and the hand. It ensures appropriate grip and also helps proper application of the hand force and minimizes fatigue.

(e) Where the work is carried out by finger the load distribution on each finger should be as per the normal capacity of the finger.

D. *Rules concerning with time conservation*

(a) Stopping of the work even at temporary level by a man or machine should be discouraged.

(b) A machine should not run idle, it is not desirable that the job is rotating but no cut is being taken.

(c) Two or more jobs should be worked upon at the same time or two or more operations should be carried out on a job simultaneously. For example, drill and turning tool may simultaneously operate on a part being manufactured on a turret lathe.

(d) Number of motion involved in completing the job should be minimized.

3.8.1. Design considerations for workplace layout

(i) Materials and tools should be available at their predetermined places and closed to the workers. These should preferably be placed in the order in which they will be used.

(ii) Wherever possible, gravity should be employed for the raw material to reach the operator and finished product to be delivered at its destination. However, it should not be too automatic to become monotonous and boring for the operator. Under such condition suitable breaks and rest periods should be provided.

(iii) The operator should have a comfortable posture and the height of the seat should position such that the work table is about 50 mm below the elbow level of the operators.

(iv) A worker should have his choice to sit and stand freely during the work i.e. it should be possible to work both while sitting or standing. Preferably, a flat foot rest should be provided for the sitting worker. Since sitting on a seat is relatively less tiring as compared to standing, the seat should have appropriate dimensions with respect to anthropometric sizes of individuals. The back of the seat should not restrict the arm movement that is essential during work.

(v) Worker should be able to operate levers and handles by changing body position.

(vi) The workplace should have enough illumination, appropriate work condition w.r.t. noise temperature, humidity, dust fumes and chemicals etc.

3.8.2. Recommended Workplace Layout

A typical workplace layout with different areas and dimension is exhibited in Figure 3.17 and Table 3.8, which shows three areas of working as follows:

(i) Actual working areas: it is most convent area for working.

(ii) Normal working area: it is within the easy reach of the operator.

(iii) Maximum working area: it is accessible with full arm stretch.

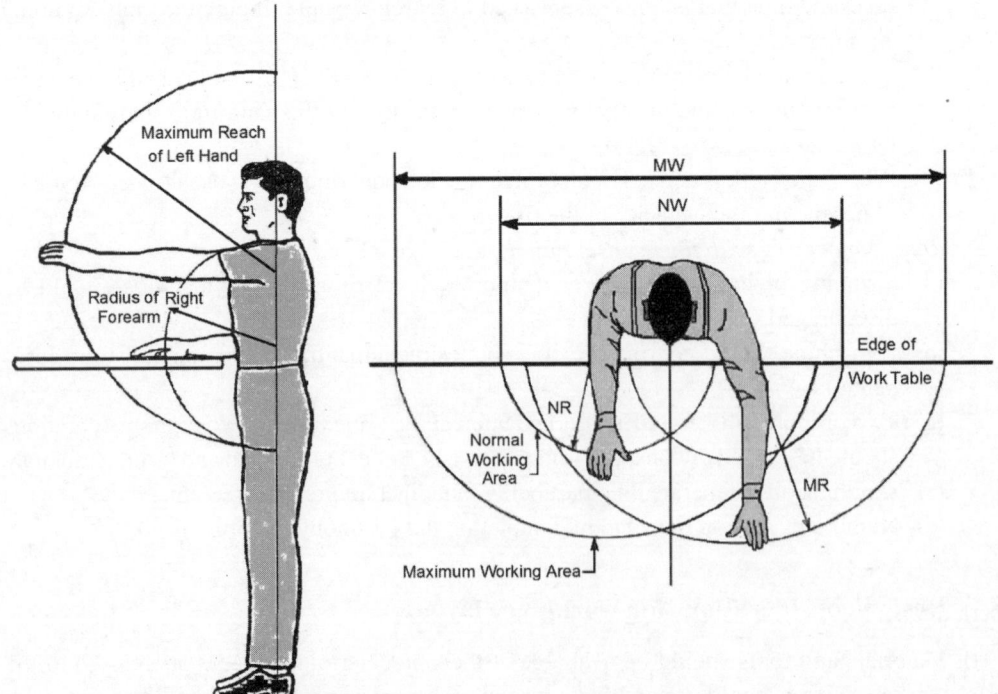

Fig. 3.17: Normal and maximum working areas in the workplace design for average user.*

Table 3.8: Normal and maximum working area dimensions as shown in Figure 3.17*

Symbol	Dimension in Working Area for Worker Seated at workplace	Male Worker in cm (inches)	Female Worker in cm (inches)
NR	Normal radius of arm reach	39 (15.35)	36 (14.17)
MR	Maximum radius of arm reach	67 (26.38)	60 (23.62)
NW	Normal width of arm reach	109 (42.91)	102 (40.16)
MW	Maximum of arm reach	163 (64.17)	147 (57.87)

Source: Work Systems and the Methods, Measurement, and Management of Work by Mikell P. Groover, ISBN 0-13-140650-7.

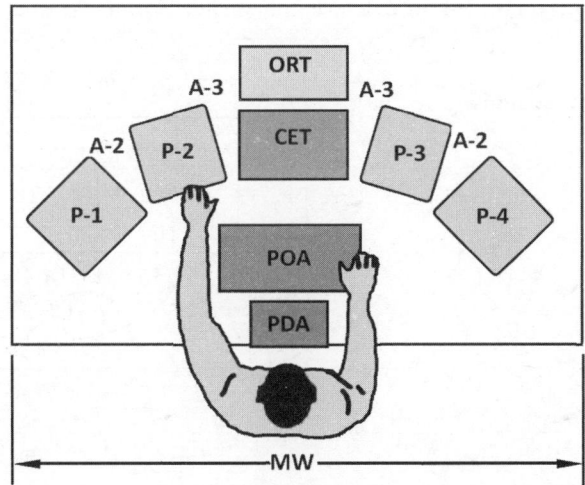

Fig. 3.18: Workplace layout for assembly of small components.

A work place layout for assembly of small component from four parts is exhibited in Figure 3.18. Areas mentioned as A-1 is the actual working area and the place of assembly (POA) where four components parts P-1, P-2, P-3, and P-4 are assembled together. Bins containing P-1, P-2, P-3, P-4 and commonly applied tools lie in the normal working area A-2. Occasionally required tools (ORT) like a hammers, etc., lie in the maximum working area A-3. After the assembly has been made at POA, it is dropped through work table at place for dropping assembly (PDA). From this area, the assembly is delivered at its destination with the help of conveyor. This workplace arrangement satisfies most of the principle of motion economy.

3.9. Operation Analysis

Operation analysis is defined as the detailed study of different operations involved in doing work. Operation analysis is necessary in order to investigate the shortcomings of the existing method and to develop the improved procedure. Operation analysis suggests whether some element should be eliminated or combined, their sequence should be altered in order to obtain effective utilization of existing man power and machinery with the minimum fatigue acquired by the workers. The analysis primarily considers the movement of limbs and intends to find a similar and economical way of doing the job.

Before the procedural steps of a task are analyzed and the motions (of an operator) are studied or eliminated, an operation chart is constructed. An operation chart of the existing method of assembly nuts and bolts is shown in Figure 3.19. As a next step the different motions involved are subjected to specific and detailed questioning with a view to eliminate unnecessary motion and to arrange the remaining motion in a better sequence. Principle of motion economy as discussed under previous section offer a very good guide in developing a better method. The chart shown in Figure 3.19 for the existing method is tested as per the rule of motion economy and the following points are observed.

(i) Distribution of work between two hands is not balanced, as the right hand is overloaded

(ii) The two hands do not follow opposite motion

(iii) Gravity has not been utilized for delivering the material to its destination, etc.

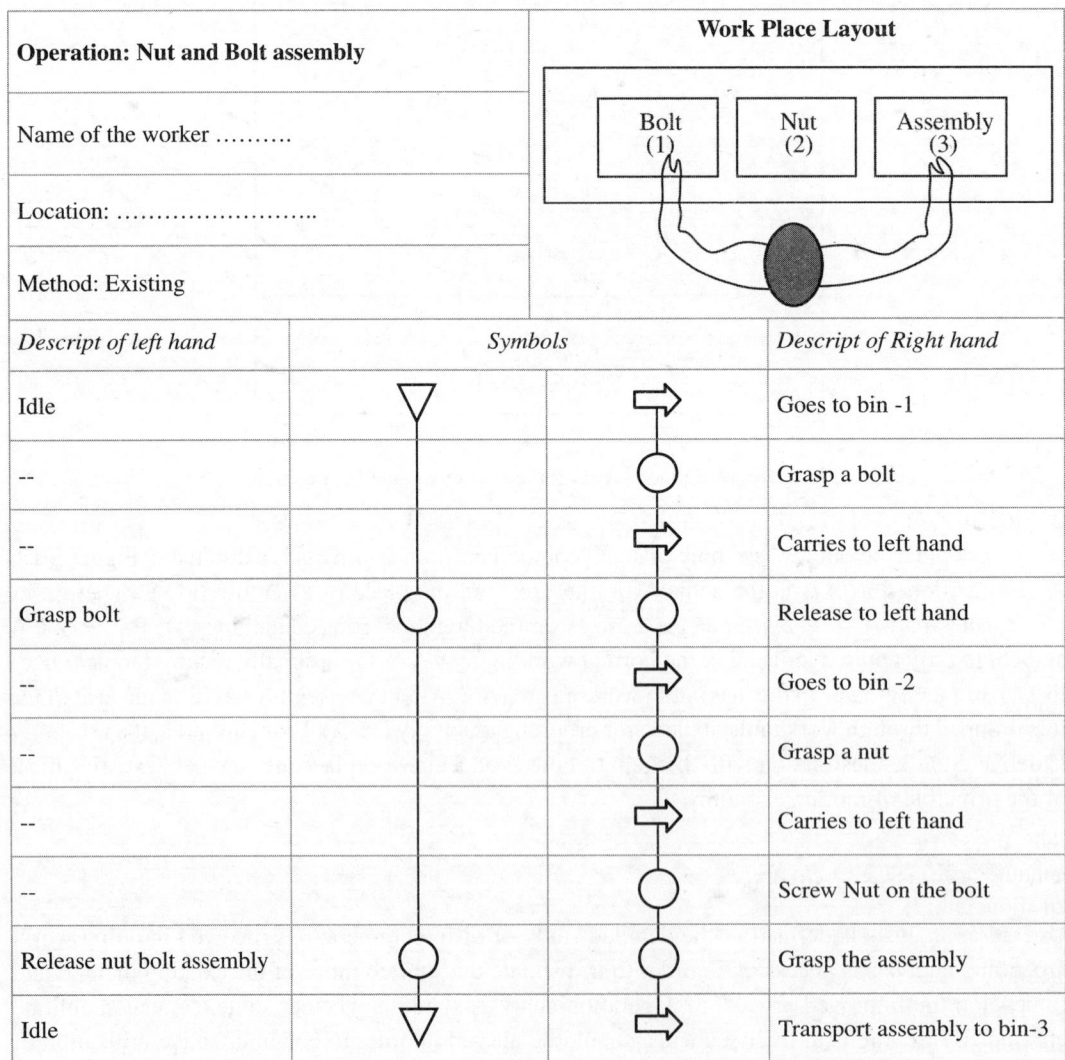

Fig. 3.19: Operation chart for the assembly of nuts and bolts by use of existing method.

Besides assisting the present method as per the principle of motion economy, it is also subjected to following question (whichever are applicable) with regard to:

(a) Questions with respect to worker
   (i) Is he mentally and physically fit?
   (ii) Does he incur unnecessary fatigue?
   (iii) Does he need training to improve?
   (iv) Does he get suitable salary?
(b) Questions with respect to set up
   (i) Are tools and other equipments readily available?
   (ii) Can the setup be modified or can the numbers of setup be decreased?
(c) Questions with respect to material
   (i) Is material of proper specification, i.e., composition, diameter, width, thickness, weight?
   (ii) Can it be substituted by a cheaper material?
   (iii) Can scrap be minimized?
(d) Questions with respect to material handling
   (i) Can material be transferred in big lots, thereby reducing the numbers of handling?
   (ii) Is it possible to avoid backtracking of the material?
   (iii) Can the distances, by which the material is moved, be cut short?
(e) Questions with respect to operations
   (iv) Can some operations be eliminated?
   (v) Can some operations be made automatic?
   (vi) What will be the effect of re-sequencing of the operation?
   (vii) Is it possible to combine some operations
(f) Questions with respect to tools and fixtures
   (i) Are they available in good condition?
   (ii) Are they suitably located in pre-position?
   (iii) Is it advantageous to modify existing jigs and fixtures for better productivity?
(g) Working condition
   (i) Are light and ventilation adequate?
   (ii) Are the operation and working condition safe?
   (iii) Are facilities of washroom, etc., adequate?

Considering the existing method in the light of motion economy principals and questions mentioned above a proposed method for the same task is given in Figure 3.20.

3.9.1. *Motion analysis*

The motion for movements of the limbs of a workers plays a major part in the fabrication or manufacture of the products by carefully observing a worker while he is doing an operation, a number of movements made by him, which appear to be unnecessary and unproductive, can be identified and eliminated. Analysis of operation when carried out in terms of individual's motion of a worker is known as motion analysis.

The purpose of motion analysis is to design an improved method which eliminates unnecessary motion and employs human efforts more productively in doing so the principle of motion economy proof to be very helpful. There are following steps involved in the motion analysis:

(a) Select the operation to be studied.

(b) List and chart various motion performed by the operators.

(c) Identify the productive and ideal motion.

(d) Eliminate the unnecessary and non-productive motion.

(e) Redesign the existing operating procedure by employing minimum number of motions in the most appropriate sequence and in accordance with the principle of motion economy.

(f) Impart necessary instruction to the workers so that they can develop good habit of cycles.

(g) Recheck the procedure in the light of step (e) above and the same may be standardized.

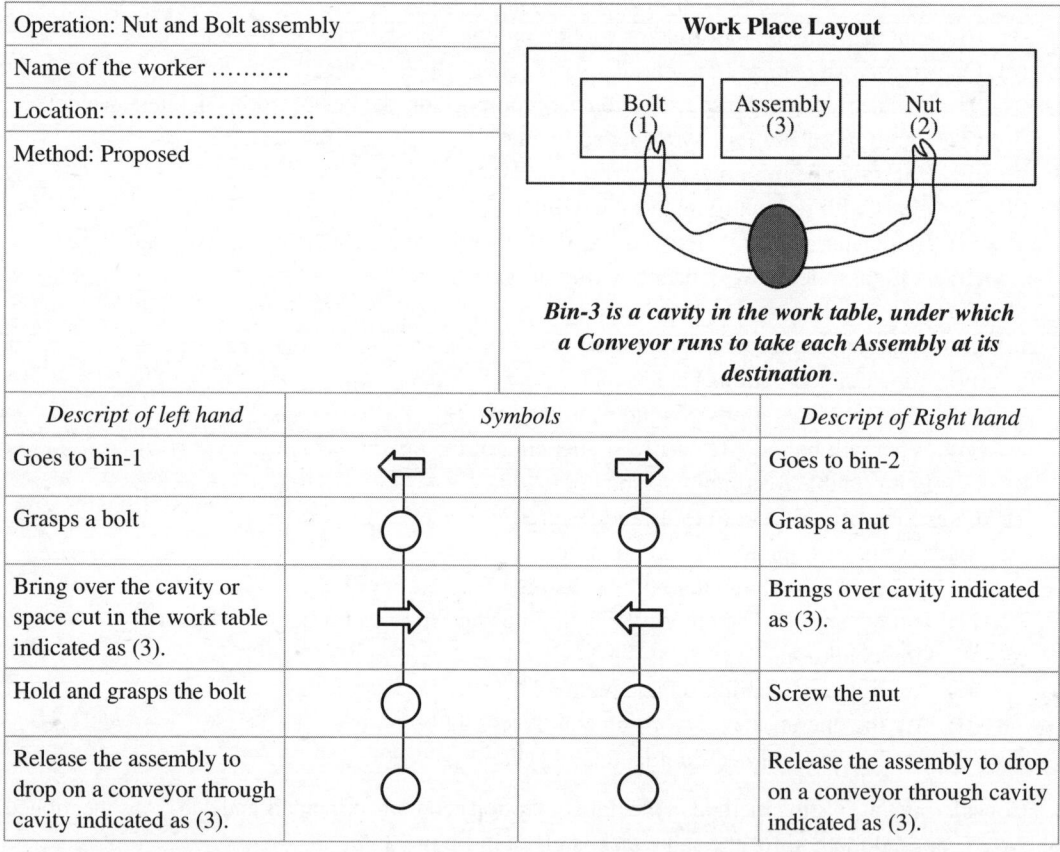

Fig. 3.20: Operation chart for the assembly of nuts and bolts by use of improved method.

3.10. Therbligs

Therbligs were suggested by Gilbreth. *As part of motion analysis, Gilbreths concluded that all work, whether productive or non-productive, is done by using combinations of 17 basic motions that they call therbligs (Gilbreth spelled backward).*

Therbligs are used to describe the basic elements of movements or fundamentals hands motion of the work cycle. Every therbligs is represented by a symbol; a definite colour with a word or two record the same. For example, therbligs grasp has symbols, red colour and is denoted by the word g. A SIMO chart employs therbligs which are of microscopic nature, whereas process chart uses symbol like operation, inspection, transportation, etc. which are macroscopic. A single operation may consist of many therbligs; for example macroscopic motion: operation of picking away a screw driver.

Microscopic motion

   (i) Reach hand for screw driver (transport empty)

  (ii) Grasp the same (grasp)

 (iii) Take away the screw driver (transport loaded)

Though it looks cumbersome to deal with a hand to chart microscopic motion, yet they possess certain advantages over macroscopic motion.

   (i) As explained above one macroscopic motion may contain a number of microscopic motion at times, it may not be possible to eliminate completely a micro motion, but an unnecessary micro motion be definitely avoided.

  (ii) Since micro system is very detailed, it is simpler to understand what precisely workers are doing.

 (iii) Therbligs colours may make the chart more meaningful.

Various therbligs along with their definition symbols and colours are given below in Tables 3.9 to 3.11.

Table 3.9: Gilbreth's seventeen therbligs.

Sr. No.	Therbligs	Symbol	Definition
1.	Transport empty	(TE)	Reach for an object with an empty hand before grasp
2.	Grasp	(G)	Grasp an object
3.	Transport loaded	(TL)	Move an object with hand and arm
4.	Hold	(H)	Hold an object
5.	Release load	(RL)	Release control of an object
6.	Use	(U)	Manipulate a tool
7.	Pre-position	(PP)	Position object for next operation
8.	Position	(P)	Position object in defined location
9.	Assemble	(A)	Join two parts
10.	Disassemble	(DA)	Separate multiple parts that were previously joined
11.	Search	(Sh)	Attempt to find an object using eyes or hand
12.	Select	(St)	Choose among several objects in a group
13.	Plan	(Pn)	Deciding on a course of action
14.	Inspect	(I)	Determine quality of object
15.	Un-avoidable delay	(UD)	Waiting due to factors beyond worker control
16.	Avoidable delay	(AD)	Worker waiting which can be avoided
17.	Rest	R	Worker is resting to overcome fatigue

Table 3.10: Classification of effective and ineffective therbligs.

Sr. No.	Effective therbligs	Symbol	Definition
1.	Transport empty	(TE)	Reach for an object with an empty hand before grasp
2.	Grasp	(G)	Grasp an object
3.	Transport loaded	(TL)	Move an object with hand and arm
4.	Release load	(RL)	Release control of an object
5.	Pre-position	(PP)	Position object for next operation
6.	Use	(U)	Manipulate a tool
7.	Assemble	(A)	Join two parts
8.	Disassemble	(DA)	Separate multiple parts that were previously joined
	In-effective therbligs		
9.	Un-avoidable delay	(UD)	Waiting due to factors beyond worker control
10.	Avoidable delay	(AD)	Worker waiting which can be avoided
11.	Hold	(H)	Hold an object
12.	Inspect	(I)	Determine quality of object
13.	Position	(P)	Position object in defined location
14.	Search	(Sh)	Attempt to find an object using eyes or hand
15.	Select	(St)	Choose among several objects in a group
16.	Plan	(Pn)	Deciding on a course of action
17.	Rest	R	Worker is resting to overcome fatigue

The basic elements of movement are fundamental hand motion of cycle. Every therbligs is presented by a definite single, colour, symbol and a word.

Table 3.11: Therbligs with symbols, word code and interpretation.

Sr. No.	Function	Symbol	Word Code	Colour	Interpretation
1.	Assembly	#	**A**	Violet	Joining of two or more components together by means of fasteners, e.g. assembly of car wheel with the hub using nuts and bolts.
2.	De assembly	#	**DA**	Light violet	Separating the two or more components, e.g. dis assembly of nut, bolt and washer.
3.	Avoidable delay	∠O	**AD**	Lemon yellow	Delay which is within the control of operator, e.g. any unwanted talk with colleagues, pause, etc.

Contd.

Contd.

Sr. No.	Function	Symbol	Word Code	Colour	Interpretation
4.	Unavoidable delay		**UD**	Yellow	Delay which is out of operator's control, e.g. error in machine, power cut, etc.
5.	Transport loaded		**TL**	Green	Movement of components using limbs, hands, etc.
6.	Transport empty		**TE**	Olive green	Movement of empty limbs; hands, e.g. reaching the hand towards a work piece or tool.
7.	Search		**SH**	Black	Searching for a tool, device, etc.
8.	Planning		**PN**	Brown	Thinking or mind prepared before starting an activity or taking an action.
9.	Rest		**R**	Orange	Any allowable idleness or pause for reviving from fatigue, i.e., rest allowances.
10.	Position		**P**	Blue	Changing position or orientation of the components or tools or jogs, etc. for assuring the proper alignment.
11.	Inspect		**I**	Burnt ochre	Checking or examining the component or tool, etc. for quality.
12.	Preposition		**PP**	Pale blue	Locating the components in a predetermined positions and place so as to make the ready and easy for use.
13.	Grasp		**G**	Red	To take hold of an object such as tool, component, etc.
14.	Use		**U**	Purple	A tool is being used or operated or manipulated to do its intended function.
15.	Hold		**H**	Gold ochre	After the grasp, retention or continuation of hold on a tool or component.
16.	Select		**S**	Light grey	Choosing one object in the middle of many.
17.	Release load		**RL**	Carmine red	Releasing the work piece, when operation or processing is over.

3.10.1. Micro-motion study

Micro motion study is concerned with the analysis of micro motions, known as therbligs. Each human activity is supposed to get divided or split into these small movements. The purpose of such a study is to find out the best pattern of movement, which involves less efforts, time and fatigue to accomplish a task for an operator. In order to arrive at the best pattern of movement, all useless motion are eliminated; remaining motions are re-sequenced and a better co-ordination of limbs' motions is obtained.

Repetitive very short operation involving quick limbs motion cannot be accurately studied and timed using operation analysis and hence cannot be improved. Under such circumstances, micro motion study is relatively more helpful. Micro-motion study involves taking motion pictures or (videography) of an activity while being performed by an operator. In this, a timing device is kept in the field of view. Later on, the film is studied by running it slowly through a projector and a frame by frame analysis is carried out.

Procedural steps involved in micro-motion study

Micro-motion study involves the following three steps:
- (a) Video recording or filming of the job being performed,
- (b) Frame by frame analysis of the film and
- (c) Graphical presentation of the data in the form of SIMO chart as a result obtained.

(a) Video recording or filming

The usual filming speed is 16 frames per second or 1000 frames per minute with a 16 mm movie camera. The general range of speed is from 1 to 64 frames per second. However, the modern digital movie cameras are used at 30 frames per second. Lower speeds are preferred for filming longperiods or crew activities whereas short cycles and very rapid hand motion are best studied at higher speeds. Normally a 30 m film capacity gives a 4 minutes run at 1000 frames per minute.

The equipment, required for filming the operation includes the following items:

- (i) A 16 mm movie camera with a provision to make film at more than one speed. The camera may be cranked, spring driven or electrically driven. Hand cranked cameras do not give uniform film speed; film speed in spring driven camera also changes as the spring unwound or loses its tension. Electric driven camera with governors is relatively better; however, electronic camera with digital video recording is much accurate. The camera should have a turret head with a number of lenses for zoom for shooting under normal condition and for taking close up views.
- (ii) A 16 mm film is normally used for recording both sound and video. A fine grain, fast black and white film usually employed. However, coloured films are more used for special purpose and which needs extra flood lighting.
- (iii) A mirco-chronometer or wink counter, which is a timing device,is kept in the field of sight while filming. A mirco-chronometer developed by Gilberth resembles a clock with its dial divided into 100 divisions. The time can be read in winks. The small hand of the clock makes 1 revolution for every 30 seconds and the long hand makes 20 revolution/minute. Wink counter as developed by Porter, is a digital timing device and thus it reads quicker

and more accurately as compared to micro-chronometer. However, in the current digital cameras timing is also recorded, which is more advantageous over the old technologies.

(iv) Exposure meter attached with the camera determines exposure accurately in addition flood lights and reflectors are also required.

(b) Frame by frame film analysis

Once the activity has been recorded in the form of video film and the film is processed, it is reviewed with the help of projector for analysis purposes. The projector runs the film very slowly and the film can be stopped or even reversed whenever desired. A frame counter counts the frames as the film is being projected. The film analysis includes the following steps:

(i) Run the film at normal speed to familiarize with the pattern of movements involved.

(ii) Select a work cycle representing the existing method for study. All repetitive work cycles should have the same beginning and the end point, thus they could be easily identified. The duration of the cycle can be found from micro-chornometreor wink counter readings.

(iii) Each frame of the work cycle is analyzed with respect to changes in limb movements. The start of each therbligs is marked. As a therbligs changes from one to another the clock reading is noted and along with it the therbligs is entered in micro-motion film analysis sheet as shown in Figure 3.21. All such changes of therbligs of right and left hand are examined and recorded either simultaneously or one after the other.

Micro-motion							
Part name:_____ Film No:_____ Operation No:_____ Operator's Name:_____				Date:____/____/___ Analyst's name:_____ Record No:_____			
Clock reading (start - end)	*Net time*	*Left hand description*	*Therbligs*	*Clock reading (start - end)*	*Net time*	*Therbligs*	*Right hand description*
60 - 65	5	To lever	TE	60 - 67	7	TL	To machine table
65 - 73	8	Lever	G	67 - 75	8	RL	Letting go the piece
And so on…	--	--	--	And so on…	--	--	--

Fig. 3.21: Micro-motion film analysis record sheet.

(d) Constructing the SIMO (Simultaneous–Motion–Cycle) chart

Once the film analysis is over and all the pertinent information has been collected in the form of film analysis sheet, the next step is to prepare a SIMO chart. SIMO chart was devised by Gilberth and it presents graphically the separable steps of each pertinent limb of the operator under study. It is an extremely detailed left and right hand operation chart. A SIMO chart for a simple chiselling operation is exhibited in Figure 3.22. It shows the simultaneous minute movements performed by the two hands of an operator a common time scale. Besides hands, the movements of other limbs of an operator may also be recorded. The time scale is represented in winks (1/2000 of minute).

SIMO chart is normally used for micro motion analysis of a short cycle repetitive jobs, high order skills jobs and finds application in job components such as assembly, packaging, repetitive use of jigs, inspection, etc. A SIMO chart shows relationship between the different limbs of an operator, for example, at any instant it can be found what one hand is doing with respect to the other, in terms of therbligs. In addition to these relationships, a SIMO chart also records the duration of micro-motions.

SIMO CHART				
Operation:_____		Date:____/____/____		
Worker's Name:_____		Film No.:_____		
Component's Name:_____		Operation No:_____		
Method: Present ☐ **Proposed** ☐				
Description of left hand	*Symbol*	*Time (winks)*	*Symbol*	*Description of right hand*
Grasp chisel	G	0,10	G	Grasp hammer
To job	TL	20	TL	To job
Position	P	30,40	AD	Idle
Hold	H	50	U	Use
				(Signature)

Fig. 3.22: SIMO chart for a simple chiselling operation.

A SIMO chart is beneficial because it permits very accurate and detailed analysis. The work cycle from the film can be studied at ease, peacefully and away from the disturbing surroundings of the actual workplace. However, the technique is limited because of the cost involved for filming and analysis.

Micro-motion analysis:

It is the process of critically examining the SIMO chart. Micro-motion analysis involves the breakdown of an operation into basic elements, known as therbligs, and this represents much finer breakdown than in the operation chart.

Improving the method:

SIMO chart is critically examined in order to explore the possibilities of improving the existing work method. The procedural steps for improvement are as follows:

(i) The elements of the work cycle having non-productive therbligs, such as search plan, select, hold and positioned, are examined with a view to eliminate these basic elements as far as possible.

(ii) Attention is centred next, towards, productive therbligs, such as transport loaded, assemble and dissemble use, which are re-sequenced in order to reduce cycle time and the fatigue incurred by the workers.

(iii) Laws of motion economy help a lot improving the existing method of doing a job or an operation.

Advantages of motion study:

A micro-motion study includes many advantages, the main are enumerated as follows:

(a) Films keep a permanent record of motion study.
(b) Films can easily reveal the difference between the present method and the proposed method.
(c) Films can be shown to large workforce at the same time and at any desired speed.
(d) Operations can be timed more accurately and conveniently than be stop watch study.
(e) Fast, short and repetitive cycles can be timed only by this process.
(f) It helps in conducting accurate and detailed analysis of the existing method.

Application of micro-motion study:

Micro-motion study can be used in a number of ways, which include the following:

(a) It helps in improving especially short and highly repetitive work cycle method.
(b) It can be used as an aid in training the operators. The improved procedures can be projected as many times, at different speeds, as desired. Different motions involved in an operation can be seen clearly and mastered.
(c) It helps in the design and development process of special machinery.
(d) It can be used for collecting motion time data for synthetic time standards.
(e) It can be used for studying man, machine and group activities.
(f) It can be used as an accurate and positive time study technique, and for carrying out research in the field of work study.

Memo-motion study:

Memo-motion study was introduced by M. E. Mundel. It is a special type of micro-motion study in which the activities are recorded at a much slower speed; say 60 or 100 frames per minute. Like micro-motion study, it involves the following procedural steps:

(a) Identify and study the operations to be filmed.
(b) Record the video film of selected operation.
(c) Analyze the video film.
(d) Depending upon the type of activities, construct an appropriate chart. For example, a multiple activity chart may be used for plotting group activities.
(e) Improve the method by using principles of motion economy.

Memo-motion study has wide applications which are listed below:

(i) It can be used to study team work, crew activities or interrelated events.
(ii) It can be used to study jobs involving long cycles and irregular sequenced events.
(iii) It can be applied to study flow of the material.
(iv) It is applied to analyze material handling activities.
(v) It is very useful where frequency and output of operation is high.

   (vi) It is applied for collecting motion time data.

  (vii) It is used for carrying out research in work study.

 (viii) It is applied for improving existing methods of doing work.

   (ix) It is used as an aid in imparting training to the workers.

Advantages of memo-motion study:

   (a) It is more economical than micro-motion study as it consumes less film.

   (b) It can record long sequences of activities.

   (c) It is a positive method of studying motions and timing operations.

   (d) It involves less time in film analysis as compared to micro motion study.

3.11. Work Simplification

Work simplification is a proprietary, simplified version of method or motion study that an individual can apply to his own work. Although in general, it may be thought of as a participation motion study approach suitable to the education level of the foreman or workers and usually involving the use of only the simple pencil and paper techniques.

Normally, it has been felt that workers and their unions dislike work study application. Perhaps they feel that if work study is applied, they will have to work more and they will be paid less. Moreover, the workers did not consider themselves to be part and parcel of work study and they offered natural reluctance to this change. Since the work study was bit the idea of the workers, they abandoned it and this thing gave birth to a new term *work simplification*.

Work simplification gives training to workers and foremen how to use basic method study techniques. They use these techniques themselves without the help of any work study engineer, which makes methods improvement.

In order to carry out work simplification, the workers collect all facts about the work being done. Subsequently, they explore the questions such as) what and why, (ii) where and why, (iii) when and why and (iv) who and why combinations. Afterwards they check various alternatives such as (a) can something be eliminated, (b) can two or more things be combined or (c) will a change in place, sequence, etc., help improvement? This in fact is a continuous improvement program conducted by the workers themselves.

Unsolved Problems

  1. What is method study? Enlist the various objectives of method study.

  2. What do you mean by process analysis? Enlist the various tools of process analysis.

  3. What are the various symbols used for making process charts?

  4. Differentiate between the functions of string diagram and travel chart.

  5. What are the various principles of motion economy?

  6. What is the use of multiple activity chart?

  7. A researcher is engaged in collecting physiological data of 3 subjects (A, B, C) in Ergonomics laboratory. The data w.r.t. setting time, recording time and unloading time is given in the table below. Make a multiple activity chart so that all the four workers are checked on each machine. Calculate the idle time for the researcher and 3 machines and waiting time for each subject.

Name of Test on Machine	HRV	BERA	BP
Setting Time (min)	5	7	2
M/C Running Time (min)	15+5 = 20	3+3 = 6	2
Un loading time (min)	3	3	1

8. Enlist the various principles of motion economy, a worker is engaged in assembly of fasteners; a bolt, two washers, one large washer and a nut. How these principles can be used to increase the efficiency of the worker? Analyze the job using therbligs.

Multiple Choice Questions

1. A management tool to achieve higher efficiency concerned with manual work is known as
   (a) Time study
   (b) Work measurement
   (c) Work study
   (d) Human study

2. The terms ILO stands for
   (a) International labour organization
   (b) Indian labour organization
   (c) International law organization
   (d) None of these

3. Work study is the technique of
   (a) Method study and work measurement
   (b) Method study only
   (c) Work measurement only
   (d) All of these

4. The father of scientific management was
   (a) Frank Gillberth
   (b) M. J. Griffin
   (c) Frederick W. Taylor
   (d) J. J. Thomson

5. The father of motion study was
   (a) Frank Gillberth
   (b) F. W. Taylor
   (c) M. J. Griffin
   (d) J. J. Thomson

6. Which of the following is not a objective of work study
   (a) Investigate and analyze the situation
   (b) To reduce rejection levels
   (c) Most efficient use of existing plant
   (d) To recommend & implement the improvements

7. Which of the following question is meant for re-sequencing the activities of the existing method? The purpose for which,
   a) What is done?
   b) Why is it done?
   c) When might it be done?
   d) How else might it be done?

8. Which of the following is an indicator of value adding activity?
   (a) Transport
   (b) Operation
   (c) Storage
   (d) Delay

9. When a number of workers work in a group or an individual operator handles two or more machines? The most appropriate chart is
   (a) Multiple activity chart
   (b) Flow process chart
   (c) Travel chart
   (d) Outline process chart

10. Which of the following is best for a situation of many repetitive paths on the shop floor?
    (a) Flow diagram
    (b) String diagram
    (c) Outline process chart
    (d) None of these

11. Gilbreth introduced a technique to record the very fast movements of limbs, which is known as
    (a) SIMO chart
    (b) Therbligs
    (c) Micromotion study
    (d) Chrono cyclegraph

12. When was the concept of scientific management came in origin?
    (a) 1800s
    (b) 1700s
    3. 1900s
    (d) 1950s

13. Who was the founder of mass production through assembly line?
    (a) Henry ford
    (b) Frank Gillberth
    (c) M. J. Griffin
    (d) Frederick W. Taylor

14. The use of study mechanism is for
    (a) Systematic recording
    (b) Critical analysis
    (c) Development
    (d) All above

15. Micro motion analysis is done by using a combination of 17 basic motions which are known as
    (a) Micro motions
    (b) Therbligs
    (c) Macro motions
    (d) None of above

16. According to the rules concerning to human body, there should be
    (a) Definite fixed and accessible place of tools
    (b) Both the hands should be engaged in the productive work
    (c) Gravity should be preferred to deliver the assembled parts
    (d) Multi purse tools should be used

Answers

1. (c)	2. (a)	3. (a)	4. (c)	5. (a)	6. (b)	7. (c)	8. (b)	9. (a)	10. (b)
11. (d)	12. (a)	13. (a)	14. (d)	15. (b)	16. (b)				

Work Measurement

4.1. Introduction to Work Measurement

It has already been described in the previous chapter that work study is mainly composed of method study and work measurement. Both these components are very closely linked with each other. Method study is mainly concerned about reducing the ineffective movements and time related to the job. Whereas work measurement is investigation of ineffective time related to a task, and subsequently establishment of time standards for the job under consideration. In broader terms '*work measurement can be defined as the application of ways or methods designed to determine the required (standard) time for a competent operator to perform a task at a distinct rate of operating*'.

Therefore, work measurement techniques are focused to find out the time required for completing an operation or an element by a qualified operator working at a standard pace and using the standard method. The time thus calculated is known as standard time. The method to do the job is normally standardized by using motion study procedure before carrying out work measurement or time study. The different commonly applied work measurement techniques are as follows:

(i) Stopwatch procedure for time study
(ii) Predetermined motion time system (PMTS)
(iii) Synthesis or synthesized time standard
(iv) Work sampling or activity sampling or ratio delay study

4.2. Stopwatch Time Study

Time study was developed by F. W. Taylor and since a stopwatch is used for making time observations; therefore it is called as stopwatch time study. It is a unique technique of work measurement as it involves direct observation of work while it is performed. Time study can be defined as follows:

'*Time study is a work measurement technique mainly comprising of recording the data with respect to time and rate of working for the elements of the specific job being carried out under the given condition and subsequently analyze the data in order to establish the basic time (normal time) and standard time required to carry out the job at a distinct level of performance*'.

The main objective of work measurement is to facilitate the management to establish the *time standards* for a particular job or its elements. The established standards can be used as a benchmark to compare the results achieved in future and exercise the necessary control. Direct time study is mainly useful for repetitive work, i.e., any job which is subsequently going to be repeated under the given circumstances and method established in method (motion) study.

The most important feature of time study is the approach of data recording, which surely affects accuracy of the results obtained. The precision results directly depend upon the number of observations taken, i.e., more the number of observations more accurate will be the results obtained from time study. Therefore, the accuracy of time study increases with the number of observation conducted to make the study.

4.3. Essentials for Time Study

The following are the four essential requirements for the time study of any job

  (i) An accurate specification of start and end of the job, and also of method of doing the operation, including details of materials equipment condition, etc.
 (ii) A system of recording the observed time taken by workers to do the job while under observation.
(iii) A clear concept of what is meant by standard rate of working.
(iv) The means of accessing the amount of rest and other allowances which should be associated with the job.

4.4. Time Study Procedure

  (i) Identify the job to be time studied and operation to be timed.
 (ii) Obtain the improved procedure of doing the job from the method study department.
(iii) Select a qualified worker for conducting the time study.
(iv) Take the worker as well as the shop supervisor into confidence and explain to them the objectives of the study.
 (v) Collect the equipment and arrange machinery, jigs, fixtures, etc., required to conduct time study and ensure the accuracy.
(vi) Explain to the worker about the improved working method/procedure and use of tools, jigs, fixtures and other attachment to the job.
(vii) Break the job into operation and operations into element and write them on the proper form, also separate constant element from variable elements.
(viii) Determine the number of observation to be timed for each element.
(ix) Conduct the observation (timing the elements) and record them on the time study form.
 (x) Also rate the performance of the worker during step (ix) above.
(xi) Repeat step (ix) and (x) for taking more number of required observations.
(xii) Compute observed time from the measure of central tendency or mean.
(xiii) Calculate normal time from observed time by using performance rating factor.
(xiv) Add appropriate allowances such as; rest/personal allowances, process and special allowances in normal time to obtain sustained time or allowed time, when policy allowance is added to allowed time, then it becomes standard time.

4.5. Applications of Time Study

Stopwatch time study is applicable to various situations in the industry, which are enlisted below:

  (i) Checking of time standard obtained by other method.
 (ii) Timing the repetitive operation employed in manufacturing of different jobs.

(iii) Where it is necessary to break down an activity in detail and analyze the same.

(iv) Determining the schedule and planning of work.

(v) Determining standard cost and as an aid in preparing budget.

(vi) Estimating the cost of a product before its manufacturing; such information is valuable in preparing bids.

(vii) Determining machine effectiveness and loading of worker, i.e., the number of machines which one person can operate; it also helps in balancing assembly lines.

(viii) Determining the time standard used as a basis for the payment of wage incentive to direct or indirect labour; it can be used as a basis for labour cost control.

4.6. Selecting the Job for Time Study

Whenever time study is to be conducted, the request to conduct stopwatch time study should preferably come from the concerned department and competent authority, such as the shop foreman; any one or more of the following reasons could be sufficient to justify the conduct of time study.

(i) If the job is newly created.

(ii) If it is required to make a change in material or method of working towards improvement and a new time standard is to be established.

(iii) If a complaint has been received from a worker or union about the unjust time standard for an operation.

(iv) In case of standard times are required prior to the introduction of incentive scheme.

(v) When cost of a particular job appears to be excessive.

(vi) When there is a necessity to compare the efficiency of two proposed methods.

(vii) To investigate the utilization of a section of plant having lower level of the output.

4.7. Selection of Worker for Time Study

The selection of a worker for the conduct of time study is a very critical decision, since the attitude affects the performance of the worker and consequently the standard time will be affected. It is also very important to take the concerned workers union and foreman into full confidence. Until and unless, they all are taken into confidence and briefed about the utility of work study, there are very less chances of success of a work study engineer. Hence, it is very important that the purpose of work study (method study/time study) is made clear to everyone, e.g., to improve the method of doing a job and the same will relief the worker from fatigue and unpleasant work. On the other side if the purpose of a time study is less obvious or is kept hidden and unless it is very carefully described to everyone concerned, its objective may possibly be completely misunderstood and defeated with consequential disturbances and even strike. Hence, the work study engineer should first get confidence of the workers representative and supervisor together and explain in simple terms what he is going to do and why. Their entire questions should be answered frankly and truthfully. It will be better to ask a foreman and workers' representatives to suggest the most suitable worker to be considered for conducting time study. Such a worker should be skilled, steady and competent enough to set a benchmark for comparing the performance of any worker in future. His rate of working should be average or slightly better than average. The worker with unsuitable temperament should be avoided in all conditions.

Usually in time study practice, a distinction is made between a representative worker and a qualified worker. *A representative worker is one whose skill and performance is the average of group under consideration.* He is not necessarily a qualified worker. *A qualified worker is one who is accepted as having the necessary physical attributes, who possesses the required intellect and education, and has acquired the necessary skill and knowledge to carry out the work in hand to satisfactory standards of safety, quantity and quality*, i.e., in time study, there is a valid reason for selecting a qualified worker. In setting time standard especially when they are used to fix up the piece rates and incentives, the standard to be aimed at is that which can be attained by the qualified worker and which can be maintained without causing him undue fatigue. The study on slow or very fast workers will tend to result in loose or tight standards; this will be unfair either towards the management or the workers and will probably be the subject of frequent complaints at a later stage. Once the proper worker has been selected, he should be carefully explained the purpose of the study. He should be asked to work at his usual pace; taking whatever rest he is accustomed to take. He should be invited to explain any difficulty which he may encounter with. In case a new/improved method has been installed, the worker must be provided sufficient time to settle down before he is timed. The work study engineer while timing a worker should remain at a place from where he can see everything the worker does without interfering with his movements or distracting his/her attention. On no account should any attempt be made to time the worker without his knowledge from a hidden position or with the watch in the pocket, as it is deceitful, and if noticed by the workers, will lead to repulsion and further problems such as protest and strike.

4.8. Time Study Equipment

In order to conduct time study certain equipments are essential. Basic time study kit consists of the following equipment:

- A stopwatch
- A time study board
- Time study forms
- Pencil, eraser, steel, rule, tachometer, micrometer, calculator, etc.

4.8.1. Stopwatch

A stopwatch is one of the principle timing devices employed for measuring the time taken by the operator to complete the job. It is also an accurate time-measuring equipment that can run continuously for one hour or half an hour normally and records time by its small hand. One revolution of the big hand of which records one minute, and the scale covering one minute may be calibrated in intervals of $1/100^{th}$ of a minute or $1/300^{th}$ of a minute. The different types of stop watch are as follows:

(i) Non-fly back
(ii) Fly back
(iii) Split hand or split second type

4.8.1.1. *Non-fly back stopwatch*

Non-fly back stopwatch is exhibited in Figure 4.1. It is mainly preferred for continuous timing. With first pressing of the winding knob the watch starts and long hand begins moving. If winding knob is pressed again, the long hand pauses and with third pressing, hands come to initial position. In the case of timing two parts, where the second part happens right after the first, the non-fly back system does not work well because it involves stopping the watch at the end of the first part, pressing the knob to bring hands back to zero, and again pressing the knob to start the hands. This consumes quite some time and it leaves less margin for timing the second element. Hence the same cannot be timed accurately. Such type of cases requires the use of fly back or split hand type of stopwatch.

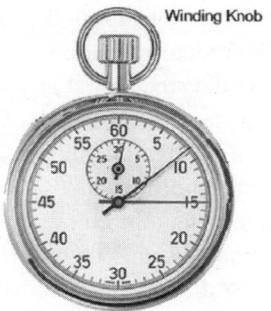

Fig. 4.1: Non-fly backstopwatch.

4.8.1.2. *Fly back stopwatch*

Non-fly back stopwatch is exhibited in Figure 4.2. In fly back system, slide is used to start and stop the watch. The hands come to zero by pressing the winding knob. However, they do not stop and begin straightaway moving forward again. The slide is employed to stop the hands at any point. This stopwatch is preferred for taking fly back timing or continuous timing observations and can easily give precise reading.

Fig. 4.2: Fly back stopwatch.

4.8.1.3. *Split hand type of watch*

This stop watch provides more accuracy in reading, particularly when two parts are to be timed and one right away follows the previous one. As one part ceases, pressing the winding knob makes one hand to stop whereas the other keeps moving. Once the reading has been taken, a second pressing on the knob stops and restarts the hand, then the two hands go along.

4.8.2. The motion picture camera and video equipment

A motion picture of an operation provides a permanent record of the method used as well as the time taken for each element of the operation. Moreover, it can be watched at any time and for a number of times if needed for purpose of analysis. When the film is projected at the exact speed at which it was recorded, it enables us to check the performance of an operator. In other words, the operator speed or tempo can be related to standard performance. The time taken for the elements of an operation can be obtained from motion pictures made with synchronous motor driven motion picture camera of known speed, or by placing a micro-chronometer in the picture when the operation is filmed. A micro-chronometer is a clock driven by a small synchronous motor. It has 100 divisions on the dial. The large hand revolves at 20 rpm and the small hand at 2 rpm each division on the dial indicate 1/2000 of a minute. The motion picture camera speed most frequently used is 1000 frames/minute that permits the measurement of time in 1000th of a minute. Whereas a normal handy cam or video camera/recorder operates at a steady speed of 30 frames/sec, the identification number indicating the scene, hour, min, sec, etc., appears at the top or bottom edge of the frame. The frame number begins with zero and each frame is numbered consecutively. Hence, it gives a direct and positive way of identifying each frame and measuring the time for an operation, a typical type of motion picture camera is exhibited in Figure 4.3.

Fig. 4.3: A motion picture camera.

4.8.3. Time study board

The time study board is simply a flat board, usually of plywood or of suitable plastic sheet or composite material, on which time study form is placed for recording the observations. A time study board has an appropriate fitting to hold the stopwatch so that the watch is in position to be read easily, and also enables the work study engineer to do his job with free hands. A strong spring clip is fitted at the centre of the upper edge of the board, to hold the time study form on which the study is recorded. The time study board should be of appropriate dimensions; otherwise either too short or too long board may be tiring to the time study engineer. A general purpose time study board is shown in Figure 4.4.

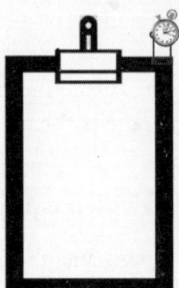

Fig. 4.4: Time study board used for making observations.

4.8.4. Proforma for time study

Although time study can be made on a plane paper but it is very awkward and time consuming to rule up a new sheet for every time study. Hence, it becomes more convenient to have a time study form of standard size in the printed format, so that they can be filled for reference, and of course it is an essential feature of well conducted time study. Printed forms are more useful because they ensure that the time study is always made in the standardized way and no essential data is lost. The time study forms illustrate a detailed formation about the operation being studied. It includes the detailed description of the operation, the name of the operator, the name of the time study observer, and the date and place of study. The form also provides space for recording stopwatch reading for each element of operation, performance rating of the worker and computation. It can also provide space for a sketch of the work place, a drawing of the part and specification of the material, jigs, gauges and tools. The form differs in size and design but a sheet 8" × 11" (215 mm × 275 mm) is widely used mainly because it fits the standard file or folder. A typical type of proforma for time study is exhibited in Figure 4.5.

TIME STUDY FORM

Product ....   Time Study Engineer ....

Operation No ....   Date ....

Operation Description ....

No. of cycles ____3____   Standard time found ....

Element description	1	2	3	--N	Average observed time	Rating factor	Normal time	Allowances	Standard time

(Signature)

Fig. 4.5: Proforma for time study.

4.9. Element Break-down in Time Study

An operation in a task may be very long and diverse. Therefore, it may become difficult to time an operation in one stretch and also may be insufficient to make an overall performance rating of the operator for the complete operation. This is for the reason that the rate of working of an operator not only may vary from one cycle to another of a job, but also varies frequently within the cycle. This requires that the job may be broken down into elements, so that an operation can be easily studied for timing and rating. Thus an element is a distinctive part of an operation selected for ease of observation, measurement and analysis. Consequently, a cycle is the sequence of the elements which is required to perform a job. A work cycle starts at the beginning of the first elements of the operation and continues to the same point in a repetition of the operation. Breaking down the elements is necessary, because of the following reasons:

  (i)   An element being a small entity makes it convenient to be observed, measured and analyzed.
  (i)   It enables the different types of element to be identified and separated.
 (ii)   It separates productive and unproductive activities, or effective and idle times.
(iii)   In order to get complete and accurate information.
 (iv)   In order to rate accurately operator's performance.
  (v)   In order to produce detailed work specification[*].
 (vi)   To select the best method by comparing the work elements of two or more given methods.
(vii)   To collect information to compile standard data.

4.9.1. Types of elements

In any organization, a number of operations takes place and each operation may consist of a number of elements. There are different types of elements which are described in the following section:

1. Repetitive element: It is the element that occurs in every work cycle of the job; e.g., picking up a part prior to an operation.
2. Occasional element: It is that element which does not occur in each work cycle of the job, but which may occur in regular or irregular intervals; e.g., cleaning the scrap from the machine tools.
3. Constant element: The reading of basic time remains consistent for this element whenever it is performed; e.g., switching on or off the machine.
4. Variable element: It is the element in which the basic time varies as per characteristic of the product, tool, or process, such as dimension, weight and quality for example, pushing a trolley of components to other section.
5. Manual element: It is an element performed by worker; e.g., manual sweeping of the floor or hand molding operation.
6. Machine element: A machine element is one, which is automatically performed by a power driven machine; e.g., turning, facing, cutting on an automatic machine tools.
7. Governing element: The element which takes longer than that of any other element which is being performed simultaneously; e.g., turning diameter on a lathe while gauging time to time.
8. A foreign element: The element is observed during a study which is not found to be a necessary part of the job after analysis; e.g., cleaning apart yet to be machined.

4.9.2. Rules concerning breaking down of the job into elements

At the time of analysis every operation is broken down into elements. Subsequently, each element is timed. Before breaking the job into elements certain rules should be considered, which are enumerated as follows:

(i)   Every element must be easily noticeable with definite start and completion points, so that once found they can be recognized again and again. The end point of an element can be noticed, as the operator lay the tool down. This is known as break point – the instant at which one element in a work cycle ends and another starts.

---

* A work specification is a document setting out the details of an operation or job – how it is to be performed, the layout of the workplace, details of machine, tools and appliances to be used and the duties and responsibility of the worker. The standard time or allowed time assigned to the job is usually included.

(ii) Elements should be short enough so that they can be conveniently timed by a trained observer.

(iii) Manual elements should be separated from machine elements. Manual time is normally completely within the control of the operator. Machine time with automatic feed and fixed speeds can be calculated and used as a check on the stopwatch data.

(iv) Constant elements should be differentiated from variable elements. Element which do not occur in every second should be timed separately from those that occurs.

The need for a precise break down of elements depends largely on a type of manufacturing the nature of the operation and the result desired. For example, punching operation in the hand tool manufacturing industries generally has short cycle operation with very short element. Elements should be checked through a number of cycles return down before timing begins.

4.10. Determination of Number of Observation

The time period needed to complete an element may vary slightly from cycle to cycle. Although the operator might have worked at an even pace, each element of successive cycles would not always be carried out in exactly the same time. The variation in time may occur due to factors like difference in the exact position or path of movement of the components and tools used by the operator, or may be from possible error in noting the particular point at which the particular reading is made.

This variation would be infinitesimally small when there is assured use of highly standardized raw materials, good tools, good working conditions and a qualified and well trained operator. But still there would be some variation. Since the time study is a sampling technique, the greater is the number of observations an element is timed; the more accurate will be the result, i.e., result will be very near to representative of the activity being observed. Also the shorter is the cycle for an element; the larger is the number of observation to be made for the desired level of accuracy. In general for an element with short cycle, say 3 sec, at least 500 observations of operation, and for relatively longer cycle, say 30 sec, at least 50-60 observations are required. However, these are only a rough estimate. Table 4.1 exhibits a guide to select the number of observations to be made for a particular cycle time.

Table 4.1: Number of cycles to be timed for a job of particular cycle time.

Cycle Time in Minutes	Recommended No. of Cycles
0.1	200
0.25	100
0.50	60
0.75	40
1.00	30
2.00	20
2 to 5	15
5 to 10	10
10 to 20	8
20 to 40	5
40 and above	3

The number of observations to be made for time study of a particular element with some cycle time can also be calculated statistically. Mathematical formulation based upon the statistical theory also provides a good channel to find out the number of observations to be taken. The main consideration in this formula is the desired precision and accuracy.

Using the characteristic of normal curve, the no. of observation or reading to be taken, N is given by

$$N = \frac{B^2}{A^2}\left(\frac{\sqrt{n\Sigma X^2 - (\Sigma X)^2}}{\Sigma X}\right)^2 \qquad \dots (4.1)$$

where, $B = 2$ for 95% confidence level

$\quad = 3$ for 99% confidence level

$\quad A = 0.05$, for $\pm 5\%$ desired precision and so on for other value of

$\quad N = \Sigma f$

$\quad \Sigma X = \Sigma fx$

$\quad \Sigma X^2 = \Sigma fx^2$

Example: The elemental time recorded in a time study sheet by snap back method for a single manual work element indicates the time in sec as shown in Table 4.2.

Table 4.2: Observations taken with snap back method.

Sr. No.	Element Time	Sr. No.	Element Time	Sr. No.	Element Time	Sr. No.	Element Time	Sr. No.	Element Time
1	13	11	14	21	15	31	18	41	12
2	12	12	17	22	14	32	16	42	13
3	18	13	19	23	20	33	14	43	14
4	15	14	13	24	14	34	13	44	15
5	16	15	15	25	15	35	19	45	12
6	17	16	17	26	20	36	14	46	13
7	14	17	17	27	18	37	13	47	14
8	16	18	19	28	17	38	14	48	15
9	18	19	14	29	19	39	17	49	16
10	15	20	17	30	18	40	12	50	20

Determine the number of observations, required precision of ± 5 per cent, with a confidence level of 95 per cent.

Solution:

The number of observations to be made, i.e., N for 95 per cent confidence level and at ± 5 per cent desired level of precession can be found by substituting the respective values in Eq. 4.1 as follows:

$$N = \left(\frac{(2)}{(0.05)}\frac{\sqrt{n\Sigma X^2 - (\Sigma X)^2}}{\Sigma X}\right)^2$$

Now substituting the values of n ΣX and ΣX^2

$$N = \left(\frac{(2)}{(0.05)} \times \frac{\sqrt{50 \times 12436 - (780)^2}}{780} \right)^2$$

$$N = \left(\frac{40 \times \sqrt{621800 - 608400}}{780} \right)^2$$

$$N = \left(\frac{40 \times 115.75}{780} \right)^2$$

$$N = 35.23$$

Table 4.3: Calculations of required parameters for calculation of number of observations.

Elemental Time (sec) x	Frequency (f)	f × x	f × x²
12	4	48	576
13	6	78	1014
14	10	140	1960
15	7	105	1575
16	4	64	1024
17	7	119	2023
18	5	90	1620
19	4	76	1444
20	3	60	1200
	$N = \Sigma f = 50$	$\Sigma fx = 780$	$\Sigma fx^2 = 12436$

Since the initial number of observations taken is 50, which is greater than the calculated value, i.e., 35, there is no need of further observations. Now, let us analyze a situation when the number of observations taken are 50 and the also frequency varies in the same trend; however, the cycle time is relatively lesser than that of the previous case. The frequency table with new cycle times is given in Table 4.4 as follows:

The number of observations can be calculated using Eq. (4.1)

$$N = \left(\frac{40 \times \sqrt{50 \times 6996 - (580)^2}}{580} \right)^2$$

$$N = \left(\frac{40 \times 115.75}{580} \right)^2 = 63.76 = 64 \text{ (approximately)}$$

Hence, more numbers of observations are required to achieve the same desired level of accuracy (± 5 per cent) at 95 per cent confidence level.

Now, if the given level of confidence is 99 per cent instead of 95 per cent at the same level of precision, i.e., ± 5 per cent, then the number of observations will increase as follows:

$$N = \left(\frac{60 \times 115.75}{580}\right)^2$$

$$= 143.004 = 143 \text{ (approximately)}$$

Whereas for original observations the number of readings will vary as follows:

$$N = \left(\frac{60 \times 115.75}{780}\right)^2$$

$$= 79.2 = 79$$

On the basis of above illustrations, it can be interpreted that lesser the cycle time more is the number of observations required to achieve the desired level of accuracy at a given level of confidence. If the level of confidence is increased from 95 per cent to 99 per cent, the required number of observations increases more than double.

Table 4.4: Calculations of required parameters when cycle time is relatively less as compared to the initial observations.

Elemental Time (sec) x	Frequency (f)	f × x	f× x²
8	4	32	256
9	6	54	486
10	10	100	1000
11	7	77	847
12	4	48	576
13	7	91	1183
14	5	70	980
15	4	60	900
16	3	48	768
	$N = \Sigma f = 50$	$\Sigma fx = 580$	$\Sigma fx^2 = 6996$

4.10.1. Conducting the observation (time study)

Once the number of readings to be made for the time study is calculated, the next step is to conduct the time study. In order to start the time study, suitable stopwatch, time study board and form, pencil, eraser, etc., are acquired by the time study officer. Then he should go to the job site and introduce himself to the worker, and must stand some distance away from the actual work place of the worker so that work is not hindered. The stopwatch and study board are positioned in line with the operation being studied, so that reading the stopwatch and recording the rating can be done while making the observation about the task. The most common methods of reading the stopwatch are as follows:

1. Continuous timing
2. Repetitive timing
3. Accumulative timing
4. Split hand watch timing

4.10.1.1. *Continuous timing*

In continuous timing method a non-fly back type of stopwatch is utilized. In this method the stopwatch is started at the beginning of first element and run continuously all through the period of the study. The reading of the watch is noted at the end of each element and the same is recorded on the observation sheet against its name or serial number. The time for each element is determined later by subtraction as illustrated in Table 4.5:

Table 4.5: Elemental times to be calculated by subtraction.

Element No.	Total Time Read at Any Instant	Actual Time for an element
1	25	25 − 0 = 25
2	45	45 − 25 = 20
3	75	75 − 45 = 30
4	100	100 − 75 = 25
5	125	125 − 100 = 25 and so on...

4.10.1.2. *Repetitive timing*

In the repetitive timing, also known as snap back method, a fly back type of stopwatch is used (Figure 4.2). At the beginning of first element the observer snaps the hand back to zero by pressing the winding knob of the watch. The hand moving forward immediately begins to measure the time for the first element. At the end of the first element, the observer reads the watch, snaps the hand back to zero and then records his reading. In the same manner the observer times the rest of the elements. This method of timing gives the direct time without subtraction and the data are recorded on the observation sheet as read from the watch. This system has one disadvantage that the observer has to read the stopwatch while its hands are moving

4.10.1.3. *Accumulative timing*

This method permits the direct reading of the time for each element by the use of two stopwatches. These watches are mounted close together on the time study board and are connected by a level mechanism in such a way that when the first watch is started, the second watch automatically stopped, and when a second watch is started the first is stopped. The watch may be snapped back to zero immediately after it is read, thus making subtraction unnecessary. The watch is read with greater ease and accuracy because its hand is not in motion at the time it is read.

4.11. Recording Observation on Time Study Form

When recording the stopwatch readings on the time study form, the observer notes only the necessary digits and skips the decimal point. Thus, in this way he is able to give as much time as possible to observe the performance of the operator. If a decimal minute watch is used, and the terminal point of the first element occurs at 0.08 minute, the observer would record only a digit 8 in the reading column of the time study form. If the decimal hour watch is used and the end point of the first element is 0.0065, the recording reading would be 65. The small hand on the watch will indicate the number of elapsed minutes and the observer can refer it periodically to verify the correct 1st digit to record after the large hands sweep past the zero. By balancing at the small hand of the stopwatch, if the observer notes that it had moved past the 5 (i.e.,5 minutes have passed in taken the given study), he will prefix the 5 and total reading may become 5.08, however the recorded reading will look like 508. All stopwatch readings are recorded in consecutive order on the reading column until the cycle is completed. If during measuring and recording, the observer fails to see a reading, then he should immediately indicate an 'M' in the reading column of a time study form. In no case should he make an approximation and try to record the missed value because this practice can destroy the validity of the standard established for the specific element.

4.12. Performance Rating

Rating and allowances are the two most debatable features of time study. Usually, in industry time studies are used to find out standard times for setting workloads and incentive plans. The procedure and policy of department have implications on the earnings of the workers as well as on the productivity and profits of the enterprise. Rating (assessed rate of working of a worker) and allowances are mainly provided for recovery from fatigue, and other purposes, these are still matters of judgement and bargain between management and labourer.

In actual meaning, rating is a comparison of the rate of working observed by the work study person with a perceived picture of some standard level in mind. This standard level is the average rate at which qualified workers will naturally work, using the correct method and fully motivated to do their work. This rate of working corresponds to the standard rating and is denoted by 100 on the rating scale recommended to readers of this book. If the standard pace is maintained and the appropriate relaxation is taken, a worker will achieve standard performance over the working day or shift.

ILO defined rating as "*an assessment of the effective speed of working of the operator relative to the observer's concept of the pace corresponding to standard rating*".

When time study engineer is conducting the study, he carefully observes the performance of the operator during the entire course of study; therefore,

- Performance rating is a method for fairly calculating the time required to do a task by a qualified/normal worker after the realistic value of the operation is recorded under study.
- Time studies should be made on a number of qualified average workers and very fast or very slow workers should be avoided.
- Different jobs require different human abilities. For example, some jobs demand mental awareness, concentration, visual perception, while others, physical strength; yet some others demand acquired skill or special knowledge.

4.13. The Qualified Worker

A qualified worker has acquired skill, knowledge and other attributes to perform work in hand to suitable standards of quantity, quality and safety.

An ordinary operator is defined as a qualified, thoroughly experienced operator working under condition as they customarily prevail at the work station, at a pace neither too fast nor too slow but representative of average. A representative is that worker whose skill and performance is the average of a group under consideration and not necessarily a qualified worker.

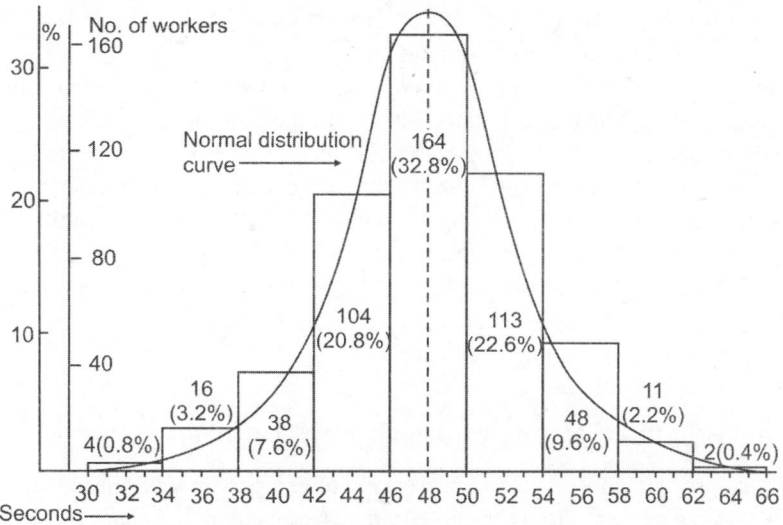

Fig. 4.6: Normal distribution of workers.

If 500 qualified workers in a given factory do the same operation by the same methods and under the same conditions, the whole operation is within the control of the workers themselves. The time taken to perform the operation would be normally distributed as shown in Figure 4.6. Once the qualified worker is selected, the next step is to make the necessary arrangements and engage the worker in the operation. The observer should make note of certain aspects of performance rating, these are described as follows.

1. Performance rating means gauging and comparing the pace rate or the performance of a worker against a standard performance level set by the time study engineer.
2. The performance rating factor is used to convert observed time into basic time.

$$\text{Basic Time} = \text{Observed Time} \times \text{Rating Factor or}$$

$$BT = OT \times RF$$

$$\text{Rating factor} = \frac{\text{Observed performance level of worker}}{\text{Standard performance level expected}}$$

3. The standard performance is a rate of output which qualified worker will achieve without overexertion, as an average over the working day, provided they know and adhere to the definite method and are motivated to apply themselves to their work. This performance is denoted as 100 on the standard rating and their performance scale.

4. A rating factor or the levelling factor is a factor by which the observed time is multiplied in order to adjust for differences in operator's performances. Performance rating becomes necessary in order to differentiate between the performances of two or more operators. Secondly, it can also be shown as day to day variation in the level of performance of the same worker.

4.13.1. Factors affecting workers' performance

Variation in actual time taken by a worker or group of workers to complete the same work element may be due to the two main types of factors; first, the factors within the control of the worker and second factors are outside the control of the worker.

The first type include variation due to worker's ability and variation due to his attitude of mind, especially towards the organizations. The factors within the control of the worker will also affect pattern of his movements and his working pace. The second type includes variation in the quality or other characteristics of the raw material used, changes in the operational efficiency of machinery and tools with the passage of time. It also includes changes in the working condition such as those of lights and temperature. These can generally be accounted for by taking a sufficient number of studies to ensure that a representative sample of times is obtained. The optimum pace at which the worker will work will depend on; the physical effort demanded by the work, the care required on the part of the worker and worker training and experience.

4.14. Comparison of Observed and Standard Ratings

It is not easy to compare the observed rate of working of an operator with the standard rating set. A work study analyst can achieve it only through long experience on different types of jobs. Assume that the job is a person has to walk from a point A to another distant point B. An analyst with almost no experience will say the person is walking slow, average, or fast. With a little practice/experience the analyst may say the person is walking at about 5 km/hour, about 8 km/hour or about 10 km/hour. With further experience the analyst may say that one person is walking at 5 km/hour and other at 8 km/hour. In order to achieve such accuracy, however, the analyst would need to have in his mind some particular rate with which to compare those which he observes.

As described above, the situation is exactly what the work study analyst does in rating, but, since the operation which he has to observe are far complex than the simple one of walking, therefore accuracy in rating can only be achieved through long experience and practice on many types of operations and obviously confidence is essential for a work study analyst.

4.14.1. Scales of rating

It is necessary to have a numerical scale if the comparison between the observed rate of working and the standard rate has to be made effectively. The rate can be used as a factor in which observed time can be multiplied to give the basic time, which a qualified worker would take, if he/she is motivated to carry out the element as standard rating.

There are various scales in use and well experienced time study analyst can obtain satisfactory result with any of them. The surveys revealed that the 0–100, standard rating scale is the most commonly used and it has been adopted as a British standard. In 0–100 scale, 0 represent no activity and 100 pertains to normal rate of working of the motivated qualified worker, i.e., is the standard rate. Therefore standard rate of performance is that, which a qualified worker will naturally achieve as an average without over exertion, through out the working day or shift. The standard rating that is usually accepted in UK and US refers to the speed of limbs movement of a man with average physique and experience walking along a straight line on a levelled floor or ground at a speed of 6.4 km/h (4 miles/hr). This is known as a brisk business like rate of walking (performance), however the person should be well accustomed to walking and able to maintain provided he take appropriate rest pauses every so often. The person walking at above cited speed is obviously moving some destination in his mind, hence he is motivated to do so. At the same time he is not in hurry unlike a person hurrying to catch a bus, because the later will walk at a significantly faster pace and not able to maintain for a longer period. It should be noted that the standard applicable to US and European population is not meant for the other Asian countries like India or so. This is because the work place conditions, physical strength, nourishment, level of training is better in UK, US and European countries as compared to the developing nations like India. This standard of walking is suitable for the outdoor conditions, however for the indoor activities, example of working at a standard rate if distribution of a pack of 52 cards in a square of one feet side in 22–23 seconds.

4.14.2. Rating factor

Referring to 0–100 standard rating scale explained above, in which 100 represents standard performance, if the work study analyst feels the operation being performed with less valuable speed than perception of standard, he may give a rating factor of less than 100 say 88 or 90 or if he feels that the effective rate of working is above the perceived standard level he may use a rating factor of 110 for the worker who is doing that operation. The rating factor is applied to observed time to calculate the normal time.

$$\text{Normal time} = \text{observed time} \times \frac{\text{Rating in percentge}}{100}$$

Assume that in a particular operation of assembling an electric switch the operator gave a consistent performance throughout the cycle and throughout the study, and that the total selected time was 55 seconds means with a rating factor for the study of 110 per cent, the normal time would be:

$$\text{Normal time} = 55 \times \frac{110}{100} = 60.5 \text{ sec}$$

This value of 60.5 sec represents the time that a qualified and well trained operator working at a normal pace would require to complete an activity or operation.

Table 4.6: Different types of rating scales.

Sr. No.	Description of Work Pace/Rate	Rating Scales		
1.	No activity i.e. operator is idle	0	0	0
2.	Very slow activity i.e. operator has no interest in the job	40	67	50
3.	Steady deliberate and unhurried or relaxed pace of the working	60	100	75
4.	Fast business/professional like performance with quality and accuracy	**80**	**133**	**100**
5.	Very fast pace with high degree of assertion, agility and coordination of movements	100	167	125
6.	Extremely fast, requires powerful efforts and concentration, cannot be sustained for longer period of time	120	200	150

4.15. Work Content

As discussed in the first chapter, the term 'work content' is the amount of work performed to complete a job or operation, excluding any ineffective time which may occur. However, in real practice of time study, the word work is understood as slightly different from its literal meaning. For example, an ordinary observer who might understand the literal meaning of word 'work' would consider in a manner, that when a worker is actually doing something he is working, and that when he is resting or doing nothing he is not working. But, in time study practice, the work is measured in numerical terms, and due to this reason the word 'work' is widened to incorporate the physical efforts along with the proper time for relaxation or rest necessary to recover from the fatigue caused by those workers. In subsequent sections, it is explained that relaxation allowances cover much higher than the revival from fatigue. Therefore, the term 'work' includes the appropriate relaxation allowance, so that the amount of work in a job also considers the extra time required for relaxation. Therefore, the work content of a job or operation can be defined as the sum of basic or normal time and relaxation/ other allowance. As the amount of allowances increases, the work content also increases; the same is described in the mathematical form as follows:

Work content (WC) = Basic Time (BT) + Relaxation Allowance (RA) + If any allowance provided for additional work like; contingency allowance (CA)

$$WC = BT + RA + CA$$

4.15.1. Allowances

It has been described in chapter 3 that during the method study examination, the emphasis is put on economizing the energy expenditure by the worker to perform a certain job, The same can be achieved through the development of improved methods, in harmony with the principles of motion economy, and also, wherever possible low cost automation and mechanization should be implemented. In spite of such improvements, still there is need of some energy expenditure in the form of human effort or the worker may need to spend some time for his personal needs like; going to toilet, drinking water, getting material or tools, or to get instructions from the supervisor of the shop floor. Hence in order to recover from fatigue, some rest allowance must be provided. Some allowance must also be provided to a worker to attend to his personal needs and alongside some contingency allowances may have to be added to the basic time in order to calculate the work content.

The establishment of allowances is perhaps the most typical part of work study, as it is very difficult to determine precisely the allowances needed for a given job. However, the attempt can be made to make an objective assessment of the allowances which can be constantly applied to a range of work elements or operations. It is very important to understand the fact that the calculation of allowances cannot be accurate under all the circumstances and also allowance cannot be used as a dumping ground for any of the missed or neglected factors during the time study. In order to arrive at fair and accurate time standards, the work study man should go to greater extent of thoroughness. Hence, these cannot be afforded to ruin by the hurried or unplanned addition of a few points here and there 'just in case'.

In practical conditions, it is very difficult to prepare a universally accepted set of precise allowances that can be applied to every working situation anywhere in the world, and there are number of factors responsible for the same; however some of the most important among them are as follows:

1. *Factors related to the individual worker*: In case every worker is to be considered individually in a particular working area, a thin, active, alert worker at the peak of physical

heath condition will require a smaller allowance to recover from fatigue than an obese, clumsy and weak worker. Likewise, each worker has its own learning curve which can affect the way in which he performs his job. There is also some reason to believe that there may be cultural/regional/racial variations in the response to the level of fatigue experienced by workers, mainly when engaged on heavy physical work load. Malnourished or weak workers usually acquire a longer time than others to recover from fatigue.

2. *Factors related to the nature of the work*: Most of the methods established for the calculations of allowances may be applicable to light and medium jobs, but simultaneously these may be found inadequate for very heavy and strenuous activities such as lifting and carrying of molten metal in a crucible from furnaces to the moulds in casting firms. Furthermore, each working situation has its own specific attributes, the level of fatigue felt by the worker which may lead to unavoidable delay in completion of job. For example, the working posture of a worker, whether he has to, sit or stand during the activity, exert force to carry loads, exposed to eye or mental strain, etc. In addition to this, there are some other in-built factors in the job, that also contribute to the need for allowances; e.g., use of protective clothing or equipment such as helmet and gloves, or when there is constant hazard, or high level of precision is required, without which there may be a risk of damaging the product.

3. *Factors related to the environment*: The development of a standardized method to meet allowances is quite difficult for every work situation affected by various work-place factors such as level of heat stress, intensity of light, level of noise, dust and fume exposure. Many of the work place environment factors are season-dependent. Now, the reader may be clear in his vision that the ILO has not adopted any allowances calculating standard. The determination of allowances under various working conditions has been mentioned in Figure 4.7. So, it is quite important in work study which must be taken as an issue of research to make some recommendations for calculating the allowances.

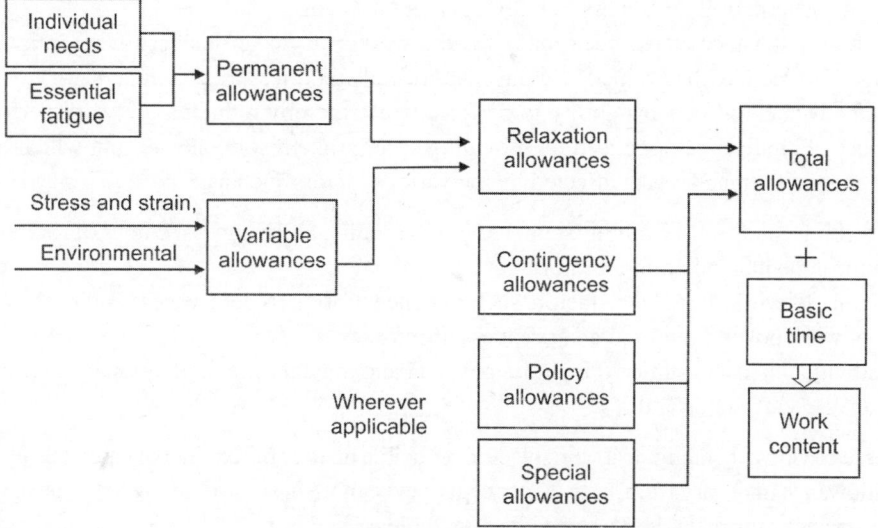

Fig. 4.7: Procedure for determination of allowances.

4.15.2. Computation of allowances

The allowance determination is represented in the form of a model as shown in Figure 4.8. Analyzing the model, it can be seen that relaxation allowances become the necessary part of the time to be added in the basic time that planned to aid recovery from the fatigue. The various other types of allowances are applicable under specific working conditions such as contingency, policy and special allowances.

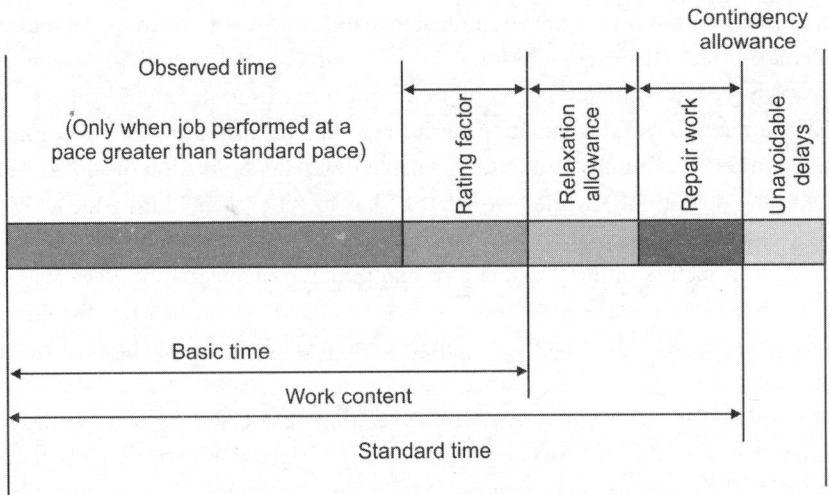

Fig. 4.8: Calculation of standard time.

4.15.3. Relaxation allowances

These are the allowances that are provided to the worker in addition to the anticipated normal time, to provide a chance of recovery from the physiological strain while performing a task under specific conditions so as to permit awareness to individual requirements.

Since these allowances have been made for the worker to make recovery from fatigue during the activity. Fatigue may be defined as a physical/mental weariness, real or imagined; existing in a person and adversely affects his ability to perform work, thereby reducing the work performance. The effects of fatigue can be reduced by providing appropriate rest breaks, during which body can revive its energy. Fatigue level is affected by the various factors such as:

1. Working conditions: Lighting, temperature, humidity, air freshness, colour of room walls, environmental and ambience noise level.
2. Nature of work: It includes factors such as concentration, repetitiveness and monotony of job, work posture required and tiredness of muscles.
3. Individual's health status: It includes physical and mental fitness, emotional stability, diet intake, living status and rest.

It is quite obvious, that fatigue at work place is reducible up to some extent but cannot be eliminated. Fatigue allowance may vary from 12 to 20 per cent or even more in some cases. A few typical personal/ fatigue allowances for male workers are given in Table 4.7.

Table 4.7: Personal/fatigue allowances for male workers.

Sr. No.	Description of Allowance	Percentage of Allowance	Remarks
1.	Personal allowance	5%	---
2.	Basic fatigue allowance	4%	---
3.	Awkward work posture	0-15%	with respect to level of risk involved
4.	Visual concentration	0-10%	--do--
5.	Poor lightening	1- 5%	--do--
6.	Work conditions; ventilation, bad smell, fumes, dust, smoke, etc.	1-5%	with respect to level of risk involved
7.	Level of noise	0-10%	--do--
8.	Protective measures/equipment	0-20%	--do--
9.	Exposure to heat stress	0-20%	--do--
10.	Tediousness or monotony	0-5%	--do--
11.	Mental strain	1-8%	--do--

Fatigue allowances are addition to normal time so as to make value of each work element separately; on the other hand for all job activities, standard times are combined to yield the standard time for the whole job/operation. This will also help to tackle with requirement of extra allowance to compensate the effect of severe climatic conditions. This is because different elements may some time be performed in different climate conditions.

Therefore, allowances for climatic conditions should be applied to full day or shift, rather than the job or work element only and it should be in such a way that the amount of work (work content) that the operator is expected to perform throughout the day is reduced. Also standard time for the job remains the same and it is irrespective of the season, whether the job is performed in summer or winter, since it is intended to be a measure of the work that the job contains.

Relaxation allowances have two major components: fixed allowances and variable allowances; the fixed allowances are composed of following two types:

1. *Allowances for personal needs*: This allowance provides for the necessity to leave the workplace to attend to personal needs such as washing, going to the lavatory and getting a drink. Common figures applied by many enterprise range 5–7 per cent.
2. *Allowances for basic fatigue*: This allowance is given to take account of the energy expended while carrying out work and to alleviate monotony. A common figure is 4 per cent of basic time. This is considered to be adequate for a worker who carries out the job while seated, who is engaged on light work in good working conditions, and who is called upon to make only normal use of hands, legs and senses.

In case of variable working conditions, variable allowances should be added to the fixed allowances, e.g., in poor environmental conditions that are inevitable, and it also involves stress and strain in performing the job in question.

As mentioned above, a number of important studies have been carried out by various research organizations to try to develop a more rational approach to the calculation of variable allowances. Most management consultants in all countries have their own tables. In Appendix 3, we give an example of relaxation allowances tables using a point system. Many of these tables appear to work satisfactorily in practice; however, recent evidence indicates that, although many of the fatigue allowances scales established empirically in a laboratory are satisfactory on physiological grounds for work involving normal or moderately intensive effort, they provide inadequate allowances when applied to very heavy operations such as those connected with furnaces, e.g. lifting and carrying of ladle of molten metal in casting process.

As already mentioned, the calculations of relaxation allowances are usually based upon element break down criteria. For example, there may be a case when the efforts spent on different elements vary widely, such as a worker engaged in a casting firm may perform an activity involving a lift and carrying of moulding box, then putting green sand into the molding box and subsequently lifting and carrying the heavy mould carrying to some pre-determined place. In this example, each of the element is involving different level of energy expenditure and variable levels of fatigue, then different amount of allowance are required to be added in the normal (basic) time. On the other hand, if it is considered that, different elements require almost the similar level of energy expenditure or causes same level of fatigue, then the easiest way is to consider the sum total of all the elemental basic (normal) time and provide the allowances as a percentage of sum total, but not a percentage of each elemental basic time.

4.15.3. Rest pauses

It is the form of allowances in which relaxation allowance are taken in a specific activity. Instead of making a standard way to indicate the required rest pauses, normally a work activity is allowed to give 10–15 minutes break time in terms of a cup of tea, coffee, soft drink and snacks. Also, a worker is allowed for discretion in the remaining of relaxation allowance. There are various reasons which show the necessity for rest pauses in a specific work activity as following:

1. The work efficiency of a worker is maintained all over the work activity and minimizes the performance variation.
2. Adequate rest pauses make the work activity boredom free throughout the day.
3. An opportunity for the recovery of fatigue level is provided to keep the work performance in a maintained way without any strain on the body.
4. The rest period leads in the reduction of time which is required to execute the work activity.

4.15.4. Other allowances

Sometimes there is a need to include some other allowances instead of relaxation allowances for the compilation of standard time, such as contingency allowance, policy allowance and special allowances. These are described as follows:

4.15.4.1. *Contingency allowances*

It is a form of small amount of time in addition to the standard time which is provided to cover up any officially permitted or expected substance of work or delays. It is very uncommon and irregular; therefore, exact calculation of the same is very complicated and too expensive.

Contingency allowances are provision to adjust some small uncontrollable delays as well as for irregular and minor extra work. These allowances are usually very short, and expressed as a percentage of the total repetitive normal time of the job. The percentage value of these allowances should not exceed more than 5 per cent, and it is to be provided only when time study man is assured that the contingencies are inevitable. It is very important to observe that such allowances should not be used as slacking factors for covering any miscellaneous gap left during the study. Moreover the duties for which the contingency allowance is provided should be specified. However, these may be necessary to be given in organizations where the production work is unorganized. Therefore, it creates the necessity for the management of the organizations to organize the work and allied activities in a well-mannered way. Hence, the need for contingency allowance will be reduced due to unorganized way of work.

4.15.4.2 *Policy allowance*

It is an increment excluding bonus or gratuity which is applied to standard time or some of their components. It is given to grant an acceptable level of income for a particular level of performance under special conditions.

These are not authentic ingredients of time study and have to be utilized with the highest vigilance and care, just under obviously defined situations. They must be handled independently from normal or basic times. They should preferably be given in addition to standard times, set by the time study. The common basis for providing a policy allowance is to match-up the standard times according to the needs of wage agreements among employers and trade unions.

In a number of firms in the United Kingdom, e.g., the standard level of performance for incentive is normally fixed so that even an average qualified worker is able to earn a bonus of at least 33 per cent of his normal time rate. However, this state of affairs can be achieved even avoiding the policy allowance. It is possible by setting up the wage rate paid per standard minute of work produced at 133 per cent of the basic time rate for each minute. It is normally easy and better to accommodate any special wage requirements by adjusting the rate paid per unit of work rather than the standard time. But still, there is some agreement between the employer and the union which enables the workers to earn higher bonuses. It may be courteous to look for an amendment of the conditions of these agreements which allow them to amend the rates paid more willingly than the time standards. Under these situations a policy allowance is given to make up the difference. It could be applied as a factor to the work content or to the standard time. It may be suitable to consider policy allowance when standard times are being introduced to only a small section of the entire workforce covered under agreement. Similarly these allowances are sometimes made as temporary additions to cover unusual situations, like the unsatisfactory functioning of a section of plant or disruption of normal working caused by rearrangements or alterations.

4.15.4.3. *Special allowances*

These are the allowances which are to be given for such activities which generally are not part of the work cycle but which are necessary to accomplish the acceptable level of work performance. In other words, it can be said that special allowances are provided for second category of activities under lean philosophy, i.e., necessary but non-value adding activities. These allowances might be permanent or temporary depending upon the situation, but should be carefully specified. Where feasible, the special allowances must be determined through time study. As soon as time standards are applied to calculate wages and make payment, it may perhaps be compulsory to formulate a start-up allowance to compensate for time taken by any work and any enforced delay time which would really occur at the beginning of a shift before actual production can start. Similarly a shut-down allowance may be provided for work or waiting time which would occur at the end of the shift, e.g., cleaning allowance is given while an operator has to give attention repeatedly for cleaning his machine or workplace. A tool allowance is given to cover up the fine-tuning and repairs of tools.

It is generally preferred to give all these allowances as a period of time per day rather than representing them within the standard times. Typically, it is better as it has an advantage of drawing the attention of the management towards the amount of time being given to these activities; consequently, there will be effort to reduce the same.

Some of the allowances are usually set per chance or per lot, e.g., set-up allowance is given to cover up the time needed to set up a machine, equipment or production process, a necessary operation to start the production of a batch of a new product or component. This set-up time is sometimes termed as make-ready time; its contrary is called as tear-down or dismantling time, which is given, to cover up the time required for doing adjustments and changes to machine or process settings after finishing a production run, e.g., changing a die. Extremely alike is the change-over allowance, which is normally provided to the operators who are not truly occupied in setting-up or dismantling. Hence it is given to reimburse the operators for the time spent on necessary activities or may be waiting time at beginning and/or the ending of a job or batch. These allowances are termed as job/batch change-over allowances.

There is another allowance, called as learning allowances, which is given to the trainee workers occupied in activities or jobs having set standard times. It is provided as a temporary benefit while the trainees can develop their skills. Similarly we have a training allowance which given to an experienced worker to compensate his training period. These are given as some minutes per hour, on a declining scale so that the allowances taper off to zero over the expected learning period.

Very similar is an implementation allowance, given to workers asked to adopt a new ways and prevent their loss of earnings by doing so. In fact, it is sometimes arranged that their earnings will actually be increased during the change-over period, so as to enable the new method to encourage them to implement enthusiastically and prevent their down earnings. For example, one system of implementation allowances credits the workers with ten minutes per hour on the first day, nine on the second, and so on, down to zero.

4.16. Calculations of Standard Time

In general practice, the time study engineer will end up with the calculation of allowed time; the allowed time is calculated by adding the percentage of rest allowance, process allowance and the special allowance, in basic or normal time; however, the addition of policy allowance is left to the decision of the management. Therefore, standard time may be defined as the amount of time required to complete a unit of work under existing working conditions, using the specified method and machinery, by a qualified worker or operator who is able to do the work in a proper manner, and at a standard pace.

The sequence wise calculations of basic/normal time, allowed time and standard time are defined as follows:

4.16.1. Basic time

It is the time required to perform a task by a normal operator working at a standard pace (rate) with no allowances for personal delays, unavoidable delays, or fatigue.

$$\text{Basic Time (Normal Time)} = \text{Observed Time} \times \frac{\text{Observed Rating}}{\text{Standard Rating}\,(100)} \quad \text{i.e.,}$$

$$BT \text{ (or } NT) = OT \times \frac{OR}{SR(100)} \qquad \text{i.e. } OT \times OR = BT \times SR$$

Basic (Normal) Time can be represented as = (Observed Time) $\times \left(\dfrac{\text{Rating factor in } \%}{100} \right)$

Rating factor (RF) is in percentage like if observed rating is 90 per cent, RF = 0.9, if rating is 110 per cent then R.F = 1.10

where, BT: Basic time

NT: Normal time

OT: the observed time with the stopwatch

OR: Observed rating

SR: Standard rating

4.16.2. Allowed time

It is the time obtained by adding the percentage of rest allowance, process allowance, and the special allowance, in basic or normal time. It can be calculated as:

$$\text{Allowed Time} = \text{(Normal Time)} \times \left(1 + \frac{\text{Allowances in } \%}{100} \right)$$

4.16.3. Standard time

When all the rest allowances are considered alone, then it is called sustained performance time, the calculation of the same is illustrated in the example given ahead. It is obtained by adding the policy allowance to the allowed time.

$$\text{Standard Time (ST)} = \text{(Allowed Time)} \times \left(1 + \frac{\text{Policy Allowances in } \%}{100} \right)$$

Thus the standard time is ultimately determined from the 'observed time' noted from the stopwatch time study or alternately it can be represented as follows:

$$\text{Standard Time (ST)} = \text{(Normal Time} \times \text{Allowance factor)}$$

where, Allowance factor $= \left(\dfrac{1}{1 - \text{Allowances in } \%} \right)$

Note: Standard time calculated by both the above mentioned methods will give almost similar results; hence any one of them can be used unless or otherwise specified.

The constituents of standard time are diagrammatically represented in Figure 4.9 as follows.

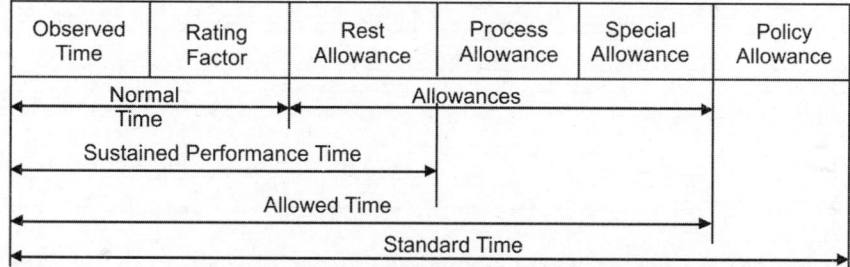

Fig. 4.9: Constituents of standard time.

In order to understand the calculation of standard time, let us consider a situation in which continuous stopwatch readings were taken in seconds. Consider fatigue allowances as 10 per cent, process allowances 5 per cent, contingency allowances 5 per cent, and policy allowance is 20 per cent. Let us calculate the normal time, sustained time, allowed time and standard time for the operation. The time recorded for the four elements is shown in Table 4.8.

Table 4.8: Showing the observations and rating for each cycle and element respectively.

Element No.	1	2	3	4
Cycle No.	Stopwatch Reading in Seconds			
1	10	30	50	70
2	90	107	125	145
3	155	175	195	210
4	225	240	260	280
5	290	310	330	350
Average Rating	110	110	120	90

Solution: The initial stopwatch reading is taken as zero, the observed time for each element is determined from the continuous readings taken at the end of each element by subtracting the initial from the final, therefore observed timings for each element is shown in Table 4.9.

Therefore normal time for the operation = 14.3 + 20.24 + 23.52 + 17.1 = 75.16 sec

The given fatigue allowance = 10%

Therefore, sustained performance time = (Normal Time) $\times \dfrac{110}{100}$ = 75.16 × 1.10 = 82.68 sec

Table 4.9: Calculation of average observed time and basic time for each element.

Element No.	1	2	3	4
Cycle No.	Stopwatch Reading in Seconds			
1	10	20	20	20
2	20	17	18	20
3	10	20	20	20
4	15	15	20	15
5	10	20	20	20
Average	$\dfrac{65}{5} = 13$	$\dfrac{92}{5} = 18.4$	$\dfrac{98}{5} = 19.6$	$\dfrac{95}{5} = 19$
Average Rating	110	110	120	90
Normal Time	$\dfrac{13 \times 110}{100} = 14.3$	$\dfrac{18.4 \times 110}{100} = 20.24$	$\dfrac{19.6 \times 120}{100} = 23.52$	$\dfrac{19 \times 90}{100} = 17.1$

Total fatigue, process, and contingency allowances = 10 + 5 + 5 = 20%

$$\text{Allowed Time} = (\text{NormalTime}) \times \left(1 + \frac{\text{Alllowances in } \%}{100}\right)$$

$$\text{Allowed Time} = (82.68) \times \left(1 + \frac{20\%}{100}\right)$$

$$\text{Allowed Time} = (82.68) \times (1 + 0.20)$$

$$\text{Allowed Time} = (82.68) \times (1.20) = 99.22 \text{ sec.}$$

Since, the given policy allowances are 20 per cent of the normal time.

$$\text{Standard Time (ST)} = (\text{Allowed Time}) \times \left(1 + \frac{\text{Policy Alllowances in } \%}{100}\right)$$

$$\text{Standard Time (ST)} = (99.22) \times \left(1 + \frac{20\%}{100}\right)$$

$$= (99.22) \times (1.20)$$

$$\text{Standard Time (ST)} = 119.64 \text{ sec or } 1.98 \text{ or } 2.0 \text{ minutes}$$

4.17. Synthesis

It is one of the work measurement techniques for setting up the standard time for a specific job using synthetic data. It is done by summing up the elemental times at a defined level of performance; the elemental times are obtained from time studies on other jobs containing the same concerned elements.

4.17.1. Synthetic data

It consists of tables and formulae derived from analysis of accumulated time study data, arranged in a suitable form for developing the standard times by synthesis. D. V. Merrick has been known as the forerunner to adapt this technique. He had compiled tables for various activities in machine shop operations. Synthetic time data are useful for determining the standard time from the drawings and specifications of the components. This method enables the time study man to estimate the time required in advance of actual implementation of the job. Therefore, it is very useful in process planning and scheduling. The development of standard time by synthesis needs to divide the elements required to perform the operation into following three categories:

(a) Constant elements: The time required for these elements is independent of the type of job.
(b) Variable elements: The time required for these elements depends upon the size, shape and weight of the component.
(c) Machining or process elements: Here the time taken by these elements are in fact the machining times, these can be estimated based upon the speed, feed and depth of cut, etc.

For the variable and machining elements, the formula, tables and graphs can be used for the calculations of time taken by these elements.

Let us consider an example of a drilling operation, which consists of elements as mentioned below:

Pick up the work piece, place and fix in the jig = 0.20 minutes

Position under the spindle = 0.06 minutes
Advance drill to the work = 0.02 minutes
Clear drill = 0.02 minutes
Unload the work piece = 0.12 minutes
Drop the work piece in the bin = 0.05 minutes
Clean jig and table = 0.01 minutes
Total Time = 0.58 minutes

The periodic activity of sharpening the drill, etc., is taken as 0.02 minutes
Therefore total handling time = 0.58 + 0.02 = 0.60 minutes

The machining element time (T) is calculated by using the formula '$T = \dfrac{L}{fN}$

where, T is the drilling time in minutes, L is the length of cut in mm, f is feed in mm/rev and N is the RPM of the spindle.

Now let us suppose that allowed time is to be calculated for drilling 2 mm diameter hole in a 5.5 mm diameter cotter pin. The machining parameters are as follows:

The spindle speed 'N' = 1200 RPM
The optimum feed 'f' = 0.05 mm / rev
The length of cut 'L' = 5.5 mm + 0.5 mm extra margin
= 6.0 mm

Therefore, $\qquad T = \dfrac{6.0}{0.05 \times 1200} = 0.1$ minutes

Thus, the total time = 0.6 + 0.1 = 0.7 minutes
The above synthetic time indicates the 'normal time' only.
The fatigue and other allowances = 20%
Therefore, allowed time = 0.7 × 1.2 = 0.84 minutes

4.17.2. Advantages of synthetic time system (synthesis)

1. The element data has been acquired by a large number of studies conducted on the shop floor; hence, it is more reliable and consistent.
2. It saves time and money to be spent in making new time study every time when needed.
3. Since it provides the estimates of time required for the various jobs, it helps the manager in production planning. Subsequently, it is very useful in line balancing of assembly work.
4. The synthetic time data could be very helpful for negotiation of wage and incentive among the management and labour union.
5. It is also very useful for quick and accurate cost estimation and fixing the price of new component or product.

4.18. Pre-determined Motion Time Standards (PMTS)

It is a work measurement technique in which time established for basic human movements are used to determine standard or allowed time for a job under the given set of conditions and at a defined level of performance. The time determined by PMTS is considered as basic time and then allowances are added to calculate the standard time. It is well recognized that every job contains some common and basic elements of motions. A very large amount of data is collected at a required level of accuracy, with the help of filming analysis, under various controlled conditions. Subsequently, time values are prepared in the form of tables for each basic motion. It should be kept in mind that the time values deployed basic elements of motions depends upon large number of actual time observations, and based upon these observations the allowed time for the job is determined. Therefore, a predetermined time standard (PTS) is a job estimation strategy through time secured for essential human movements (characterized by nature of movement and conditions under which it is made) are used to develop the time for an occupation at a characterized level of execution.

4.18.1. Characteristics of PMTS

There are main five characteristics of PMTS which are enumerated as follows:

(a) It includes mainly the elements of the basic human motions.

(b) The elements are classified according to the nature of motions and the conditions under which they are to be used.

(c) Identification of basic elements and their times are established.

(d) The job under consideration is analyzed based upon the proper sequence of elements for which time data is available.

(e) Normal time is determined for the job under consideration using the established data.

There are numerous PMTS established by various organizations; however, some of them are briefly described in the following sub sections.

1. Motion Time Analysis (MTA)

   It was developed by A. B. Segur in 1924. It is considered as the fore-runner of all other PMTS. It is based upon the fundamental motions, developed by Gilbreth, and it is called as 'therbligs'. Segur used the motion pictures, micro-motion analysis and kymograph to establish the time values for various therbligs. It is used for the estimation of time values of human motions, by dividing into 'therbligs' and build up by the addition of the time values developed.

2. Basic Motion Times

   This was developed by Palph Presgrave, Bailey and their associates of J. B. Woods and Gordon Limited of Toronto during 1945–1951. This system was developed through laboratory study and experimentation. It has the following concepts:

   (a) Basic motion: The basic motion is defined as a motion in which the body member at rest moves and again come to rest; the same is depicted in Figure 4.10.

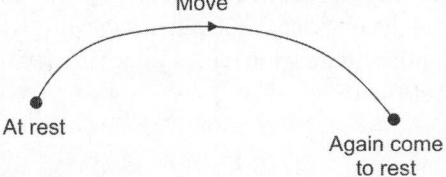

Fig. 4.10: Basic motion of a body: at rest (begin), movement and rest at end point.

   (b) Categories of basic motions: Basic motions can be classified into three categories; A, B and C depending upon the nature of stopping of basic motion and then further into two sub-categories based upon the visual direction required.

(i) Basic motion 'A': When the motion is stopped without muscular efforts or control.

(ii) Basic motion 'B': When the motion is stopped with muscular control.

(iii) Basic motion 'C': When the motion is slowed down by muscular efforts, e.g., before the object or tool is grasped or positioned.

Basic motion 'B' and 'C' can further be sub categorized as 'BV' and 'CV', i.e., when motions are visually directed. The categories of basic motions are exhibited in Figure 4.11.

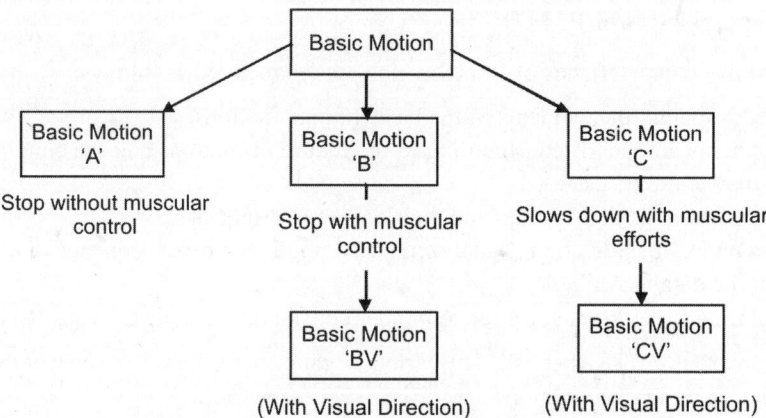

Fig. 4.11: Classifications of basic motions.

(c) There are following five aspects that are taken under consideration for estimation of time for basic elements.

(i) The distance moved in mm or cm

(ii) The visual attention required

(iii) The degree of precision required in terms of tolerance in mm

(iv) The force required to overcome the resistance in kg

(v) Simultaneous motions[*]

(d) The time values given in the BMT tables are expressed in terms of 0.0001 minutes (10^{-4} minutes), the following two variables are used.

(i) Basic motion like reach (R) or move (M) for the A, B, C, BV and CV classes for different distances in mm. For example, the symbol used such as; 'R 200 B' indicates that reach component 200 mm away with 'B' class motion.

(ii) Precision (P) for the different tolerances values in rows for the distances in columns as shown in Table 4.10.

---

[*] Simultaneous motions are considered when both arms move together in case of assembly. The time required is given depending upon the distance of separation (in mm) of assembly points and the required precision (in terms of tolerance in mm) within which the part is to be assembled. The required time is estimated in three parts, the basic motion with distance and visual direction, grasping the part with precision required for grasp and the time with effect of simultaneous motion depending upon separation distance and precision required for assembly.

Table 4.10: Different tolerances values and respective distances.

Distance (mm)	25	50	75	100	...	750
12 mm tolerance	3	4	6	7	...	...
...	...	...	...	...	...	...
...	...	...	...	...	...	...
0.75 mm tolerance	90	97	102	106	...	165

For example, symbol 'P 12 mm' indicates precision of tolerance 12 mm, the distance is same for the reach, a move used for the basic motion, and both should be added as given below:

$$R\ 100\ C = 64$$

$$P\ 12\ mm = 07$$

$$Total = 71,\ i.e.,\ 0.0071\ minutes$$

This can be interpreted as reach an object placed 100 mm away to grasp the component with a limit of 12 mm.

(iii) Simultaneous motions (Simo): Here the values are provided for separation distances (mm) in the column and for the tolerances limits in the rows, for example, simo 1.5 mm simo 200 mm indicate that a precision of 1.5 mm tolerance and the separation distance of 300 mm.

(iv) Turn (T): This refers to the turning of forearm along the axis through the angle in degrees and the class of motion. For example, T 90 C indicates that turning through 90° with C class motion.

(v) Force (F): This refers to the weight of the component or the force applied. The force timings are added for the start, move and stop cases separately. For example, when a component of 4.5 kg is picked, moved through a distance of 800 mm and positioned (or placed carefully), it can be noted as follows:

For motion M 300 C = 96
For weight 3 F 4.5 = $3 \times 14 = 42$ (three times the weight conditions are added)
Therefore, total = 126 = 0.0128 minutes

3. Methods Time Measurement (MTM): It can be defined as the procedure that analyzes any manual operation or method into the basic motions required to perform it. It assigns a pre-determined time standard to each motion; these times are determined by the nature of the motion and the conditions under which it is made. MTM was developed by Maynard, Stegmerten and Schwab during the Second World War. This is known to be the most popular method over the others. Since significant work has been made in each and every field of human activity; therefore, special types of MTMs have been developed for predetermining the time in maintenance work as well as for the measurement of non-repetitive indirect work. The MTM International Directorate has been setup and under its guidance various MTM associations at national levels are formed with a view to protect MTM against in correct use, coordinate research in the field of MTM and supervise the correct training for instructors and practitioners through approved courses.

MTM was developed from the analysis of motion pictures taken from various industrial operations. Thus the original data is based on frame by frame micro motion film analysis. The data collected were grouped under eight basic motions and predetermined time values are given in separate tables. The basic motions are move hand or fingers to a destination, the time values depend upon:

(i) Distance of movement.

(ii) Type of reach like; A, B, C, D and E depending upon the location of reach, speed, visual attention and control needed.

(iii) Type of motion: Types-I, II and II depending upon whether the hand is at rest or in motion either in the beginning and end of the reach.

4.19. Work Sampling Process

It is the method of statistical sampling and random observations to determine the proportion of time spent on a precise activity by workers. Within the work sampling process, the study engineer obtains a large number of observations of a worker or machine randomly all through the working shift or day. The work study engineer observes and record precisely about the working or non-working of machine or worker; it also includes specific information if a machine is being set or maintained and what exactly the worker is doing, i.e., whether setting the machine, waiting for material or talking with the foreman regarding the job, etc. It is also very important to note that in work sampling study stopwatch is not used. It is due to the reason that the objective is to find the frequency of occurrence of every work element. The work sampling technique is entirely based upon the law of probability. Work sampling is supported by the statistical principle that 'the incidence in an adequate random sample observation of an activity will follow the same distribution pattern that may be found in a lengthy, continuous study of the same activity,.

Mathematically, the probability 'p' of occurrence of an activity is calculated as:

$$p = \frac{x}{n} = \frac{\text{no of observation of the activity}}{\text{total no of observation}}$$

Therefore, the work sampling technique includes taking a number of intermittent, randomly-spaced, instantaneous observations about an activity and determining the percentage of time spent on each part of the process.

It is also required to rate the performance of worker and calculate the actual number of units produced during the period. Finally, it may be assured that the precision of the approach depends on the number of observation made; higher the number of observations, greater will be the accuracy of result obtained regarding the job under observation.

4.19.1. The need for work sampling

Work sampling (also known as 'activity sampling', 'ratio-delay study', 'random observation method', 'snap-reading method' and 'observation ratio study') is, as the name implies, a sampling technique. Let us first see why such a technique is needed.

In order to get an entire and precise image of the productive time and idle time of the machines in a specific production area, it would be essential to examine constantly all the machinery in that area and make a record for the disturbance of the machines. It would be some what unachievable to do this unless a large number of employees spent a long time on this particular task alone,which is perhaps an unrealistic proposal.

However, it might be found that, say 80 per cent of the machines were working and 20 per cent stopped. As this action repeated 20 or more time at varying times and if each time the proportion of machines working was always 80 per cent that would be possible to say at any one time there were always 80 per cent of the machinery is running.

As it is not generally possible, the other best method has to be adopted; that is making tours of the factory at arbitrary intervals. This is the basis of the work sampling technique. When sample size is large then observations made are indeed at random, there is quite a high probability that will reflect the real situation, plus or minus a certain margin of error.

4.19.2. Basic of sampling

Unlike the costly and impractical method of continuous observation, sampling is mainly based on probability. Probability has been defined as 'the extent to which an event is likely to occur'. A simple and often-mentioned example that illustrates the point is that of tossing a coin. When we toss a coin there are two possibilities of outcome; either head or tail. The law of probability says that we are likely to have 50 heads and 50 tails in every 100 tosses of the coin. Note that we use the term 'likely to have'. In fact, we might have a score of 55-45, say, or 48-52, or some other ratio. But it has been proved that the law becomes increasingly accurate as the number of tosses increase. In other words, the greater the number of tosses, the more chance we have of arriving at a ratio of 50 heads to 50 tails. This suggests that the larger the size of the sample, the more accurate or representative it becomes with respect to the original 'population', or group of items under consideration.

We can therefore visualize a scale where, at one end, we can have the complete accuracy achieved by continuous observation and, at the other end, very doubtful results derived from a few observations only. The size of the sample is therefore important, and we can express our confidence in whether or not the sample is representative by using a certain confidence level.

4.19.3. Principle of work sampling

The principle of work sampling is statistical hypothesis of random sampling and probability of normal distribution with associated confidence level. It is elaborated by following example.

Let x = number of observation of an activity i.e. machine/operator is working.

N = total no of observation in the activity of the pilot study, then the proportion of the activity is

$$p = \frac{x}{N}$$

The proportion of activity = $1 - p = q$

The both states which are cohesively exclusive is 1

$$\text{i.e. } p + q = 1$$

where, p = probability of an occurrence and q = probability of non-occurrences.

This may be extended to take account of a number of observation (n) and become $(p + q)^n = 1$.

If this term is extended by the binomial theorem, the first term of the expression will have the probability that $x = 0$, the second term $x = 1$ and so on ....

The distribution of these probability will follow the binomial distribution and will have a mean value = Np, and standard deviation = $(Npq)^{1/2}$. As N becomes higher, the binomial distribution approaches the normal distribution as a satisfactory approximation. In order to use the normal distribution, we need normal distribution and we may use the normal distribution as satisfactory approximation.

4.19.4. Steps in work sampling

The work sampling study consists of essentially the following steps:

1. Establish the goal of study which contains explanation of the state of action to be noticed.
2. Arrange the sampling process including:
    (a) an approximation of the percentage of time devoted to every face of the activity precision limit.
    (b) situation of precision limit.
    (c) judgement of number of observation needed.
    (d) selection of the duration of study and programming of the number of reading during this period.
    (e) the foundation of the apparatus for making observations, route and recording of data.
3. Gathering of data as considered.
4. Analyze the data and present result.

4.19.5. Benefits and limitations of work sampling

Work sampling is a very useful tool of work study. It has many advantages along with the limitations which are enumerated as follows:

Advantages:

1. Inexpensive to apply and usually cost is significantly less than continuous time study.
2. Used to determine various impractical actions which are difficult to measure by time study.
3. Trained work measurement analyst are not essential to make observation.
4. Work sampling measurement can be made with a predetermined assigned level of consistency.
5. Directly measure the working/engagement of individuals and equipment.
6. Eliminate the need of stopwatch for measurement.

Limitations:

1. Does not recommend several of the opportunity for methods study that accompanies time; study and less improved work methods.
2. The worker may be unable to understand statistical work sampling.
3. The results may be unfair if random sampling is not done.

4.19.6. Establishing confidence levels

Let us go back to our previous example and toss five coins at a time, and then record the number of times we have heads and the number of times we have tails for each toss of these five coins. Let us then repeat this operation 100 times. The results could be presented either in table form or graphically as shown in Figure 4.12. If we considerably increase the number of tosses and in each case toss a large number of coins at a time, we can obtain a smoother curve, such as that shown in Figure 4.13.

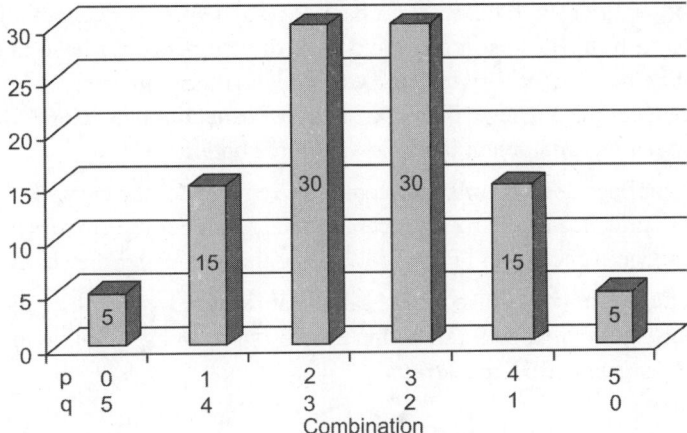

Fig. 4.12: Proportional distribution of 'heads and tails' after 100 tosses of coins at a time.

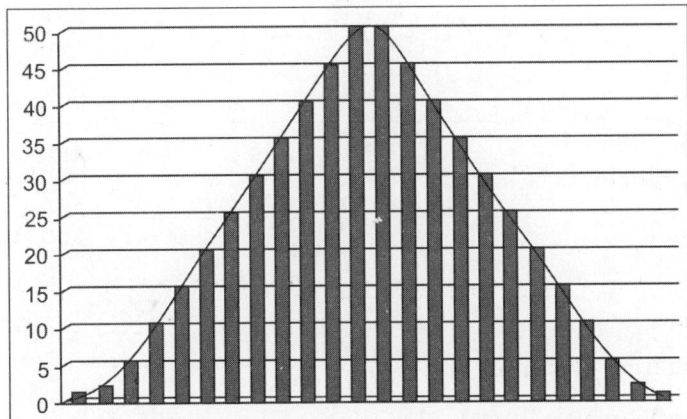

Fig. 4.13: Distribution of 'heads and tails' after a huge number of tosses
of coins at a time to make a normal distribution of data.

The curve in Figure 4.13 is called the curve of normal distribution. Basically, this curve tells us that, in the majority of cases, the tendency is for the number of heads to equal the number of tails in any one series of tosses (when $p = q$ the number of tosses is a maximum). In a few cases, however, p is considerably different from q due to mere chance.

Curves of normal distribution may be of many shapes they may be flatter, or more peaked. To describe these curves we use two attributes; one is average or mean, denoted as x bar, which is the average or measure of central dispersion; it depicts the measure of central dispersion and second attribute is the deviation from the average, referred to as standard deviation, which is denoted by 'σ'. Since in this case we are dealing with a proportion, we use σp to denote the standard error of the proportion.

The area under the curve of normal distribution can be calculated. In Figure 4.14 one σp on both sides of mean x bar gives an area of 68.27 per cent of the total area; two σp on both sides of x bar gives an area of 95.45 per cent and three σp on both sides of x bar gives an area of 99.73 per cent. We can put this in another way and say that, provided that we are not biased in our random sampling, 95.45

per cent of all our observations will fall within *x* bar ± 2 σ*p* and 99.73 per cent of all our observations will fall within *x* bar ± 3 σ*p*. This is in fact the degree of confidence we have in our observations. To make things easier, however, we try to avoid using decimal percentages; it is more convenient to speak of a 95 per cent confidence level than of a 95.45 per cent confidence level. To achieve this we can change our calculations and obtain the following three options:

- 95 per cent confidence level or 95 per cent of the area under the curve = 1.96 σ*p*;
- 99 per cent confidence level or 99 per cent of the area under the curve = 2.58 σ*p*;
- 99.9 per cent confidence level or 99.9 per cent of the area under the curve = 3.3σ*p*;

In this case we can say that if we take a large sample at random we can be confident that in 95 per cent of the cases our observations will fall within ± 1.96 σ*p*. In work sampling the most commonly used level is the 95 per cent confidence level.

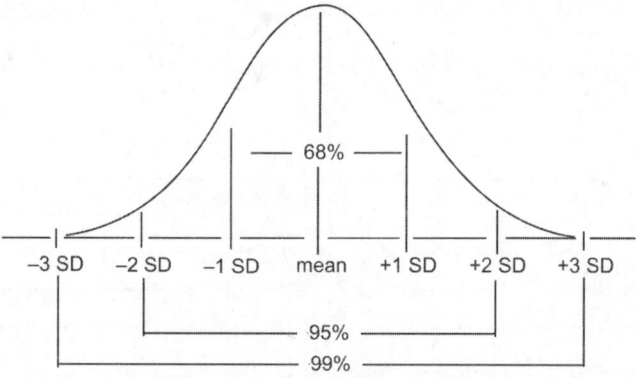

Fig. 4.14: Curve of normal distribution.

4.19.7. Determination of sample size

In order to determine the sample size as well as defining the confidence level for our observations, we have to decide on the margin of error that we can allow for these observations. We must be able to say that we are confident that for 95 per cent of the time this particular observation is correct within ± 5 per cent or 10 per cent, or whatever other range of accuracy we may decide on. Let us now return to our example about the productive time and the idle time of the machines in a factory. There are two methods of determining the sample size that would be appropriate for this example: the statistical method and the monogram method, the statistical method is explained, where as monogram method is beyond the scope of this text.

4.19.7.1. *Statistical method*

The formula used in this method is:

$$\sigma p = \sqrt{\frac{pq}{n}}$$

where, σ*p* = standard error of proportion

     p = percentage of idle time

     q = percentage of working time

     n = number of observations or sample size we wish to determine.

Before we can use this formula, however, we need to have at least an idea of the values of p and q. The first step is therefore to carry out a number of random observations in the working area. Let us assume that some 200 observations were carried out at random as a preliminary study, and showed that machine was idle in 20 per cent of the cases ($p = 20$) and working 80 per cent of the time ($q = 80$). We thus have approximate values for p and q; now in order to determine the value of n, we must find out the value of σp.

Let us choose a confidence level of 95 per cent with a 10 per cent margin of error (that is, we are confident that in 95 per cent of the cases our estimates will be within $\pm 10\%$ (i.e. $\pm 2\,\sigma p$) of the real value).

At the 95 per cent confidence level:

$$1.96\;\sigma p = 10$$

$$\sigma p = 5 \text{ (approx.)}$$

We can now go back to our original equation to derive n:

$$\sigma p = \sqrt{\frac{pq}{n}}$$

$$5 = \sqrt{\frac{20 \times 80}{n}}$$

Squaring both sides

$$5^2 = \left(\sqrt{\frac{20 \times 80}{n}}\right)^2$$

or,

$$25 = \frac{20 \times 80}{n}$$

$$n = \frac{20 \times 80}{25} = 64$$

Now if we reduce the margin to ± 5 percent, then we have

$$1.96\;\sigma p = 5$$

$$\sigma p = 2.5 \text{ (approximately)}$$

i.e., $$2.5 = \sqrt{\frac{20 \times 80}{n}}$$

Squaring both sides

$$(2.5)^2 = \left(\sqrt{\frac{20 \times 80}{n}}\right)^2$$

or

$$n = \frac{20 \times 80}{6.25} = 256 \text{ observation}$$

Hence, it can be interpreted that if the margin of error is reduced by half, the number of observations or sample size will have to increase by a multiple of four.

4.19.8. Making random observations

Our previous conclusions are valid provided that, we can make the number of observations needed to attain the confidence level and accuracy required, also provided that these observations are made at random. To ensure that our observations are in fact made at random, we can use a random table such as given in Table 4.11. Various types of random table exist and these can be used in different ways. In the present case let us assume that we shall carry out our observations during a day shift of

Table 4.11: Random table for making observations.

53 74 30 77 40	44 22 78 84 26	04 33 46 09 52	68 07 97 06 57
59 29 97 68 60	71 91 38 67 54	13 58 18 24 76	15 54 55 95 52
48 55 90 65 72	96 57 69 36 10	96 46 92 42 45	97 60 49 04 91
66 37 32 **12** 30	77 84 57 03 29	10 45 65 04 26	11 04 96 67 24
68 49 69 20 82	53 75 91 93 30	34 25 20 57 27	40 48 73 51 92
57 24 55 **06** 88	77 04 74 47 67	21 76 33 50 25	83 92 12 06 76
49 54 43 54 82	17 37 93 23 78	87 35 20 96 43	84 26 34 91 64
78 64 56 **07** 82	52 42 07 44 38	15 51 00 13 42	99 66 02 79 54
16 95 55 67 19	98 10 50 71 75	12 86 73 58 07	44 39 52 38 79
09 47 27 **96** 54	49 17 46 09 62	90 52 84 77 27	08 02 73 43 28
96 83 50 87 75	97 12 25 93 47	70 33 24 03 54	97 77 46 44 80
40 33 20 **38** 26	13 89 51 03 74	17 75 37 13 04	07 74 21 19 30
88 42 95 45 72	16 64 36 16 00	04 43 18 66 79	94 77 24 21 90
33 27 14 **34** 09	45 59 34 68 49	12 72 07 34 45	99 27 72 95 14
50 27 89 87 19	20 15 37 00 49	52 85 66 60 44	38 68 88 11 80
06 09 19 **74** 66	02 94 37 34 02	76 70 90 30 86	38 45 94 30 38
83 62 64 11 12	67 19 00 71 74	60 47 21 29 68	02 51 37 03 31
33 32 51 **26** 38	79 78 45 04 91	16 92 53 56 16	02 75 50 95 98
42 38 97 01 50	87 75 66 81 41	40 01 74 91 62	48 51 84 08 32
96 44 33 **49** 13	34 86 82 53 91	00 52 43 48 85	27 55 26 89 62
64 05 71 95 85	11 05 65 09 68	76 83 20 37 90	57 15 00 11 66
75 73 88 **05** 90	52 27 41 14 86	22 98 12 22 08	07 52 74 95 80
33 96 02 75 19	07 60 62 93 55	59 33 82 43 90	49 37 38 44 59
97 51 40 **14** 02	04 02 33 31 08	39 54 16 49 36	47 95 93 13 30
15 06 15 93 20	01 90 10 75 06	40 78 24 89 62	02 57 74 17 33
22 35 85 **15** 33	92 03 51 59 77	59 56 78 06 83	52 91 05 70 74
09 98 42 99 64	61 71 52 99 15	06 51 29 16 93	58 05 77 09 51
54 87 66 **47** 54	73 32 08 11 12	44 95 92 53 16	29 56 24 29 48
58 37 78 80 70	42 10 50 67 42	32 17 55 85 74	94 44 67 16 94
87 59 36 22 41	25 78 63 06 55	13 08 27 01 50	15 29 39 59 43
71 41 61 50 72	12 41 94 96 25	44 95 27 36 99	02 96 74 30 83
23 52 23 33 12	96 93 02 18 39	07 02 18 36 07	25 99 32 70 23
31 04 49 69 96	10 47 48 45 88	13 41 43 89 20	97 17 14 49 17
31 99 73 68 68	35 81 33 03 76	24 30 12 48 60	18 99 10 72 34
94 58 28 41 35	45 37 59 03 09	90 35 57 29 12	82 52 54 55 60
44 17 16 58 09	79 83 86 19 62	06 76 50 03 10	55 23 64 05 42
82 97 77 77 99	83 11 46 32 24	20 14 85 88 45	10 93 72 88 71
82 97 77 78 81	07 45 32 14 08	32 98 94 07 72	93 85 79 10 75
50 92 26 11 97	00 56 76 31 38	80 22 02 53 12	86 60 42 04 53
83 39 50 08 30	42 34 07 96 88	54 42 06 87 98	35 85 29 48 39

eight hours, from 8 am to 4 pm. An eight-hour day has 480 minutes. These may be divided into 48 ten-minute periods. We can start by choosing any number at random from the random number Table 4.11. For example let us close our eyes and place a pencil point somewhere on the table. Now, let us assume that in this case we pick number 20, by mere chance, which is in the first block, fourth column, and fourth row as shown in Table 4.11.

We now choose any number between 1 and 10. Assume that we choose the number 2; we now go down the column picking out every second reading and noting it down, as shown below (if we had chosen the number 3, we should pick out every third figure, and so on).

12   06   07   **96**   38   34   **74**   26   **49**   05

Looking at these numbers, we find that we have to discard 96, 74 *and* 49 because they are too high (since we have only 48 ten-minute periods, any number above 48 has to be discarded). Similarly, the numbers which are repeated will also have to be discarded. We therefore have to continue with our readings to replace the three numbers that we have discarded. Using the same method, that is choosing every second number after the last one (05), we now have 14, 15 and 47. These three numbers are within the desired range and have not appeared before. Our final selection may now be arranged in ascending order and the times of observation throughout the eight-hour day worked out. Thus our smallest number (05) represents the fifth ten-minute period after the work began at 8:00 am. Thus our first observation will be at 8:50 am, and so on, the same is exhibited in Table 4.12.

Table 4.12: Work sampling schedule on basis of random observations.

Usable number Selected from Random table	Number Arranged in Ascending order	Lapsed time in minutes (hours) from the start of shift (i.e. 8:00 am)	Time of Observation w.r.t start of shift (8:00 am)
12	05	05 × 10 = 50 (0:50 hr)	8:50 am
06	06	06 × 10 = 60 (1:00 hr)	9:00 am
07	07	07 × 10 = 70 (1:10 hrs)	9:10 am
38	12	12 × 10 = 120 (2:00 hrs)	10:00 am
34	14	14 × 10 = 140 (2:20 hrs)	10:20 am
26	15	15 × 10 = 150 (2:30 hrs)	10:30 am
05	26	26 × 10 = 260 (4:20 hrs)	12:20 pm
14	34	34 × 10 = 340 (5:40 hrs)	01:40 pm
15	38	38 × 10 = 380 (6:20 hrs)	02:20 pm
47	47	47 × 10 = 470 (7:50 hrs)	03:50 pm

4.19.9. Conducting the study

Before making our actual observations, it is very important to decide the objective of work sampling. The simplest objective is that of determining whether a given machine is idle or working. In such a case, our observations aim at detecting one of the two possibilities only. Although we may wish to get an idea of the percentage distribution of time when the machine is working and when it is idle, in which case we combine the last two models.

We may also be interested in the percentage time spent by a worker or groups of workers on a given element of work. If a certain job consists of ten different elements, by observing a worker at the defined points in time we can record on which element he or she is engaged and hence turn up at a percentage distribution of the time he or she has been spending on each element. A general procedure for conducting the work sampling is exhibited in Figure 4.15. The objectives to be achieved by the study determine the design or format of the recording sheet to be used in work sampling, as can be seen from Tables 4.13, 4.14 and 4.15.

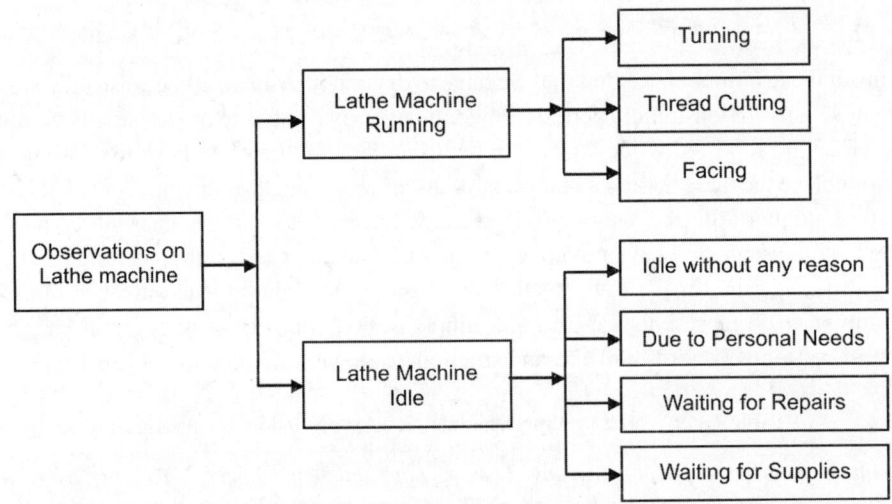

Fig. 4.15: Work sampling procedure for making observations.

Table 4.13: Work sampling record sheet showing observation when machine is running and idle.

Date	Name of Observer:		Study No:	
Number of Observations: 85			Total	Percentage
Machine Running	ЦНІ ЦНІ ЦНІ ЦНІ ЦНІ ЦНІ ЦНІ ЦНІ ЦНІ ЦНІ ЦНІ ЦНІ ЦНІ ЦНІ		70	82.35
Machine Idle	ЦНІ ЦНІ ЦНІ		15	17.65

Table 4.14: Work sampling record sheet showing observation for machine running and distribution of machine idle time.

Date:	Name of Observer:		Study No:	
Number of Observations: 85			Total	Percentage
Machine Running		ЦНІ ЦНІ ЦНІ ЦНІ ЦНІ ЦНІ ЦНІ ЦНІ ЦНІ ЦНІ ЦНІ ЦНІ ЦНІ ЦНІ	70	82.35
Machine	Idle Without Reason	ЦНІ	5	05.88
	Due to Personal Needs	III	3	03.52
	Waiting for Repairs	II	2	02.35
	Waiting for Supplies	ЦНІ	5	05.88

Table 4.15: Work sampling record sheet showing observation on distribution of machine idle time of nine elements of operation performed.

Date:				Name of Observer:				Study No:
No. of Observations:								
Worker ID	Elements of Work							
W-1	1	2	3	4	5	6	7	8	9
W-2									
W-3									
W-4									
W-5									

4.19.9.1. *Making the observations*

So far we have taken the first five logical steps in conducting a work sampling study. Thus, to summarize these consist of the following steps:

1. Selecting the job to be studied and determining the objectives of the study.
2. Making a preliminary observation to determine the approximate values of p and q.
3. Determining the number of observations 'n' needed in terms of a chosen confidence level and accuracy range.
4. Determining the frequency of observations, using random tables.
5. Designing record sheets to meet the objectives of the study.
6. Observing and recording the observations and analyzing the results.

In making the observations, it is essential from the outset that the work study person is clear about what is to be achieved and why. Ambiguity should be avoided when classifying activities. For example, if the engine of a fork-lift truck is running while the truck is waiting to be loaded or unloaded, it should be decided beforehand whether this means that the truck is working or idle. It is also essential for the work study person to contact the persons he or she wishes to observe, explaining to them the purpose of the study, indicating to them that they should work at their normal pace and attempting to gain their confidence and cooperation.

The observation itself should be made at the same point relative to each machine. The work study person should not note what is happening at the machines ahead, as this tends to falsify the study. For example, in a weaving department, the observer may notice a loom that has stopped, just ahead of the one he or she is observing. The weaver may have it running again by the time the observer reaches it. The observer would, by noting it as idle, be giving an untrue picture.

The recording itself, as can be seen, consists simply of making a stroke in front of the appropriate activity on the record sheet at the proper and predetermined time. No stopwatches are used.

The analysis of results can be conducted readily on the record sheet. It is possible to find out the percentage of effective time compared with that of delays, to analyze the reasons for ineffective time and to ascertain the percentage time spent by a worker, groups of workers or a machine on a given work element. These, in themselves, provide useful information in a simple and reasonably quick way.

4.19.10. Rated work sampling

We have discussed the problem of rating a worker's performance relative to a perceived standard pace. Thus qualified workers who work according to a specified method and who are motivated to apply themselves to work briskly but naturally without over-exertion, are said to be working at 100 per cent standard rating on the performance scale. As rating is an important factor in deriving a time for an operation since not all workers work at the same pace. As a result, a work study person has to take the pace of work into consideration when timing a study.

This raring of pace can equally be combined with work sampling to give what is known as rated work sampling or rated activity sampling.

In this method, observations are taken at fixed intervals rather than at random times. When using fixed interval sampling, care must be taken to ensure that the fixed interval selected does not coincide with a natural cycle in the work. Such a coincidence would distort the results, but generally if the interval is short enough when compared to the overall job cycle time, normal variations in the work will avoid such a problem occurring.

During the sampling study, in addition to the activity being undertaken at the instant of the observation, a recording is also made of the pace of the worker using a performance rating scale. This rating can be used to modify the results of the study through the process of extension (converting observed times to basic times) which is already discussed.

4.19.11. Group sampling techniques

As the name suggests, these are designed for the measurement of work carried out by group of workers. The techniques are sometimes referred to by the term 'high-frequency sampling' since when used for the measurement of short-cycle work, they use fixed short-time intervals with the observer in constant attendance. They are thus very close to time study but have the advantage that the observer can cover the work of the group. Group sampling techniques may make use of rating.

Consider a very simple example of three workers, each producing the same parts by a process that involves only hand tools. The sampling is carried out at 30 sec (0.5 minutes) intervals and involves the categories of 'working' and 'not working' only. The sampling observations have been rated and this is thus an example of both rated activity sampling and group sampling. The sampling sheet would look as shown in Table 4.16 and followed by calculations of basic time.

Total Time of observations	= 300 minutes
Total Observations on each operator	= $300 \times 2 = 600$
Total number of observations for 3 operators	= $600 \times 3 = 1800$
Total number of observations (working)	= 1420
Total number of observations (not working)	= 180
Average rating of the worker	= 95% (based upon 100% standard performance)
Total number of components produced	= 75
Then total working time (observed time)	= $1420 \times 0.5 = 710$ minutes

Therefore, total basic time $\quad = $ Observed Time $\times$ Rating factor

$$= 710 \times \frac{95}{100} = 674.5 \text{ minutes}$$

Hence basic time per piece $\quad = \dfrac{674.5}{75} = 8.99 \text{ or } 9.0 \text{ minutes per piece}$

Table 4.16: Group sampling sheet with rating factors.

Time	Operator-1		Operator-2		Operator-3	
	Working	Not Working	Working	Not Working	Working	Not Working
8:000	90				90	
8:005			85			
8:010	97		95		97	
8:015			100		85	
8:020	80		80			
8:025	90		90		90	
8:030	95		95		95	
...	...	...	...	...	...	...
...	...	...	...	...	...	...
4:580	95		95		95	
4:585	85				85	
4:590			95			
4:595	95		95		95	
4:600	85		85		85	

4.19.12. Using work sampling

Work sampling either individually or by group, with or without rating, is widely used. It is a relatively simple technique that can be used beneficially in a wide variety of situations, such as manufacturing, service and office operations. Apart from providing a quick result, it is a fairly low-cost method and one that is less controversial than time study. The information derived from work sampling can be used to provide for a more reasonable distribution of work in a group and, in general, to provide the management with the percentage of ineffective time and also reasons behind the same. As a result it may indicate where method study needs to be applied, materials handling improved or better production planning methods introduced, as may be the case if work sampling shows that a considerable percentage of machine time is spent idle, waiting for supplies to arrive.

Solved Problems

1. In a hand tool manufacturing unit, for a definite aspect of job as; punching the basic (normal) time established is 20 sec, if for 3 observations a time study observer records ratings of 110, 125 and 90, respectively on a 0–100 normal scale. What are the observed timings?

Solution:

Given: standard rating = 100, Observed rating = 110, Basic Time = 20 sec

Observed Time × Observed Rating = Basic Time × Standard Rating

For observation no. 1

$$\text{Observed time} \times 110 = 20 \times 100$$

$$\text{Observed time} = 20 \times \frac{100}{110} = 18.18 \text{ sec}$$

For observation no. 2

$$\text{Observed time} = 20 \times \frac{100}{125} = 16 \text{ sec}$$

For observation no. 3

$$\text{Observed time} = 20 \times \frac{100}{90} = 22.22 \text{ sec}$$

2. An eight hour work measurement study in a plant reveals that number of units produced = 320, ideal time =15 per cent, performance rating =120 per cent, allowance = 12 per cent of normal time. Determine standard time/unit produced.

Solution:

$$\text{Observed time for 320 unit} = (\text{working time - ideal time}) \text{ hours}$$

$$= (8.0 - 8.0 \times 0.15) = 6.8 \text{ hours}$$

$$= 6.8 \times 60 = 408 \text{ minutes}$$

$$\text{Observed time per unit} = \frac{408}{320} = 1.275 \text{ minutes}$$

$$\text{Normal time per unit} = \frac{(\text{observed time per unit} \times \text{observed rating})}{\text{standard rating}}$$

$$= \frac{1.275 \times 120}{100} = 1.53 \text{ minutes}$$

Standard Time per unit (ST) = (Normal Time + Normal Time × Allowance Factor)

$$= \left(1.53 + 1.53 \times \frac{12}{100}\right) = 1.714 \text{ minutes}$$

3. Calculate the standard production per shift of 8 hours duration with the following data. Observed time per unit = 5 minutes, rating factor=120 per cent total allowance =33 1/3 per cent of normal time.

Solution: Normal time/unit = observed time per unit × rating factor

$$= 5 \times \frac{20}{100} = 6 \text{ minute}$$

Allowances = 33.33% of normal time = 2 minute
Standard time/unit = (Normal time per unit + Allowances)

$$= 6 + 2 = 8 \text{ minutes per unit}$$

Standard production in shift of 8 hour = $8 \times \dfrac{60}{8}$ = 60 units

4. In a time study, the stopwatch observations (expressed in hundredths of a minute) are as follows for the most variable element of the study: 10, 9, 11, 10, 9, 11, 10, 11, 12, 11, 13, 10, 12, 11, 12, and 13.

Elemental Time (sec) x	Frequency (f)	f × x	f × x²
9	2	18	162
10	4	40	400
11	5	55	605
12	3	36	432
13	2	26	338
	$N = \Sigma f = 16$	$\Sigma fx = 175$	$\Sigma fx^2 = 1937$

Using Eq. (4.1)

$$N = \left(\frac{(2)}{(0.05)} \frac{\sqrt{n \Sigma X^2 - (\Sigma X)^2}}{\Sigma X} \right)^2$$

$$N = \left(\frac{40 \times \sqrt{16 \times 1937 - (175)^2}}{175} \right)^2$$

$$N = 19.17 \text{ or } 19 \text{ (Approx.)}$$

Since the number of observations taken are 16, which is less than 19, therefore there is need to conduct more number of observations.

5. A work study engineer conducted a work sampling study and found that 20 per cent of a work week of 48 hours was consumed by unavoidable delays. If each time work sampling study was made the operator was rated. The average of such rating was 105 per cent. If 100 units were produced by the operator in that period, determine the standard time.

Solution:
Duration of work in a week = 48 hours
Percentage working time = 80%
Therefore, actual time worked by the operator = (48 × 60 × 0.80) = 2304 minutes

$$\text{Actual time taken per piece} = \frac{2304}{100} = 23.04 \text{ minutes}$$

$$\text{Standard time} = \frac{\text{Actual time per piece} \times \text{Rating \%}}{100}$$

$$= \frac{23.04 \times 105}{100} = 24.192 \text{ minutes}$$

6. Calculate the standard time per article produced from the data obtained by a work sampling study; total number of observation = 2500, working observation = 2100, number of units produced in 100 hours duration = 6000. Proportion of manual labour = 2/3, proportion of machine time = 1/3, observed ratting factor = 115 per cent and total allowances = 12 per cent of normal time.

Solution:

Actual working time in the duration of 100 hours $= \frac{2100}{2500} \times 100 = 84$ hours

Time taken per article $= \frac{84 \times 60}{6000} = 0.84$ minutes

Observed manual labour time per article $= 0.84 \times \frac{2}{3} = 0.56$ minutes

Observed machine time/article $= 0.84 \times \frac{1}{3} = 0.28$ minutes

Normal labour time/unit = observed time per unit × rating factor

$$= 0.56. \times 1.15 = 0.644 \text{ minutes}$$

Standard labour time/unit $= (\text{Normal Time}) \times \left(1 + \frac{\text{Allowances \%}}{100}\right)$

$$= 0.644 \times \left(1 + \frac{12}{100}\right) = 0.721 \text{ minutes}$$

Standard time per unit of article produced = 0.721 + 0.28 = 1.0 minutes

7. The time for manufacture of 4 pieces of the item was observed during time study. The manufacture of the item consist of 4 elements a, b, c, d the data collected during the time study are as under. Time observed (minutes) during the various cycles are as below:

Element	Cycle Time (minutes)				Element Rating on B.S. Scale (0–100)
	1	*2*	*3*	*4*	
A	1.2	1.3	1.3	1.4	85
B	0.7	0.6	0.65	0.75	120
C	1.4	1.3	1.3	1.2	90
D	0.5	0.5	0.6	0.4	70

Calculate the production cost per piece if data reveal a direct material of Rs. 2 per piece, the wage rate Rs. 2000 per month consisting of 25 working days and 8 hours per day, overhead expressed as 200 per cent of direct labour cost. The personal, fatigue and delay allowances may be taken as 25 per cent of basic time.

Solution:

Step 1. Calculate the standard time for the job based on data given

Element	Average Observed Time (O.T) in Minutes	Normal time = $O.T \times \dfrac{\text{Observed Rating}}{\text{Std. Rating}\,(100)}$
A	$\dfrac{1.2 + 1.3 + 1.3 + 1.4}{4} = 1.3$	$= 1.3 \times \dfrac{85}{100} = 1.105$
B	$\dfrac{0.7 + 0.6 + 0.65 + 0.75}{4} = 0.675$	$= 0.675 \times \dfrac{120}{100} = 0.81$
C	$\dfrac{1.4 + 1.3 + 1.3 + 1.2}{4} = 1.3$	$= 1.3 \times \dfrac{90}{100} = 1.17$
D	$\dfrac{0.5 + 0.5 + 0.6 + 0.5}{4} = 0.5$	$= 0.5 \times \dfrac{70}{100} = 0.35$

Normal time for the job = 3.435 minutes

Standard time for the job = normal time + allowances

$$= 3.435 + \frac{25}{100} \times 3.435 = 4.3 \text{ minutes}$$

As this time is a time taken for producing 4 pieces

Therefore, the standard time per piece = $\dfrac{4.3}{4}$ = 1.075 minutes

Step 2. Calculation of production cost

Direct labour cost of the job = standard time/job in hours × labour rate/hour

Labour rate/hour = $\dfrac{2000}{25 \times 8}$ = Rs. 10

Direct labour cost for the job = $1.075 \times \dfrac{10}{60}$ = Rs. 0.18

Direct material cost per piece = Rs. 2

Overhead cost at the rate of 200% of labour cost = $\dfrac{200}{100} \times 0.18$ = Rs. 0.36

Total production cost per piece = 0.18 + 2 + 0.36 = Rs. 2.54

8. A work sampling study is to be made of a typist pool. It is felt that typist is ideal 30 per cent of time. How many observations should be made in order to have 95 per cent confidence so that accuracy is within ±4 per cent.

Solution:

No of observation required for work sampling study = $N = C^2 \times p \times \dfrac{q}{E^2}$

Where C = constant depending on confidence level
P = % of idling, q = % of activity, E = absolute error
$C = 2$ for 95.5% confidence level, $p = 0.3$, $q = 1 - 0.3 = 0.7$, $E = \pm 4\% = 0.04$

$$N = 4 \times 0.3 \times \frac{0.7}{0.04^2} = 525$$

Hence, $N = 525$

9. A work study engineer conducted stopwatch time study on a job, for taking the observations the job was divided into 5 elements. The observations made on 4 cycles (in minutes) of all the 5 elements are shown in the table given below:

Element	Time (minutes) for Cycle				Performance Ratting
	1	2	3	4	
1	1.246	1.328	1.298	1.306	90
2	0.972	0.895	0.798	0.919	100
3	0.914	1.875	1.964	1.972	100
4	2.121	2.198	2.146	2.421	110
5	1.253	1.175	1.413	2.218	100

Calculate the normal time and standard time for the job. If a relaxation allowances of 12 per cent, contingency allowance of 3 per cent and an incentive of 20 per cent are applicable for the job.

Solution:

The normal time is calculated on the basis of data excluding the outliers as below:

Element	Mean Actual Time (minutes)	Performance Rating	Normal or Basic Time (minutes)
1	1.295	90	$1.295 \times 90/100 = 1.165$
2	0.896	100	$0.896 \times 100/100 = 0.896$
3	1.681	100	$1.681 \times 100/100 = 1.681$
4	2.222	110	$2.222 \times 110/100 = 2.444$
5	1.515	100	$1.515 \times 100/100 = 1.515$

Normal time for total job which include all the five elements = 7.701 minutes

$$\text{Standard time for the job} = (\text{Normal Time}) \times \left(1 + \frac{\text{Allowances \%}}{100}\right)$$

$$= (7.701) \times \left(1 + \frac{15 \%}{100}\right)$$

$$= 8.856 \text{ minutes}$$

Since 20 per cent incentive allowance are given,

Hence, total time allowed under incentive scheme = $8.856 + \frac{20}{100} \times 8.856 = 10.63$ min

10. A company 'ABC' is working 8 hours a day and 6 days a week, a work sampling study was conducted for two weeks and the following observations are made. Machine controlled elements 480, manual controlled elements 240, personal needs 50, tool grinding and other contingencies 90, unavoidable delays 66. Total observations were made 926. Number of pieces produced 828. If rating index is 110 per cent, determine the standard time per piece.

Solution: Given is:

M/C control elements	$= 480$
Manual control elements	$= 240$
Personal needs	$= 50$
Tool grinding and other contingencies	$= 90$
Total no. of observations	$= 926$

Total production activity elements $= 480 + 240 = 720$

$$\text{Percentage of production time} = \frac{720}{926} \times 100 = 77.75\%$$

Total clock time for the period under study $= 2 \times 6 \times 8 \times 60 = 5760$ minutes

$$\text{Total production Time} = \frac{5760 \times 77.75}{100} = 4478.40$$

No. of pieces produced $= 828$

$$\text{Actual time per piece} = \frac{4478.40}{828} = 5.41$$

$$\text{M/C control portion of production time} = \frac{480}{720} \times 100 = 67.67\%$$

$$\text{M/C time per piece} = \frac{5.41 \times 67.67}{100} = 3.66$$

Manual control time (per piece) $= 5.41\text{-}3.61 = 1.75$

Rating Index is applicable only for manual control elements (operations)
Therefore,
Normal Time = M/C time + [Rating Index × Manual Control Elements (Operation time)]

$$= 3.61 + (1.1 \times 1.80)$$

Normal Time $= 5.59$ minutes

The personal need, contingency and unavoidable delays are eligible for allowances. Hence they can be clubbed together to determine the total allowance.
The total number of observations $= 50 + 90 + 66 = 206$
The percentage of allowances are determined based on the portion of these elements w.r.t the total production activity elements.
Therefore;

$$\text{Total allowances} = \frac{206}{720} \times 100 = 28.60\%$$

$$\text{Std. time} = \text{Normal Time} \,[1 + \text{Allowances}]$$

$$= 5.59 \,[1 + 0.286]$$

$$= 7.21 \text{ minutes}$$

11. In an auto parts manufacturing company a production study was conducted for a period of one shift (eight hour). It was observed that; production time constitutes of 360 minutes. Time consumed for; personal needs = 15 minutes, tool grinding =15 minutes, cleaning of work place = 5 minutes, preparation of work ticket = 3 minutes, talk with foreman = 5 minutes, talk with colleagues (not connected with production) = 20 minutes, the numbers of parts produced = 100. Normal time per piece from stopwatch time study= 3.45 minutes. The standard time given by the time study personnel = 4.05 minutes. What would be (a) fatigue, (b) personnel needs and (c) contingency allowance according to the production study? Does the total allowance compare favourably with the stopwatch time study?

Solution: As the reader may be thinking about; what production study is? Therefore before solving the problem, it is mandatory to define the term production study. Hence a production study is a continuous study of comparatively long period of one or more shifts; the main objective of production study is to examine an existing or proposed standard time, and to obtain information about other aspects which affects the rate of output.

According to the Production Study

(a) Fatigue allowance: it is defined as the ratio of difference between production time and normal time to the normal time; it can be given as follows:

$$\text{Fatigue Allowance} = \frac{\left(\text{Production Time} - \text{Normal Time}\right)}{\text{Normal Time}}$$

It is given that:

The number of piece produced = 100

As per the stopwatch time study the normal time = 3.45 minutes per piece.

Therefore, the normal time for 100 pieces produced = 3.45 × 100 = 345 minutes.

Production time = 360 minutes.

Therefore, fatigue allowance = $\dfrac{\left(360 - 345\right)}{345} \times 100 = 4.35\%$.

(b) Time for personal need, during the study = 15 minutes.

Normal time for items produced during the study = 345 minutes.

Therefore, personal need allowance = $\dfrac{15}{345} \times 100 = 4.35\%$.

(c) Contingency refer to the sum total time required for non-cycle preparation times and other unavoidable delay, excluding personal needs.

$$\text{Contingency allowance} = \frac{\left(\text{Contingency Time}\right)}{\text{Normal Time}}$$

In this problem, contingency time is taken as sum total of:

Time for tool grinding	= 15 minutes.
Time for clean the work place	= 5 minutes.
Time for preparation of work ticket	= 3 minutes.
Time for talk with foreman	= 5 minutes.
Total contingency time	= 28 minutes.

Note: Talk with foreman is considered as getting instructions from foreman, which is unavoidable.

Therefore, contingency allowance $= \dfrac{28}{345} \times 100 = 8.12\%$.

Now, the total percentage allowances according to the production study = (a) + (b) + (c)
Therefore, total percentage allowances according to the production study

$$= 4.35 + 4.35 + 8.12$$

Total percentage allowances = 16.82%

The allowed time is given by time study personal = 4.05 minutes.

Therefore, Total % allowance $= \dfrac{\left(\text{Allowed Time} - \text{Normal Time}\right)}{\text{Normal Time}} \times 100$

$$= \dfrac{\left(405 - 345\right)}{345} \times 100 = 17.40$$

Since the total percentage allowances as per production study is less than the total percentage allowance as per the time study, thus the production study compares favourably with time study.

Unsolved Problems

1. In a stopwatch time study, the elemental time observed in sec is; 10, 9, 10, 9, 10, 10, 11, 10, 10 and 11. Examine whether the number of observations are enough at ± 5 per cent accuracy with a 95 per cent confidence level.

2. The time recorded select time for a job is 6 minutes. The mean rating is 110 per cent. The rest allowance is 5 per cent. Process & contingency allowance have been given as 5 per cent. How many workers are needed to produce 300 pieces per day?

3. A job is consisting of 5 motion elements. The observed times for these elements made by a work study trainee are 0.15, 0.23, 0.07, 0.10, 0.52 minutes respectively. The motion time is calculated using MTM tables are 0.16, 0.25, 0.05, 0.10, and 0.50 minutes respectively. The trainee had given 110 per cent rating for the job. Is it high or low?

4. Production study has been conducted during an eight hour shift. It has been noted that production time = 353 minutes. Time for personal needs = 12 minutes. Time for tool grinding =15 minutes. Clean work place = 5 minutes, preparation for work ticket = 3 minutes, talk with foreman = 5 minutes, talk with colleagues (not connected with production) = 22 minutes, Numbers of parts produced = 110. Normal time per piece from stopwatch time study= 3.38 minutes. The allowed time is given by the time study personnel = 3.96 minutes. What would be (i) fatigue, (ii) personnel and (iii) contingency allowance according to the production study? Does the total allowance compare favourably with the stopwatch time study?

5. At the end of a 25 day study, 200 total hours had been checked or logged in by the machine operator, and 500 parts had been produced. Work sampling study had shown that out of 400 observations the operator had been found working during 300, if the company has a policy of 15 per cent allowance of personal, fatigue and delay requirements. Determine the standard time for the job.

6. During a work sampling study 200 observation were made. The number of observations recorded as non-working is 48. For 95 per cent confidence level, at what limits of accuracy the percentage of non-working could be reported? How many observations are further needed for accuracy of ± 5 per cent?

7. The following observations are made during work sampling study conducted during two weeks. Machine controlled elements 448, manual controlled elements 228, personal needs 42, tool grinding and other contingencies 75, unavoidable delay 52. Total observation made 845. Number of pieces produced 768. Rating index is 110 per cent. The firm is working 8 hours a day and 5 days a week. Determine the Standard time per piece.

Multiple Choice Questions

1. Work measurement is the technique is used to determine the
   - (a) Work load
   - (b) Work content
   - (c) Standard time
   - (d) Type of work
2. The term PMTS stands for
   - (a) Pre-determined motion time system
   - (b) Post-determined motion time system
   - (c) Provisional motion time system
   - (d) Predetermined method time system
3. The term EMTS termed as
   - (a) Elemental motion times system
   - (b) Elemental method times system
   - (c) Elemental motion times sequence
   - (d) None of these
4. BMTS stands for
   - (a) Basic motion time system
   - (b) Base motion time system
   - (c) Basic method time system
   - (d) Basic motion time sequence
5. Which of the following is a work measurement technique?
   - (a) Synthesis-synthesized time standard
   - (b) Analytical estimating
   - (c) Work sampling
   - (d) Activity sampling
   - (e) All of these
6. Time study was developed by
   - (a) M. J. Griffin
   - (b) A. S. Thomson
   - (c) F. W. Taylor
   - (d) None of these
7. Time study is the basic technique for
   - (a) Time measurement
   - (b) Work measurement
   - (c) Motion analysis
   - (d) All of these
8. Time study of any job is a straight forward approach of
   - (a) Standard rate of working
   - (b) Partial rate of working
   - (c) Predetermined rate of working
   - (d) Final rate of working
9. Which of the following is not a part of time study procedure?
   - (a) Identification of job
   - (b) Selection of qualified worker
   - (c) Determine number of observations
   - (d) Critical examination of operation
10. Nominal time is calculated with the help of
    - (a) Predetermined rating factor
    - (b) Performance rating factor
    - (c) Post determined rating factor
    - (d) All of these
11. The formula to calculate the standard time is
    - (a) Standard time = Basic time + Allowances
    - (b) Standard time = Observed time + Allowances

(c) Standard time = Basic time + Rest pauses

(d) Standard time = Predetermined time + Allowances

12. The basic time is calculated as

(a) Basic time = Observed time × Observed rating/Standard rating (100)

(b) Basic time = Predetermined time× Observed rating/Standard rating (100)

(c) Basic time = Time × Observed rating/Standard rating (100)

(d) Basic time = Observed time × Observed rating/Predetermined rating (100)

13. Basic time is also termed as

(a) Rating time (b) Predetermined time

(c) Normal time (d) Recorded time

14. which of the following is not a part of stop watch time study?

(a) Stop watch (b) Time study forms

(c) Pencil (d) Time study board

(e) None of these

15. Which of the following element is a part of element breakdown in time study?

(a) Repetitive element (b) Occasional element

(c) Constant element (d) All of these

16. If a worker working at 110 per cent rating complete his job in 10 minutes, then the basic allowed time is.

(a) 8 minutes (b) 10 minutes

(c) 11 minutes (d) 12 minutes

17. If an operation has a cycle time of 45 sec, then the required number of observations is approx.

(a) 30 (b) 50

(c) 40 (d) 60

18. In a garment manufacturing company the operators are engaged on sewing machines, and need to concentrate visually on job, what should be the range of concentration allowance.

(a) 0–5% (b) 0–15%

(c) 0–10% (d) 0–20%

19. In a work sampling study 200 observations were carried out as a preliminary study and these showed the machine to be idle in 20 per cent of the cases ($p = 20$) and to be working 80 per cent of the time (q = 80). The required number of observations at a confidence level of 95 per cent with a ± 5 per cent margin of error will be.

(a) 300 (b) 250

(c) 340 (d) 256

20. An 8 hour work measurement study in a plant reveals that; number of units produced = 408, ideal time = 15 per cent, performance rating = 120 per cent, allowances = 12 per cent of normal time, determine standard time /unit produced.

(a) 1.2 minutes (b) 2.1 minutes

(c) 1.34 minutes (d) 3.2 minutes

Answers

1. (c)	**2.** (a)	**3.** (a)	**4.** (a)	**5.** (e)	**6.** (c)	**7.** (b)	**8.** (a)	**9.** (d)	**10.** (b)
11. (a)	**12.** (a)	**13.** (c)	**14.** (e)	**15.** (d)	**16.** (c)	**17.** (c)	**18.** (c)	**19.** (d)	**20.** (c)

Wages and Incentive Plans

5.1. Introduction

In any organization no work can be accomplished without involving manpower; hence, the efforts and time devoted by the manpower are payable in terms of wages and incentives. Before we discuss the various wages and incentive schemes, let us first define these terms. The term wages refers to the payment or remuneration made by the employer for the efforts put in by the worker, employee or staff. According to the Payment of Wages Act, *'wages refer to all remunerations capable of being expressed in terms of money, which would, if the terms of the contract of employment, expressed or implied, were fulfilled; be payable to a person employed in respect of his or her employment or of work done in such employment and includes any bonus or other additional remuneration of the nature aforesaid which should be so payable'*. Thus it may be observed that the wage is a contractual obligation of the management to be paid to the persons employed by them. Wages are tangible and have to be paid in terms of money. The basic wage of a worker, once fixed by the management, must not be altered and has to be paid irrespective of the quality or quantity of work.

5.1.1. Incentive

Incentive may be defined as an influence which tends to produce an increased effort on the part of the employee above a standard. It provides the stimulus to motivate the worker towards higher productivity. Incentives can be negative or positive. The disciplinary actions taken against non-compliance are termed as negative incentives. Psychologically negative incentives are not conducive to human-relations. Positive incentives refer to both extra financial and non-financial benefits provided by the management. Positive incentives would lead to increased production levels and productivity. Non-financial incentives are generally more welcome by the executives. Financial incentives are more effective with the workers.

5.1.2. Incentive plan

Incentive plan refers to financial incentive based on the wage structure. Incentive plan is a scheme of arrangement, which lays down a systematic procedure of payment to the employee based on the performance of the task set. It is developed in such a manner as to produce benefit to both the employee and the employer. The earnings of the worker bear a direct relationship to the quantity of output by the individual or by the group in which he/she belongs.

5.1.3. Requirements of a good incentive plan

1. The incentive should be significant and there should be no restrictions on the earnings of the workers.
2. The minimum wage should be guaranteed.
3. The reward should be at least proportional to the extra efforts put forth by the worker.
4. It is essential that proper work study has been conducted, job methods are standardized and the standard times are available for the same, before implementing the Wage Incentive Plan.
5. Adequate precautions should be taken to ensure that quality does not suffer for quantity.
6. The plan should aim at higher labour earnings and lower production costs.
7. The plan should be simple to understand by the workers and easy to operate by the management.

5.2. Types of Wage Incentive Plans

Wage incentive plan cannot be same for all industries, therefore a range of plans may be prevailing among them. A wage or incentive plan depends upon the nature of the industry and conditions of production. The policies of the management and union have a substantial role on the choices and implementation of the incentive plan.

The wage incentive plans are classified as follows:

1. Time Rate Schemes (Day Work Scheme/Time Wage System)
2. Piece Rate System
3. Differential Piece Rate System
4. Task and Bonus Plan
5. Premium Bonus Schemes
6. Standard Time Incentive Schemes (Work Study Incentive Schemes)
7. Group Incentive Schemes
8. Profit Sharing
9. Co-partnership

A firm may adopt any one or more types of incentive plan. Let us consider the types of incentive plans and their implications.

5.2.1. Time rate schemes

According to this scheme, a worker is paid for the period of work at rate fixed per hour/day consideration of output in terms of production. Time rate wage schemes do an appreciable incentive. It requires close and active supervision.

(a) *Straight Day Work Scheme*: Worker's rate per hour is fixed and the daily wage depends upon the hours of work, he/she had worked during the day. If the rate is Rs 30/hour and one has worked for 6 hours only during the day, he/she would get Rs 180 for that day. This scheme is not practised generally.

(b) *Graded Day Work (Graded Time Rate) Scheme*: In this scheme, higher rate of payments are offered whenever better performance is achieved.

5.2.2. Piece rate system

Here the worker is paid on the basis of output rather than on the time spent by him on the job. This may be in the form as:

(a) *Straight piece rate system*: Piece refers to a measurable unit of production, it may be in the terms of the number of items or specific unit of area. The rate per piece in rupees, (R) is fixed. If N number of pieces are produce per day, then the earnings per day (E) = Number of pieces produced × Rate per piece, i.e., $E = N \times R$

Example: A worker is working at a standard piece rate of Rs 0.95 per piece, what will be his daily earning if he produces 200 pieces in 8 hour day.

Solution:

Standard piece rate = Rs 0.95/piece

Number of pieces produced = 200

Earning per day $E = N \times R$

$E = 200 \times$ Rs 0.95

$E =$ Rs 190/day

(b) *Piece rate with minimum wage guarantee*: In this case, the basic wage per day (W) is fixed. As before, the rate per piece (R) is also fixed. If a person had produced number of pieces in a day, then his wage for that day is calculated as Max ($W, N \times R$). In this manner the worker is assured of his wage rate (W). If he could produce more pieces, and then he would be paid on piece rate system.

   (i) This system is applicable, where the work is standardized with the establishment of standard times and the work is repetitive in nature and the output is capable of being measured in the form of piece (units).

  (ii) The workers earnings are proportional to the output.

 (iii) It is easy to calculate the wages as well as labour cost per unit of production.

 (iv) The worker is rewarded based on the efforts put in by him.

  (v) There is no need for close supervision.

Example: In a small scale unit, the minimum guaranteed wage is fixed at Rs 185 per day. The standard piece rate is Rs 0.95 per piece. Two workers A and B are producing 200 and 190 pieces per day respectively, what would be their daily wages?

Solution:

The given minimum guaranteed wage 'W' = Rs 185/day

Standard piece rate = Rs 0.95/piece

Number of pieces produced by worker 'A' = 200

Earning of A per day $E =$ Max ($W, N \times R$)

$E =$ Max (185, 200 × Rs 0.95)

$E =$ Max (185, Rs 190)/day

Therefore, earning of 'A' per day $E =$ Rs 190/day

Now, since number of pieces produced by worker 'B' = 190

Earning of B per day $E =$ Max ($W, N \times R$)

$E =$ Max (185, 190 × Rs 0.95)

$E =$ Max (185, 180.5)

Hence earning of 'B' per day $E =$ Rs 185/day, worker B will get minimum guaranteed wage.

5.2.3. Differential piece rate systems

In this system, two or more rates per piece are being followed; instead of a single uniform rate up to a particular level of production a rate per piece is fixed. For higher rates of outputs by the worker, higher rates per piece are paid.

(a) *Taylor's differential piece rate system*

This has been originally developed by F. W. Taylor as an outcome of his time study technique. A standard output per hour for every job is fixed based on the time study. When the number of pieces per hour is equal to or greater than the standard output, a higher rate per piece is fixed. When the number of pieces produced is less than the standard output, and then a lower rate per piece is paid.

Example: The standard time per piece of production is 2 minutes. The differential rates are Rs 0.80 and Rs 0.70 per piece. Workers A and B are producing 30 and 25 pieces per hour respectively. If both of them keep the same tempo for the whole day (of 8 hours), what would be their daily wages?

Solution: The standard time per piece = 2 minutes

Therefore, the standard output per hour = $\dfrac{60}{2}$ = 30 pieces/hour.

The higher piece rate (0.80) is for outputs equal to or greater than standard output, i.e., 30 pieces per hour. The lower piece rate (0.70) is for outputs less than 30 pieces/hour (standard output).

Worker 'A' produces 30 pieces/hour and his piece rate = Rs 0.80/piece
Hence, worker A's wage/hour = 30 × 0.80 = Rs 24.
A's daily wage = 24 × 8 =Rs 192.0

Worker 'B' produces 25 pieces per hour and hence his piece rate = Rs 0.70/piece
Hence, worker B's wage/hour = 25 × 0.70 = Rs 17.5
B's daily wage =17.5 × 8 = Rs 140.0

In this system, the slow worker is penalized for not achieving the expected standard rate of production.

(b) *Merrick's differential piece rate system*

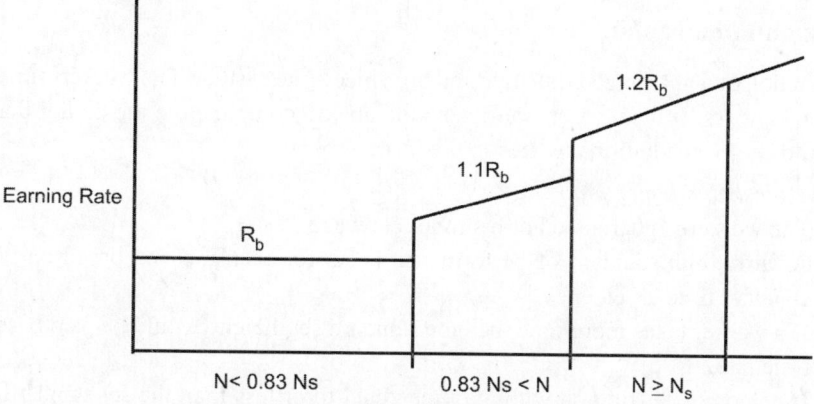

Fig. 5.1: Merrick's differential piece rate system.

In this system, the slow and moderate workers are not penalized. The standard output is fixed based on work measurement. If a person is able to achieve up to 83 per cent of standard output, he is assured of the basic piece rate being paid. If the output is equal to 83 per cent of the standard output, but less than the standard output, the piece rate is 10 per cent more than the basic price rate. For output equal to or more than the standard output, he would be paid at a piece rate of 20 per cent more than the base rate.

Assume that N_s is the standard output piece per hour and R_b is the basic piece rate per piece, and if the number of piece produced by a worker is N pieces per hour, then Merrick's differential piece rate can be presented as under:

If $0 \leq N < 0.83$ Ns, then the piece rate is R_b.

If 0.83 Ns $\leq N < N_S$, then the piece rate is $1.1\ R_b$.

If $N \geq$ Ns, then the piece rate is $1.2\ R_b$

Example: In a hand tool manufacturing industry, two workers A and B are engaged in grinding operation. If the standard output is 20 pieces per hour, and the basic piece rate is Rs 0.80 per piece, what would be the daily wage for the workers A and B who are producing at a rate of 20 and 19 pieces per hour under Merrick's Scheme?

Solution:

Let us first calculate the minimum level of output for entitlement of basic wage rate.

$$0.83\ N_s = 0.83 \times 20 = 16.6 = 17 \text{ pieces (Approx.)}.$$

The output rate for A = 20 pieces/hour which is equal to the standard output N_s.
Hence, the piece rate is $1.2 \times R_b$.
As the basic piece rate is $R_b = 0.80$,
Hence, A's piece rate = 1.2×0.80 = Rs 0.96 per piece
Hence, A's daily wage = $20 \times 0.96 \times 8$ = Rs 153.6
The output rate for B = 19 pieces/hour. It is greater than $0.83\ N_s$, (17) and less than N_s (=20). Hence the piece rate for B is $1.1 \times R_b$

$$= 1.1 \times 0.80 = \text{Rs } 0.88/\text{piece}$$

Hence, B's daily wage = $19 \times 0.88 \times 8$ = Rs 133.76

5.2.4. Task and bonus plan

It is a combination of a guaranteed time rate and high piece rate system. The workers time rate wages are guaranteed always. If the worker is efficient and able to complete the job earlier than the given time, he would be given additional wages in the form of bonus.

(a) *Gantt's task and bonus plan*
- The worker is guaranteed of his time rate wage.
- A fairly high standard of performance is set for each job and based on which the standard time is set.
- If a worker takes more time than the standard set then he would be paid based on the time rate.
- If he completes the job within a time equal to or less than the set standard, then he would get extra bonus (usually 20 per cent).

Example: The standard time for producing a piece is fixed as 180 minutes. A, produces 3 pieces per day and B could produce only 2 pieces per day. For an eight hour working day and 20 per cent bonus, what would be the daily earnings of A and B under Gantt's Plan? The hourly rate is Rs 15 per hour.

Solution:

The standard time is 180 minutes, i.e., $\dfrac{180}{60}$ = 3 hours/piece.

A produces 3 pieces per day, i.e., at rate of $\dfrac{60 \times 8}{3}$ = 160 minutes/piece.

Since time taken by worker 'A' is lesser than the standard time, hence he is eligible for the 20% bonus. Therefore his, rate per piece produced = $(3 \times 15) \times \dfrac{120}{100}$ = Rs 54.

As the number of pieces produced by worker 'A' is 3/day.
Hence worker A's daily earnings is 3 × 54 = Rs 162

B produces 2 pieces per day, i.e., at a rate of $\dfrac{60 \times 80}{2}$ = 240 minutes per piece. This is more than than the fixed standard time and he is eligible for time rate only.
Hence B's daily earning is 15 × 8 = Rs 120.

(b) *Emerson's efficiency plan*

This is similar to Gantt's plan; but instead of a fixed bonus, there is a graded scale for the bonus based on the efficiency of the worker.

- Worker is guaranteed of the time rate.
- A standard of performance is set for each task.
- Worker starts earning bonus even if he reaches 2/3 (66.67 per cent) efficiency.
- From 66.67 per cent to 80 per cent efficiency, the bonus is 4 per cent.
- From 81 per cent to 90 per cent efficiency, the bonus is 10 per cent.
- From 91 per cent to 100 per cent efficiency, the bonus is 20 per cent.
- For more than 100 per cent, for every increase of 1 per cent efficiency, the bonus is further increased by 1 per cent.
- Thus a moderate slow worker is also eligible for bonus.

Examples: The standard time per piece is fixed at 10 min. The daily wages is fixed as Rs 190 per day of 8 hours . What would be the daily earnings of workers A, B, C, D and E, if they produce 30, 38, 42, 48 and 55 pieces per day according to Emerson's plan?

Solution:

As the standard time is set at 10 minutes/piece,

$$\text{Therefore, the standard output} = \frac{8 \times 60}{10} = 48 \text{ pieces/day}$$

$$\text{Efficiency of a worker} = \frac{\text{Actual output}}{\text{Standard output}}$$

$$\text{Efficiency of Worker 'A'} = \frac{30}{48} = 62.5\%;$$

Hence, worker 'A' is eligible only for daily wages of Rs 190
Hence A's daily earning = Rs 190

Efficiency of Worker 'B' = $\dfrac{38}{48}$ = 79.17%;

Hence, worker 'B' is eligible for a bonus of 4%
Hence B's daily earning = 190 × 1.04 = Rs 197.6

Efficiency of Worker 'C' = $\dfrac{42}{48}$ = 87.5%;

Hence, worker 'C' is eligible for a bonus of 10%
Hence C's daily earning = 190 × 1.10 s Rs 209/ -

Efficiency of Worker 'D' = $\dfrac{48}{48}$ = 100%;

Hence, worker 'D' is eligible for a bonus of 20%
Hence D's daily earning = 190 × 1.2 = Rs 228

Efficiency of Worker 'E' = $\dfrac{55}{48}$ =114.5%;

Hence, worker 'E' is eligible for a bonus of 20% +14% = 34%
Hence E's daily earning = 190 × 1.34 = Rs 254.6

5.2.5. Premium bonus schemes

In this scheme, a worker is given wages on basic hourly rate and extra bonus is given for the increase in productivity in the form of time saved in completing the job.

(a) *Halsey's premium bonus plan*
- This plan, originated in the USA, was formulated by F. A. Halsey.
- The standard time required for the completion of the job is fixed and agreed upon.
- Operator is paid for the hours worked the job at the basic hourly rate (R).
- If he completes the job in less than the standard time, he is paid an extra bonus based on the portion of the time saved multiplied by the basic rate.
- The portion may vary from 1/3 to 3/4 of the time saved.

Let us assume that T_s is the standard time in hours, T_a is the actual time taken for the job and α is the portion (P) to be taken on the time saved, and R is the basic hourly rate in Rs/hour, then

The earnings for the job, $E = RT_a + \alpha\,(T_s - T_a)\,R$.

The value of $\alpha\,(= p/100)$ is usually taken as 50 per cent. It is known as 50–50 Halsey bonus plan or split bonus plan. When $\alpha = \dfrac{1}{2}$, the earning for a job can be determined as follows:

$$E = RT_a + \frac{(T_s - T_a)}{2}\,R$$

$$= \frac{1}{2}\,(T_a + T_s)\,R.$$

$$E = 0.5\,(T_a + T_s)\,R$$

When the actual time taken for the job (T_a) is greater than or equal to the standard time (T_s), the worker is paid only the basic hourly rate and no bonus would be given, i.e., $E = RT_a$ (when $T_a \geq T_s$).

(b) *Weir scheme (Halsey–Weir premium plan)*

It is similar to Halsey plan. In this case, the bonus is 30 per cent of the time rate for the time saved; i.e., $\alpha = 0.3$.

The earning for the job, $E = RT_a + 0.3 \, (T_s - T_a) \, R$

$$E = R \, (0.7T_a + 0.3 \, T_s)$$

In this case the earning by the productive worker is less in comparison with Halsey's plan.

(c) *Rowan's premium plan*

This plan was developed by James Rowan and widely used in England. This is only a modification of Halsey's plan. The main difference is in the manner in which the bonus (premium) is given. Here the premium (bonus) is based on the proportionate (percentage) of the time saved. The premium hourly rate is base rate multiplied by the percentage time saved.

The earnings for the job, $E = RT_a + \dfrac{(T_s - T_a)}{T_s} \times RT_a$

$$E = RT_a + \dfrac{(T_s - T_a)}{T_s} \times RT_a = RT_a \left[2 - \dfrac{T_a}{T_s} \right]$$

(d) *Barth's premium plan*

Here the minimum wages are not guaranteed. It is mainly more useful for new entrants who have not developed their full capacity of higher rates of output.

The standard time (T_s) is fixed for the job. If R is the basic wage rate in rupees per hour, the actual hourly rate for the worker is taken as $\sqrt{\dfrac{T_s}{T_a}} \times R$

The earning for the job $(E) = \sqrt{\dfrac{T_s}{T_a}} \times RT_a = \left(\sqrt{T_s \times T_a} \right) \times R$

Example: The basic rate for an operator is fixed Rs 30 per hour. The standard time for the job is set as 8 hours. The actual time taken by an operator is 6 hours. The firm follows Halsey plan (50–50). Determine each of the following:

  (i) The operators earning for the job,

 (ii) Operators earning rate per hour,

(iii) Direct labour cost per unit,

(iv) The saving in the direct labour cost,

 (v) The ratio of actual to standard labour cost,

(vi) The ratio of actual to standard output,

(vii) The ratio of actual to standard time rate.

Solution: The standard time for the job $(T_s) = 8$ hours.

Actual time taken by the operator $(T_a) = 6$ hours.

Basic Wage rate $(R) = $ Rs 30/hour.

Since 50 – 50 Halsey Plan is followed in the company, therefore;

  (i) Operator's earning for the job $(E) = 0.5 \, (T_s + T_a) \, R$

$$= 0.5 \, (8 + 6) \times 30 = \text{Rs } 210.$$

(ii) Operator's earning rate = Earnings for the job/Actual time

$$= \frac{E}{T_a} = \frac{210}{6} = \text{Rs 35 per hour.}$$

Operators earning per day = 35 × 8 = Rs 280

(iii) Direct labour cost per unit = Operators earnings for the job (E) = Rs 210.

(iv) The standard direct labour cost per piece

= Standard time per piece (T_s) × Standard labour rate (R)

= $T_s \times R = 8 \times 30 = $ Rs 240.

The actual direct labour cost per piece = Rs 210

The saving in the direct labour cost = 240 – 210 = Rs 50

(v) The ratio of actual/standard labour cost = $\frac{210}{240} \times 100 = 87.5\%$

(vi) The standard time per unit = T_s.

The standard output per hour = $\frac{1}{T_s}$ units

Similarly the actual output per hour = $\frac{1}{T_a}$

The ratio of actual/standard output per hour = $\frac{T_s}{T_a} = \frac{8}{6} \times 100 = 133\%$

(vii) The actual time rate is the (actual) earning rate of the operator = Rs 35/hour

The standard time rate = Rs 30 /hour.

Actual time rate/standard time rate = $\frac{35}{30} \times 100 = 117\%$

Example: For the data given in the previous example determine for which premium bonus scheme the operator would get the maximum wage rate?

Solution: The wage rate 'W' = Earnings for the job/Actual time taken in hours.

Therefore,   'W' = $\frac{E}{T_a}$

(i) The operator's wage rate (earning rate) for Halsey plan has already been found as Rs 35 per hour.

(ii) According to Weir scheme:

Operator's earning $E = R\,(0.7\,T_a + 0.3\,T_s)$

= 30 [0.7 × 6 + 0.3 × 8] = Rs 198

Wage rate 'W' = $\frac{E}{T_a} = \frac{198}{6} = $ Rs 33 per hour

(iii) Similarly, for Rowan Scheme,

The Wage Rate 'W' = $E/T_a = R\,T_a\left[2 - \dfrac{T_a}{T_s}\right]/T_a$

$$= R\left[2 - \frac{T_a}{T_s}\right] = 30 \times \left[2 - \frac{6}{8}\right] = \text{Rs } 37.5$$

(iv) For Barth scheme,

$$\text{Wage rate} = \frac{E}{T_a} = \frac{\left(\sqrt{T_s \times T_a}\right) \times R}{T_a} = R\sqrt{\frac{T_s}{T_a}} = 30 \times \sqrt{\frac{8}{6}} = \text{Rs } 34.62$$

Hence, the maximum wage rate earned by the operator is for the Rowan Scheme.

Example: Under what condition of production rate, the earning rate by the worker is same in both the Halsey and Rowan Plan?

Solution: Since the earning rate (wage rate 'W') = E/T_a, therefore the earning rate

under Halsey = $\dfrac{R}{2}\left[\dfrac{T_a + T_s}{T_a}\right]$

Similarly, the earning rate under Rowan = $R\left[2 - \dfrac{T_a}{T_s}\right]$

Let $\dfrac{T_s}{T_a} = \rho$ If the earning rates of both the systems are same,

then, $\dfrac{R}{2}\left[\rho + 1\right] = \dfrac{R}{2}\left[2 - \dfrac{1}{\rho}\right]$

i.e. $\dfrac{\rho + 1}{2} = \dfrac{2\rho - 1}{\rho}$, i.e., $\rho^2 - 3\rho + 2 = 0$

Solving the quadratic equation, we have $\rho = 1$ or 2.

When $\dfrac{T_s}{T_a} = 1$, then $T_s = T_a$ and the question of bonus does not arise.

Hence, when $\dfrac{T_s}{T_a} = 2$, the earning rates of both plans are same.

The actual production rate\standard production rate = $\dfrac{T_s}{T_a} = \rho = 2$. *Hence when*

actual production rate is twice the standard production rate, the earning rates of both these plans are same.

5.2.6. Standard time incentive schemes (work study incentive schemes)

In the standard time incentive schemes, every minute saved by the worker would give him the money worth of it. As it is essential that work study has to be conducted properly before implementing the scheme, it is also called as work study incentive scheme. The jobs are subjected to method study. The optimum method is chosen and specified along with tools, equipment and work place layout. Then the jobs are accurately time studied (by any form of work measurement), and the time standards are well set. The average level of production in the shop is to be about 80 per cent of total capacity. One has to be guarded against loose ratings as otherwise the labour cost would increase due to unwanted high bonus rates and the efficiency would go down. Separate as well as accurate records of time of the operator are to be maintained. The important concept in this system as against others is that in any job a portion of time would be spent towards (unauthorized) rest, waiting time, and non-repetitive type of work, etc. This is called as unmeasured work. The other portion of the time is only made for the actual repetitive work of the job for which time standards are available. This is called as measured work. Moreover, a minimum wage rate is fixed, which is 75 per cent of the standard wage rate. The wages and the bonus are given for the measured work portion of the time, based on standard wage rate. The standard time incentive schemes are of following types:

(a) 75–100 straight proportional scheme.
(b) 50–100 geared scheme.
(c) Bedaux point scheme.

(a) 75–100 straight proportional scheme
In this scheme, the minimum wage rate is given up to 75 per cent of the standard performance and also for the unmeasured portion of the time. The incentive starts from 75 per cent of the performance and increases linearly (i.e., proportionately to performance) afterwards for higher percentage of performance.

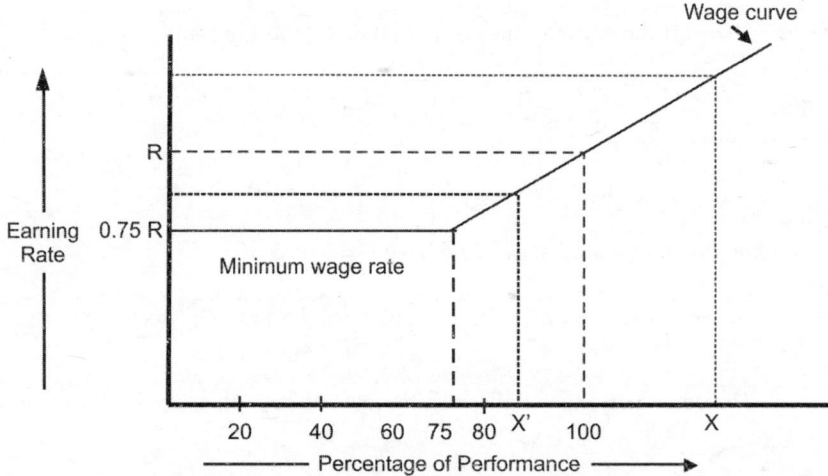

Fig. 5.2: Earning rate versus percentage of performance.

R = Standard wage rate.
Minimum wage rate = 0.75R

[**Note:** Standard wage rate = $\frac{4}{3}$ × minimum wage rate

From Figure 5.2, it may be noticed that if X is the percentage of performance ($X > 100$), earning rate for the measured portion of work can be found as:

$$E_R = R + \frac{R(X - 100)}{100}$$

where, $X > 100$

If the percentage of performance (X) is between 75 per cent and 100 per cent, then the earning rate is

$$E_R = \text{Min. Wage Rate} + \frac{R(X - 75)}{100}$$

where, $75 < X \le 100$

The proof is left as an exercise for the student. Prove also that in both the cases the earning rate, $E_R = \dfrac{RX}{100}$ when $X \ge 75$, and thus it is directly proportional to the percentage of performance.

Example: The standard time for producing one unit is fixed as 10 minutes. A worker produces units 50 pieces per day. The waiting time is 20 minutes and other unmeasured work takes 1 hour, the minimum guaranteed rate is Rs 20/hour, determine the earnings per day of the worker 75 –100 proportional scheme.

Solution: The minimum wage rate = Rs 20 per hour.

Hence, the standard wage rate $(R) = \dfrac{4}{3} \times$ Minimum Wage Rate

$$= \frac{4}{3} \times 20 = \text{Rs } 26.68 \text{ per hour}$$

The unmeasured work time = Waiting time + other unmeasured work time

$$= 20 \text{ minutes} + 1 \text{ hour} = 1 \text{ hour } 20 \text{ minutes}$$

$$= 80 \text{ minutes}$$

Assuming 8 hours working, the attendance time = $8 \times 60 = 480$ minutes.

The measured work time = $480 – 80 = 400$ minutes.

The number of units produced during the period = 50.

Actual time taken per unit= 400/50 = 8 minutes.

The standard time per unit = 10 minutes.

The percentage of performance $(X) = \dfrac{\text{Standard time}}{\text{Actual time}} = \dfrac{10}{8} = 1.25 = 125\%$

The earning rate for measured work $(E_R) = \dfrac{RX}{100}$

$$= 26.68 \times \frac{125}{100} = \text{Rs } 33.35 \text{ per hour}$$

The wages for the unmeasured work period $= \dfrac{80}{60} \times 20$

$$= \text{Rs } 26.68 \text{ per day}$$

The wages for the measured work period $= \dfrac{400}{60} \times 32.5$

$$= \text{Rs } 216.67 \text{ per day}$$

Total earnings per day = 216.67 + 26.68 = Rs 243.35

(b) 50–100 Geared scheme

This is similar to the previous one except that, (i) the incentive starts from 50 per cent performance level itself; (ii) the minimum wage rate is fixed at 75 per cent of the standard wage rate, which is given up to 50 per cent of performance level; and (iii) the wage curve increases proportionately (linearly from 50 per cent of performance level. The unmeasured work is paid at the minimum wage rate. It may give a satisfaction to the low performance worker, but the high performance worker would get less incentive when compared with 75–100 system, the system is exhibited in Figure 5.3.

The earning rate 'E_R' (for the measured work), is found as under:

$$E_R = \frac{R}{2}\left[\frac{X}{100} + 1\right]$$

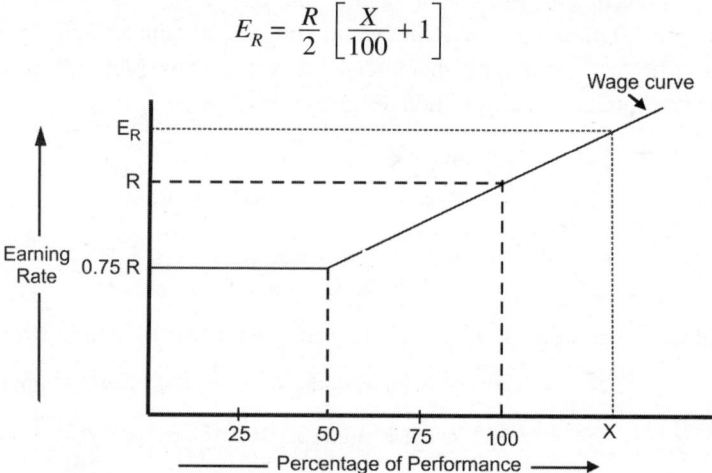

Fig. 5.3: Earning rate versus percentage of performance.

Example: For the same data given in the previous example, what would be the earnings per day based on 50 – 100 Geared scheme?

Solution: Since, there is no change in the wages for unmeasured work.
Hence the earnings for unmeasured work period = Rs 26.6/day.
The percentage of performance (X) = 125%
The standard wage rate (R) = Rs 26.68/hour

The earning rate for measured work (E) = $\frac{R}{2}\left[\frac{X}{100} + 1\right]$

$$\frac{26.68}{2}\left[\frac{125}{100} + 1\right] = \text{Rs 30.01 per hour}$$

Measured work period = 400 minutes

The wages (earnings) for the measured work $\frac{400}{60} \times 30.01 = \text{Rs. 200.067 per day.}$

Total earnings per day = 200.07 + 26.68 = Rs 226.75 per day

(c) Bedaux point scheme

In Bedaux point scheme, the standard time is given in the form of Bedaux Points (called as 'B' values). In fact it is only the standard time for a job in minutes. [Example: If the standard time for a task is 3.5 hours then the 'B' value for the task (Job) is 3.5 × 60 =

210, sometimes it is stated as 210 B's]. This plan also guarantees basic hourly rates (i.e., minimum wage rates), there is no difference between minimum and standard wage rates. A saving in time in terms of B's (i.e. in terms of minutes) calls for a premium.

- This method is easy to calculate and could be understood easily by the workmen.
- It has a wide operational scope and due to incentives for the foreman; also it can bring better production control.
- The main feature of this scheme is that, it provides incentive to the foreman (or the supervisor who supervises and gives instructions towards the work) to enable him to provide the necessary cooperation to the worker.
- The extra time saved is paid in the form of premium, at the base rate in which 25 per cent goes to the foreman (supervisor) and 75 per cent goes to the worker.

If T_s is the standard time in hours (i.e., $60\ T_s$ of B values) and T_a is the actual time for the job in hours, then;

the time saved is $= T_s - T_a$.

If R is the base rate in rupees per hour, then;

the premium for the worker $= 0.75\ (T_s - T_a)\ R$.

the premium for the foreman is $0.25\ ((T_s - T_a)\ R$.

The time saved is usually taken in minutes and called as Bedaux Points (B value base rate of wages is expressed as $\dfrac{R}{60}$ rupees per minute, i.e., as premium per Bedaux.

Example: In a small scale casting firm, a worker engaged in moulding section has completed six jobs during a week. The standard times for jobs are 5, 8, 6, 10, 20 and 15 hours respectively. If the base wage for the operator is Rs 20 per hour, what would be his earnings for the week under Bedaux Point Scheme?

Solution:

The total standard time for the six jobs $= 5 + 8 + 6 + 10 + 20 + 15 = 64$ hour

Assuming 6 days a week and 8 hours of work per day,

The actual time taken for the week $= 6 \times 8 = 48$ hours (T_a).

The time saved $= T_s - T_a = 64 - 48 = 16$ hours

The premium for the week $= 0.75\ (T_s - T_a) \times R = 0.75 \times 16 \times 20 = $ Rs. 240.

The time rate wage for the week $(W) = RT_a = 48 \times 20 = $ Rs. 960.

Hence the total earning during the week $= RT_a + 0.75\ (T_s - T_a) \times R$

$$= 960 + 240 = \text{Rs. } 1200$$

5.2.7. Group incentive scheme

These are generally applicable to a group of workers working as a team and given for the assembly work. A fixed percentage of their basic earning is given as a bonus when they reach the target. There are various methods through which the bonus can be fixed.

(a) Nunn–Bush plan

It is based on the concept that 20 per cent of the total factory sales revenue constitutes total direct labour cost. Thus 20 per cent of the sales revenue is credited to labour fund. The regular wage (or salary) payments to the workers are paid out of this fund. At the end of 6 months or one year the balance amount remaining (after payment of wages) apportioned as bonus, based on the basic wages, are paid as bonus to the workers.

(b) Scalon plan

It is generally applied to all the company employees. A total normal cost is established as a percentage of sales and it is credited to wage and salary account, against which the actual wages and salaries are charged. About two-thirds (67 per cent) or three-fourths (75 per cent) of the balance amount is distributed as bonus. The leftover fund is withheld as a safety measure towards possible deficits. This method helps in bringing about cooperation of all the members towards improvement of the firm. An increase in sales volume of the products manufactured by the firm would lead to increased bonus amount.

(c) Rucker's 'share of production' plan

In many organizations it is limited to hourly wage earners. It is based on the value added by manufacture (rather than sales revenue). The value 'Red' is calculated as a difference between the sales revenue and the amount spent on materials, supplies and allied costs. Usually 40 per cent of the value added is considered as the share of the workers. The difference between worker's share and actual wages paid is distributed as bonus, with about one-fourth as reserve. This method is helpful for preventing wastage and improves the quality of work.

(d) Collective bonus system

Collective bonus system is being experienced in almost all the industries. A collective bonus in the form of extra salary of one or more months is given, if throughout business year the profits are fairly good. The bonus is declared at the end of the year and will be payable at the time of declaration.

(e) Group bonus system

Here every department or section is provided a separate fixed bonus per unit time or per unit of job, if the standard time is saved or production exceeds the predetermined quantity. Such bonus is shared between foreman and workers as agreed upon. It must be noted that it is not given on the basis of individual's performance. It is given to the whole group and hence helpful for bringing out better cooperation among the workers and the supervisor.

5.2.8. Profit sharing

It is based on the idea that the employees who have put in their hard work are entitled for a share in the profits gained by the organization. Employees receive a certain percentage of the recognized profits in addition to their normal wages. This payment is made after specified intervals based on the audit report. The share of the profit is determined by mutual agreement of the management and union.

5.2.9. Co-partnership

It is an upgrade over the profit sharing system. The part of the profit is not given in the form of money. Instead it is distributed in the form of share certificates of the company. The workers would become shareholders of the company. The workers' interest in the organization is enhanced as it is connected with the capital. The workers not only get the dividends over their share but also benefit with the appreciation of the share values. It offers better relationship between labour and management. It gives the workers a psychological feeling as or partners of the company. It brings increased productivity. As the status of the workers has been changed, their attitude with the capital and management are changed. Thus it increases the morale of the workers. It improves quality of work. There will be better cooperation and higher team spirit. However, there are also certain limitations such as, there may be disputes over the determination on the value of the profit. The union may blame the management stating that the values are manipulated. As the reward of payment is not based on individual's effort, the efficient workers may lose interest and begin to put in effort as that of average worker. However, the slow workers may be motivated to grow up.

Multiple Choice Questions

1. Which of the following is not a form of financial incentives
   - (a) Bonus
   - (b) Profit sharing
   - (c) Both bonus and profit sharing
   - (d) Job security
2. Which of the following is considered as non-financial sharing
   - (a) Job satisfaction
   - (b) Chance of promotion
   - (c) Training
   - (d) All of these
3. Operator's performance may depends upon the estimates of
   - (a) Standard output
   - (b) Job security
   - (c) Job satisfaction
   - (d) Product quality
4. The various aspects of job performance are
   - (a) Quantity
   - (b) Product quality
   - (c) Utilization of tools
   - (d) Efficiency
   - (e) All of these
5. An incentive scheme should provide
   - (a) Improvements in utilization of tools and plant
   - (b) Recognition to a worker for good contribution
   - (c) Improve relations between workers and management
   - (d) All of these
6. The earning of a worker can be calculated as
   - (a) Earning of a worker = No. of pieces produced × rate per piece
   - (b) Earning of a worker = No. of workers × rate per piece
   - (c) Earning of a worker = No. of pieces produced × production rate
   - (d) All of these
7. If a worker brazes 20 heat exchangers per day and each having wage rate Rs 4, in straight piece rate system the earning would be
   - (a) Rs 80
   - (b) Rs 20
   - (c) Rs 60
   - (d) Rs 30
8. Consider an output standard of 20 pieces per day with a wage rate of Rs 40 per hour under 8 hour a day, calculate guaranteed wage rate.
   - (a) 320
   - (b) 360
   - (c) 380
   - (d) 300
9. The system in which a worker who exceeds more than standard output is paid higher wage rate per piece and another who fails to do so get earning at low piece rate is
   - (a) Straight piece rate system
   - (b) Differential piece rate system
   - (c) Halsey plan
   - (d) None of these

10. The differential piece rate system was suggested by
    (a) F. W. Taylor
    (b) J. J. Thomson
    (c) M. J. Taylor
    (d) F. J. Taylor

11. In Halsey plan
    (a) Minimum wage is guaranteed
    (b) Bonus is given to worker
    (c) Output standard are based on production record available
    (d) All of these

12. According to the Halsey Paln, the wage rate of a worker is given by
    (a) $W = RT_a + (P/100) \times (T_s - T_a) \times R$
    (b) $W = RT_a + (R/100) \times (T_s - T_a) \times R$
    (c) $W = RT_a + (P/100) \times (T_a - T_s) \times R$
    (d) $W = 8 + (P/100) \times (T_s - T_a) \times R$

13. The disadvantage of Halsey plan is
    (a) Minimum wage is guaranteed
    (b) Quite simple
    (c) Management also shares bonus
    (d) Less expensive
    (e) None of above

14. In Rowan plan
    (a) $W = RT_a \times (2 - T_d/T_s)$
    (b) $W = RT + (8 - Ta/Ts) \times R.T$
    (c) $W = RT + (T_s - T_s/T_a) \times RT$
    (d) $W = S.T + (T_s - T_a/T_s) \times RT$

15. In Bedaux plan wage rate of a worker is calculated as
    (a) $W = RT_a + 0.75 (T_s - T_a) \times R$
    (b) $W = R.T + [NS - NT/60] \times 65/100 \times R$
    (c) $W = R.T + [N_S - N_T/60] \times 55/100 \times R$
    (d) $W = R.T + [N_S - N_T/60] \times 85/100 \times R$

16. As per Emerson's efficiency plan, the efficiency of a worker is calculated as
    (a) Output time × 100/Actual time taken by worker to complete job
    (b) Standard time × 100/Actual time taken by worker to complete job
    (c) Standard time × 50/Actual time taken by worker to complete job
    (d) None of these

Answers

1. (d) **2.** (d) **3.** (a) **4.** (e) **5.** (d) **6.** (a) **7.** (a) **8.** (a) **9.** (b) **10.** (a)
11. (d) **12.** (a) **13.** (e) **14.** (a) **15.** (a) **16.** (b)

Introduction to Ergonomics

6.1. Background of Ergonomics

Ever since human first began to interact with his environment in any complex way, the tendency to produce conformity by aggressive or illogical means has been widespread and same is called procrustean approach. Particularly the industrial worker has been 'customized to fit the demands of his physical work, consequently with most sufferers usually accepting a significant degree of discomfort and disability without too much argument. The procrustean approach explained through a situation like the arms of an operator have elongated to reach inaccessible controls and perceptual abilities stretched to be able to hear or to see virtually inaudible or invisible signals. At the other end of the procrustean range, legs have cut down to fit confined workplaces and cognitive capacities shrink to fit tedious tasks.

The problem has become increasingly important since the industrial revolution, particularly with the expansion in the complexity of both work and machines. Because of a poor 'fit' between the human operator and his environment, lives have been lost, productivity reduced and errors have been vital, the needs and abilities of the man and compatibility with in the environment have been as secondary importance.

Training the operator in tasks is no doubt difficult to carry out, but also a costly procedure. In most of the cases, training and production time can be reduced to a significant extent by designing the machine in harmony with the operator's capability. Even though it is possible to train an operator to a high level of competency and skill, training alone will not solve their problems and increase their output. Moreover with the increasing complexity of industrial setup, it is impractical to assure operator's effectiveness without creating the appropriate environment. Further, he needs help in the form of suitable designed controls, information displays and other aspects of his environment for maximizing the effects of training. Chaney and Teel (1967) performed a study and compared the detection efficiency of skilled inspectors of machine parts, either after a four-hour training programme or after giving a set of specially designed visual aids and display to help them for defect detection. The result of the study revealed that, the training programme resulted in a 32 per cent hike in defect detection, whereas the use of the suitable visual aids resulted in 42 per cent increase.

In another laboratory study done by Ellis and Hills (1978), it has been established that numbers formed from the common seven segment liquid crystal display are more difficult to read under short viewing times (i.e., they lead to more reading errors), than that the numbers formed from conventional displays. An appropriate training could help to cease these difficulties.

One of the most important and ultimate limitation of the procrustean approach, has been suggested by Taylor and Garvey (1966). This stated, *"no matter how well the operator has been trained, his behaviour can break down under stress and this lead him making irrelevant response to a situation"*. For instance, Murell (1971) citied the case of hydraulic press which was broken during an emergency action (that is, when the operator was under stress). For normal operations, a lever needed to be pushed down to lift the press up, and the operator was trained enough to perform this action at the required level of efficiency. However, the normal and natural anticipation of a human would be to lift a lever upward to lift some part of a machine. When the emergency occurred, the operator desiring to lift the press forgot his training and pulled the lever up. This causes the press to move down and the same had ultimately broken. Hence, in the majority of cases, no amount of training will overcome the tendency of an operator to do what comes naturally when he is put under stress. It might of course be argued that some types of training systems, e.g., in the military, are designed to overcome this breakdown in behavior, which occurs under stressful conditions. This may be so but in most working environment the cost of such training would possibly be excessive and the type of training would not be tolerable for civilian trainees.

It is obvious from the above example that merely training is not sufficient to utilize the full potential of an operator. The higher performance level and efficiency of an operator is realizable, if the training schedules connected with a full consideration of a task carried out, and the work designed is in harmony to the physical, cognitive and emotional capacities of an operator. Therefore, it is the function of *ergonomics* to enterprise and highlight this harmony between the environment and the man. Hence, *ergonomics is a discipline, which attempts to restore the balance, which in the past been biased towards the procrustean approach*. Therefore, the basic aim of ergonomics is to measure the capabilities of the man and then to arrange the environment to fit such abilities. As Rodger and Cavangh (1962) illustrated it as follows:

> "Ergonomics attempts to fit the 'job to the man' rather than 'fit the man to the job".

6.2. Historical Evolution of Ergonomics

The birth date of ergonomics can be pin-pointed accurately to 12 July 1949. A meeting was held at the Admiralty in which an interdisciplinary group was constituted for those interested in human work problems (Edholm and Murrell, 1973). Later at a meeting on 16 February 1950, the term ergonomics was adapted and the discipline was born. (*The word ergonomic has evolved from a Greek; Ergon; means work and Nomos; means natural laws i.e. natural law of working.*)

Although the birth date of ergonomics could exactly been defined but the evolution and development period of this new discipline was long and twisting, and certainly no precise date can be given for its conception. However, the initial rise of interest in the relationship between man and his working environment believed to have originated at the time of the First World War. Workers in ammunition factories were required in maintaining the war effort, but with the drive for a higher output of arms, a number of unforeseen complications occurred. The attempt to resolve some of these problems led to the establishment of the committee for health of munitions workers in 1915, which includes some individuals, trained in physiology and psychology among its investigators. At the end of the war, the committee was reconstituted and approved at the Industrial Fatigue Research Board (IFRB), primarily to carry out research into fatigue problems in industry.

In 1929, the IFRB renamed as Industrial Health Research Board and its scope widened to investigate general conditions of industrial employment particularly about the preservation of health among the workers and to industrial efficiency. The committee had investigators, trained as psychologists, physiologist, physician, engineers, and they worked both separately and together on problems covering a wide area. These include posture-carrying loads, the physique of working men and women, rest pauses, inspection, lighting, heating, ventilation and music-while-you-work. With the outbreak of the Second World War, there occurred a rapid development in the military field. However, as if the stresses of battle were not enough, but the military equipment had become so complex and the required operating speeds so high, it resulted in additional stresses which caused men either to fail to get the best out of their equipment or to suffer operational breakdown. Therefore, it became essential that much more to be known about individual's performance capabilities and limitations. Thus, the physical, mental/cognitive capabilities and limitations were considered at the time of design and development of weapon and war equipment. It is obvious that starting from the beginning till date ergonomics has evolved at a great pace and consequently human became much more efficient and precise at the work places. A brief of historical evolution of ergonomics is exhibited in Figure 6.1, and portrays that the ergonomics discipline took birth in the form of classical ergonomics in 1950.

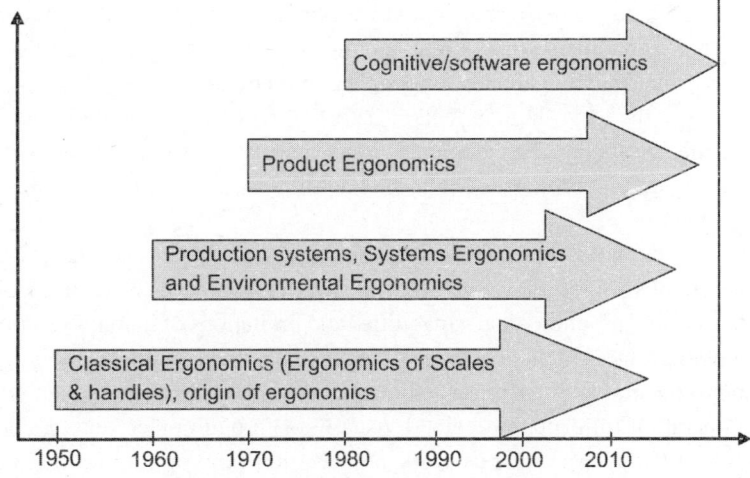

Fig. 6.1: Historical evolution of ergonomics.

About 1960 onwards, the concept of production systems, environmental ergonomics, and system ergonomics inside the factory came into practice. However, outside the factory, i.e., ergonomics in products was introduced from 1970 onwards. With the passage of time after 1980, the industrial and product design advanced and also software became an integral part of the manufacturing and service industry, the cognitive and software ergonomics played a vital role. In the year 2000 onwards, ergonomics has been an inbuilt feature in software and computer systems. In the present scenario of Laptops, Tablets and smart cell phones, etc., the software ergonomics now a day is commonly incorporated.

6.3. Introduction

As mentioned previously that ergonomics has developed as a known field at the time of the Second World War, enforced by arranging technology and human sciences in a synchronized way. Physiologists, psychologists, anthropologists, doctors, work scientists and engineers, working in a well-defined way, figured out the complications. The auspicious outcomes of this inter-disciplinary access were benefitted by pursuing after the war in industries. The importance grew very fast particularly in Europe and the United states, leading to the setting up of National Ergonomics Society in 1949, and then Ergonomics term was adopted. After that in 1961 foundation of International Ergonomics Association (IEA), representing ergonomics societies took place actively in 40 countries of the world with total membership of 15000 people, thereafter the count of membership has significantly increased.

6.4. Definitions of Ergonomics

The word 'Ergonomics' is derived from the Greek words *(Ergon 'Work' + Nomics 'Law')* i.e. *Ergonomics* is the application of scientific information concerning humans to the design of objects, systems and environment for work place use, i.e., *Ergonomics is the laws of work that define the limits to human capability.*

However, the modern definition of ergonomics can be narrated as: *the science of fitting workplace conditions and job demands to the capabilities of majority of the working population. In other words, Ergonomics is 'fitting the job to the person engaged in the tasks'.* Therefore, Ergonomics is the science of improving employee performance and well-being in relation to the job tasks, equipment and the environment. Ergonomics is a continuous improvement effort to design the workplace for what people do well, and design against what people could not do well.

In the United States, the term 'human factors' is often used. A concise definition would be that ergonomics aims to design appliances, technical systems and tasks in such a way as to improve human safety, health, comfort and performance. The formal definition of ergonomics, approved by the IEA, reads as follows:

In the design of work and everyday-life situations, the focus of ergonomics is human. Unsafe, unhealthy, uncomfortable or inefficient situations at work or in everyday life are avoided by taking account of the physical and psychological capabilities and limitations of humans. A number of aspects play a role in ergonomics; these include body posture and movement (sitting, standing, lifting, pulling and pushing), environmental factors (noise, vibration, illumination, climate, chemical substances), information and operation (information gained visually or through other senses, controls, relation between displays and control) as well as work organization (appropriate tasks, interesting jobs). These factors determine to a large extent safety, health, comfort and efficient performance at work and in everyday life. Ergonomics draws its knowledge from various fields in the human sciences and technology, including anthropometry, biomechanics, physiology, psychology, toxicology, mechanical engineering, industrial design, information technology and industrial management. It has gathered selected and integrated relevant knowledge from these fields. In applying this knowledge, specific methods and techniques are used. Ergonomics differs from other fields by its inter-disciplinary approach and applied nature. The inter-disciplinary character of the ergonomic approach means that it relates to many different human facets. Due to its applied nature, the ergonomic approach results in the adaptation of the workplace or environment to fit people, rather than the other way round

6.5. An Ergonomist

In some countries, it is possible to graduate as an ergonomist. Other people who are trained in one of the relevant basic technical, medical or social science fields can also acquire knowledge and capabilities in ergonomics through training and experience. In several countries, independent certifying bodies can certify professional ergonomists. For example, in Europe, the Center for Registration of European Ergonomists (CREE) decides on candidates for registration as European Ergonomists. Expert ergonomists can work for the authorities (legislation), training institutions (universities and colleges), research establishments, the service industry (consultancy) and the production sector (occupational health services, personnel departments, design departments, research departments, etc.). Many professional ergonomists who are active in business (company ergonomists) practise their profession mainly by being an intermediary between the designers and the users of production systems. The ergonomist highlights the areas where ergonomic knowledge is essential, provides ergonomic guidelines and advises to designers, purchasers, management and employees, regarding the more acceptable systems. Other experts, besides professional ergonomists, make use of ergonomic knowledge, methods and techniques. These include industrial designers, company doctors, company nurses, physiotherapists, industrial hygienists and industrial psychologists.

6.6. Significance of Ergonomics in Society

Ergonomics can contribute to the solution of a large number of social problems related to safety, health, comfort and efficiency. Daily occurrences such as accidents at work, in traffic and at home, as well as disasters involving cranes, aircrafts and nuclear power stations can often be attributed to human error. From the analysis of these failures it appears that, the cause is often a poor and inadequate association between operators and their task. The probability of accidents could be reduced by taking better account of human capabilities and limitations when designing work and everyday life environments.

Many of the daily life situations and workplaces are hazardous to health. In Western countries, diseases of the musculoskeletal system (mainly lower back pain) and psychological illnesses (due to stress) constitute the most important cause of absence due to illness and occupational disability. These conditions can be partially attributed to poor design of equipment, technical systems and tasks. Here also, ergonomics can help to reduce the problems by incorporating the advanced working conditions. So in numerous countries, occupational health services are obliged to employ ergonomists.

Ultimately, it can be stated that ergonomics can offer convenience to some level. However, in the architecture of complicated technical systems such as process installations, nuclear power stations and aircraft, ergonomics has become one of the most important design factors introducing operator error. Some ergonomic knowledge has been compiled into official standards, whose objective is to stimulate the application of ergonomics. A range of ergonomic subjects is covered by international standards of the International Standardization Organization (ISO), European EN-standards of the Committee European de Normalization (CEN), as well as national standards, for example in the United States (ANSI) and Britain (BSI). In addition, there are specific ergonomic standards, which are applied in individual companies and in industrial sectors.

6.7. The Scope of Ergonomics

Since ergonomics evolved through the interest of a number of different professionals and it remains a multidisciplinary field of study, therefore, it crosses the different boundaries between many scientific and professional disciplines and draws on the data finding and principles of each.

Presently ergonomics can said to be an amalgam of mainly three areas: physiology, physics and engineering. The physiological and biological science provides information about the structure of the body; the operators' physical capabilities and limitations; the dimension of his body; how much one can lift; the physical pressure he can endure, etc. Finally, physics and engineering provide similar information about the machine/systems and the environment with which the operator has to interact and contend. These areas enable an ergonomist acquire and integrate data to maximize the operator's safety, efficiency and reliability of performance to make his task easier to learn and to increase his feeling of comfort.

These criteria however are independent; for example, an operator's efficiency is highly dependent on his accuracy but accuracy is not the only component of efficiency, others include reliability, speed and reduction in efforts/fatigue. Arguing in the same way, it is ascribed that ergonomics seeks to increase safety. Therefore, consequently this should result in a reduction of time lost due to illness and a corresponding increase in worker's efficiency. However, safety itself depends upon efficiency certainly. There are many examples to illustrate the fact that the margin of safety, which left in operations, is largely a function of the operator's speed or reliability.

A further aim of ergonomics is to attempt to reduce the unpredictability of operator performance and increase his (or her) reliability. Thus, the human operator should not only be fast and efficient, but should also be reliable. No doubt, reliability relates to accuracy but they both may be independent. An operator may perform his task accurately most of the time but because of some intermittent action of his work situation, he may be unreliable in his accuracy. There is also a question of ease of learning. Thus a system, which has been designed to produce a series of tasks, which are easier to learn will reduce training time and costs and assure less errors under stress. The final aspect comfort is a subjective criterion that is becoming increasingly important in present day situations and refers to a sense of well-being and ease induced by the system. The concept of comfort and the controversies surrounding its definition will be discussed in later chapters but it is sufficient to point out here that an uncomfortable operator is prone to errors and is likely to perform less efficiently.

In summary, therefore it can be stated that the task of the ergonomics is first to determine the capabilities of the operator and then to attempt to build a work system around these capabilities. In this respect, ergonomics often referred as the science of fitting the environment to the man. The spirit of procrustes should finally said to be driven out, only when this approach is fully accepted.

6.8. Ergonomics and Related Disciplines

It is pertinent at this point to question that where ergonomics stands in relation to the associated disciplines such as operation research, work-study and time and motion study. Each tries to maximize the worker's effectiveness and certain areas of overlap are bound to exist. Despite this similarity of objectives, it is also possible to perceive differences between the disciplines. As its title suggests time and motion study is concerned primarily with increasing performance by measuring and then minimizing the time taken to perform various motions of an operations. The fundamental philosophy of this discipline suggests that (a) although there are usually numerous ways to perform any task, one method will be superior to other and (b) the superior method can be determined by observing and analyzing the time taken to carry out parts of activity. The integration of Ergonomics with the other disciplines is shown in Figure 6.2. Human–Computer Interaction (HCI) is concerned with the design, evaluation and implementation of interactive computing systems for human use and the study of major phenomena surrounding them.

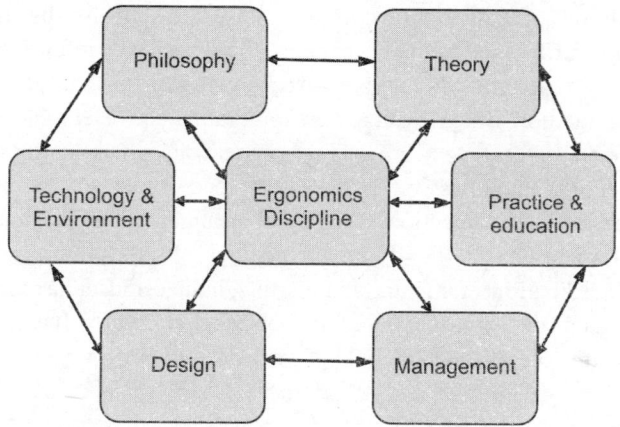

Fig. 6.2: Ergonomics and interaction with other disciplines.

6.9. Aspects of Ergonomics

There are five aspects of ergonomics: safety, comfort, ease of use, productivity/performance and aesthetics. Any product/process or system having all the five aspects can be considered as ergonomically designed. In present scenario of competition, there is a big challenge to make a product, which is ergonomically designed at competitive price. At the same time, the durability of the product is making the companies to strive more for developing alternative materials, which proves to be durable.

6.10. Application Areas of Ergonomics

The principles developed by Ergonomics study and research have wide applications; some of the areas are enumerated as follows:

1. Study and measure of manual work
2. Physical and psychological aspect of fatigue and stresses
3. Compatibility of man–machine system
4. Design of work-place layout
5. Design of tools and equipment
6. Design of display and controls
7. Design of furniture
8. Product design
9. Study of manual operations
10. Safety in work conditions
11. Study of job stresses and measures of strain for job evaluation
12. Work postures and lifting, etc.

6.11. Man and Machine Interaction

The performance of a worker in an industry depends upon the interaction and interrelationships of the worker with the machine/equipment and the environment. A production unit is usually a man–machine integrated system performing under an environment envelop. The environment refers to not only the ambient conditions of temperature, humidity, noise, etc., but also the arrangement of layout, facilities, display and controls. The human capacities and limitations refer to the physical, sensory and psychology parameters of human beings (called as human factors) that are studied to synchronize effectively with the production system with a view to obtain a consistent and efficient output from the system. Thus, human factors engineering is that endeavour which seeks to match human with the machine, equipment and facilities so that their combined output will be efficient, comfortable, and safe. The emphasis is on the selection, design and arrangement of the components taking into account the capabilities and limitations of human personnel. It takes into account of psychomotor co-ordination (i.e., coordination of mental actions that induce muscular contractions), complex information processing, and decision-making efforts for effective control. It calls for the interaction of equipment designers with specialists in the field of work design, physiology and biological sciences.

6.12. Man-Machine Closed Loop System

The man–machine closed-loop control system comprises a machine and an actively participating human operator. An operator, while operating (or working with) the machine will be continuously processing mentally the stimulus received by him through the display and taking the necessary control actions. This produces the stress on the operator. In the absence of proper compatibility of the components of the system (operator–machine–display–control) the stress (tension/fatigue) increases with the consequent impairment of productivity, quality and safety. The *Man–Machine System* (*MMS*) is depicted in examples like worker–working tool, socio-technical system (factory, ship, aircraft, etc.), sportsman–sport device, driver–vehicle, operator-controlled system, user–product, soldier–weapon, housewife–household appliances, human–computer system, etc. A general type of man machine closed loop system is shown in Figure 6.3.

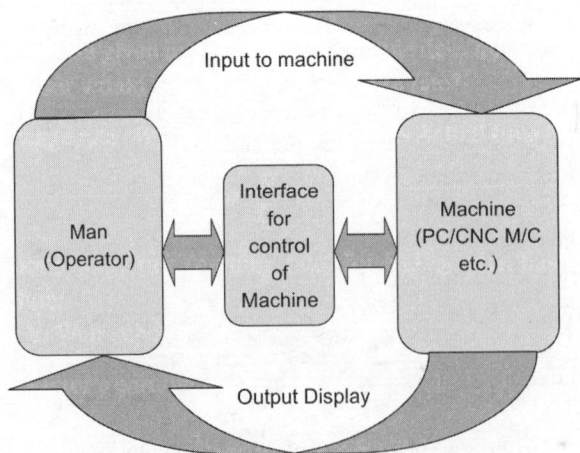

Fig. 6.3: Closed loop of man-machine system.

6.13. Man-Machine System (MMS)

Man-Machine System (MMS) always has a human subsystem, a technical subsystem and a user interface (UI). These subsystems can further be divided into smaller and even smaller elements as necessary, depending on the particular aim of the analysis. If the human subsystem and the technical subsystem are not compatible, the particular activity may not be safe, comfortable and efficient and therefore the user may experience increased stress. The characteristics of human and technical subsystems user interface are shown in Figure 6.4. The *human subsystem*, for example, consists of anthropometric, physiological, perceptual, cognitive, emotional, etc., sub-subsystems that can further be divided into even smaller elements if necessary.

The *technical subsystems*, on the other side, can have a very big variety, and therefore we cannot give here even a general level description. The technical subsystems of the following MMSs, e.g., have very different characteristics and working behavior such as a pilot and the aircraft, a designer and the CAD system, a bank official and his information system, a bank system administrator and the system itself, client and the ATM and a tennis player and his racket.

The *user interface* (UI) is the machine, as the human perceives it. The user gets into touch with the perceptible surface of the machine and creates a general judgement about the whole system based only on this perceptible surface. Therefore, the same technical equipment/device/tool, etc., may appear differently for different users. Because of learning during usage, the perceived quality of the same UI may change even within the same person.

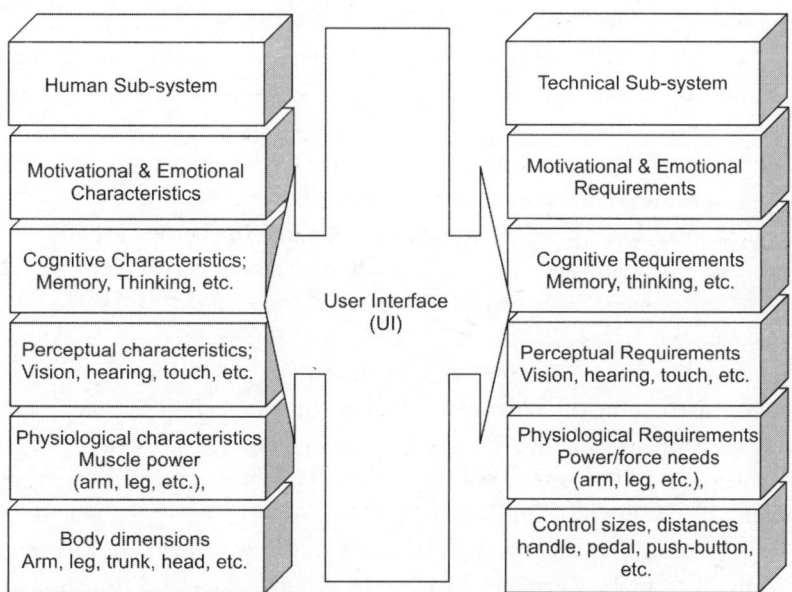

Fig. 6.4: Human and technical subsystems user interface.

6.14. Determinants of MMS System Performance

The objective of every organization to introduce ergonomics is to increase the performance at required level of quality and safety. There are several factors that determine the man–machine system performance, which are briefly described as under.

(a) Operator's ability: It is based on two factors – selection and training of the personnel.

(b) The system design: It is also based upon two things – equipment and display & control.

(c) The system operation: It is based on three parameters that include workload, communication channel and maintainability.

(d) Environment: This is based on two factors – compatibility, layout and environmental conditions like illumination, noise, temperature, humidity, etc.

6.15. Limitations of Human Beings

The basic definition of human factors engineering is based upon the two prime factors: first is the capabilities and second is the limitations of human being. Therefore, the limitations are more critical to consider while designing a new system. The limitations of human can be categorized as: physiological, sensory, analytical, motor, stress and environmental. These are described here:

1. Physiological limitations

   This pertains to limits of an individual to perform a physical activity during any given task. It is understood that in practical situations, there may be a need of exerting great force, doing continued work, repetitive and precise motions, and more number of simultaneous motions at equal pace, postures etc. However, human being is capable of performing all these within a limit; hence, it is very important to consider the physiological limitations of population who will be engaged to the systems, which are under design considerations, i.e. physiological requirements of the system must be designed according to the physiological capabilities and limitations of the operator or user.

2. Limitations of sensory perception

   (a) Limited channel capacity: A person can receive the stimulus for a group of stimuli only one at a time, before one goes to the next one (or group).

   (b) Absolute thresholds: There are certain superluminal (higher levels) and subliminal (lower levels) beyond which the stimulus will have no effect on the human being. For example: eyes cannot see beyond the visible spectrum; ears cannot sense beyond the audible octave bands.

   (c) Differential threshold: This indicates the threshold of consciousness, wherein the amount of change in stimulus, below which is not perceptible. Human mind fails to note small changes in the stimulus. For example: movie projection, projection on TV screen, where the reference frame to consider as stationary, and objects seem continuously as moving.

   (d) Sense of equality: Even though two stimuli are different many times, they are taken as equal; for example: when luminosity is decreased, red and green spots are taken as same as that of black spot.

3. Analysis of information

   Human being has limited capacity for analysis of large amount of information. It has short term of human memory. It lacks the full knowledge of information.

4. Sensory motor response

   It refers to the response of the motor nerves and consequent reaction to stimulus. There is a delay between stimulus and response, the reaction times for simultaneous operation have been found to be different, and it is called as differential delay.

5. Stress

   Human being is subjected to mental and physical stress, feeling of tiredness, fatigue, mental strain, etc., beyond which he can not work but these are not applicable for the machines.

6. Environmental factors

Thermals, lighting, colour, noise, vibrations, radiation, toxic fumes, etc., pose restrictions that are more complicated to human beings.

7. Psychological factors

These are not applicable for machines. One has to consider the capabilities of the human being separately.

A human being is better for different cognitive and psychological factors that are listed below.

1. Discrimination: Discrimination of relevant from irrelevant signals
2. Selection: Selecting his own inputs
3. Recall: Able to have selective recall of old information
4. General pattern perception: Perceiving the patterns and generalizing
5. Innovation: Innovating new ideas in problem solving
6. Adaptation: Improvising and adapting flexible procedure
7. Vide variety: Stimuli sensing
8. Learning: Profiting from experience

Machine is more suitable for work, which is beyond the capabilities of a human and tackles with his limitations.

1. Exerting great force
2. Receptive work
3. Precise operations
4. Routine processing
5. Storing and retrieval of information
6. Rapid response to signals
7. Simultaneous operation
8. Accurate rapid and complex work
9. Sensing stimuli beyond the range of human thresholds
10. Sustained work
11. Working under conditions that are stressful or intolerable for human being

Multiple Choice Questions

1. The ergonomics origins on
   (a) 12 July 1949
   (b) 14 July 1949
   (c) 12 June 1949
   (d) 14 June 1949
2. The term IFRB stands for
   (a) Industrial fatigue research board
   (b) International fatigue research board
   (c) Industrial fatigue revolution board
   (d) None of these
3. The word ergonomics is made up of two words which are
   (a) Erg + Nomos
   (b) Ergo + Nomos
   (c) Ergon + Nomos
   (d) None of these

4. Ergonomics is the science of defining work limits to
   - (a) Work measurement
   - (b) Work capability
   - (c) Human capability
   - (d) Human postures

5. The term IEA stands for
   - (a) Industrial ergonomics association
   - (b) Indian ergonomics association
   - (c) International ergonomics association
   - (d) None of these

6. CREE referred as
   - (a) Center for registration of European ergonomists
   - (b) Center for registration of European economics
   - (c) Center for registration of engineering ergonomists
   - (d) Center for renowned European ergonomists

7. The term CEN stands for
   - (a) Committee European de Normalization
   - (b) Constitute European de Normalization
   - (c) Combined European de Normalization
   - (d) Creative European de Normalization

8. HCI stands for
   - (a) Human computer inputs
   - (b) Human computer interaction
   - (c) Human computer interference
   - (d) None of these

9. The application area of ergonomics is
   - (a) Compatibility of man–machine system
   - (b) Design of work-place layout
   - (c) Design of tools & equipment
   - (d) Design of display & controls
   - (e) All above

10. Man and machine interaction consist of
    - (a) Worker and machine
    - (b) Worker and display
    - (c) Equipment and environment
    - (d) Worker and environment
    - (e) All above

11. Human factors engineering is that endeavour which seeks to match human personnel with
    - (a) Machine
    - (b) Equipment
    - (c) Facilities
    - (d) All above

12. MMS stands for
    - (a) Man machine system
    - (b) Machine man system
    - (c) Man material system
    - (d) Material man system

13. MMS consists of
    - (a) Human technical subsystem
    - (b) Human technical system
    - (c) User interface
    - (d) Human technical subsystem and user interface

14. The system operation is based on
    - (a) Work load
    - (b) Communication channel
    - (c) Maintenance
    - (d) All above

15. Certain superluminal (higher levels) and subliminal (lower levels) beyond which the stimulus will have effect on the human being is termed as
    - (a) Absolute thresholds
    - (b) Differential thresholds
    - (c) Sense of equality
    - (d) None of these

Answers

1. (a)   **2.** (a)   **3.** (c)   **4.** (c)   **5.** (c)   **6.** (a)   **7.** (a)   **8.** (b)   **9.** (e)   **10.** (e)
11. (d)   **12.** (a)   **13.** (d)   **14.** (d)   **15.** (a)

Work Physiology

7.1. Introduction

This chapter focuses upon the physiological aspects of muscular work, and it describes how various physiological systems work together to meet the energy expenditure requirements of work load and how these requirements can be measured quantitatively and considered in the analysis of physical work. Whereas in subsequent chapters the focus is on the mechanical aspects of physical work and awkward postures and heavy exertion forces that can lead to severe musculoskeletal problems, such as low-back pain and upper-extremity disorders.

This chapter begins with a depiction of the physiological structure of muscles and how energy is generated and made available for use by the muscles. Subsequently, it describes how the raw materials for energy production are supplied and its waste products removed by the circulatory and respiratory systems. The human body has a musculoskeletal system of bones, muscles and connective tissues which enable it to uphold the body posture, walk and run and lift and carry objects. Physical work is possible only when there is sufficient energy/vigour to support muscular contractions. Energy expenditure requirements of various types of activities are then described, together with a discussion about how the levels of energy expenditure can be measured quantitatively. Clearly, there are upper limits of energy production and muscular work for each individual. The implications of these work capacity limits for ergonomic job design are discussed in the last section of the chapter.

7.2. Muscle Structure

The primary role of muscle is to generate force and produce movement by exerting forces such as pull and push. The human body has mainly three types of muscle cells, also called as muscle fibres. However, this chapter is primarily focused upon the skeletal muscle, since it is directly responsible for physical work, these are described below.

- (a) Smooth muscles
- (b) Cardiac muscle
- (c) Skeletal muscles

- (a) Smooth muscles: Smooth muscles are found in the stomach and intestines, blood vessels, urinary bladder, and uterus, etc. These muscles basically help in the digestion of food and are responsible for regulating the internal environment of the body. The contraction of smooth muscles is controlled by autonomic nervous system but not by consciousness.

(b) Cardiac muscles: Cardiac muscle is the muscle of the heart, and like smooth muscle it is not controlled by consciousness, but under direct control of autonomic nervous system.

(c) Skeletal muscles: Skeletal muscles are the main muscle tissues in the body and account for around 40 per cent of the body weight. These are attached to the bones of the skeleton through tendons, and these act as puller as their contraction facilitate the bones to act like levers. The contraction of most skeletal muscles is under direct conscious control. It is the movement of skeletal muscles that makes physical work to happen. Each skeletal muscle is composed of thousands of elongated cylindrical muscle fibres. Each fibre consists of many cylindrical elements arranged in parallel which are called as myofibrils, each of which is further divided longitudinally into a number of sarcomere that are arranged in series and form a repeating pattern along the length of the myofibril. These sarcomeres are the contractile unit of skeletal muscle. The sarcomere comprises two types of protein filaments; one thick filament, called myosin, and another one is thin called as actin. The two types of filaments are layered over each other in alternate dark and light bands. The layers of thick filaments are found in the central region of the sarcomere, forming the dark bands, known as the A bands. The layers of thin filaments are connected to either end of the sarcomere to a structure called the Z line. Two successive Z lines define the two ends of one sarcomere, the structure of a muscle is exhibited in Figure 7.1.

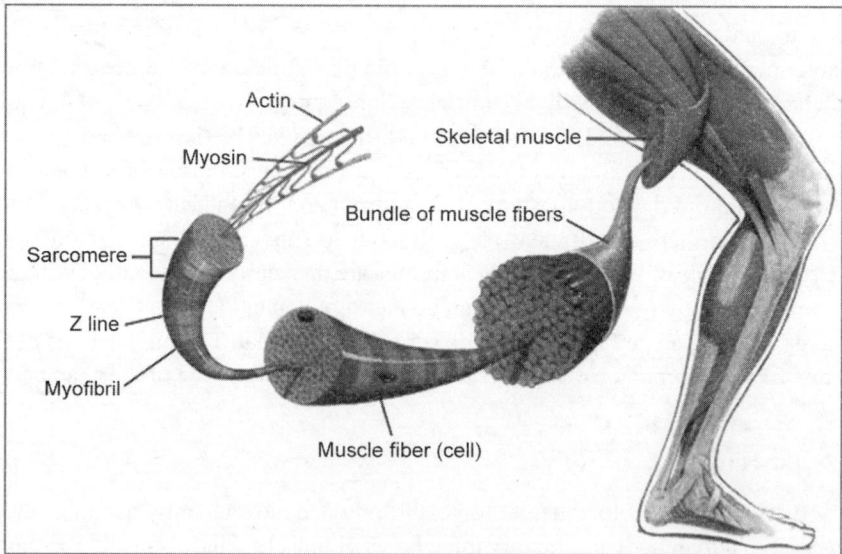

Fig. 7.1: Detailed structure of muscle.

7.3. Metabolisms (Aerobic and Anaerobic)

Physical work is possible only when there is some energy to support muscular contraction. The energy required for muscular contraction and various other physiological functions of the body itself is built up in the form of two high-energy phosphate compounds, called as adenosine triphosphate (ATP) and creatine phosphate (CP). These compounds are generated from aerobic or anaerobic metabolism of nutrients. *The metabolism which occurs in the presence of oxygen is called as aerobic metabolism and the one in absence of oxygen is known as anaerobic metabolism.* This process of creating high-energy phosphate compounds is known as phosphorylation. The ATP and CP compounds are basically the energy carriers and are found in all body cells. They are involved to perform energy activities of the body and thus sustain life. Whenever there is energy demand for muscle contraction and relaxation, ATP is converted to ADP (adenosine diphosphate) by splitting off one of the phosphate bonds, and energy becomes available for use. In this respect, ATP performs like a rechargeable battery, which provides a short-term storage of directly available energy (Astrand and Rodahl, 1986). Figure 7.2 demonstrates the various physiological systems that work together to meet the energy expenditure demands of work.

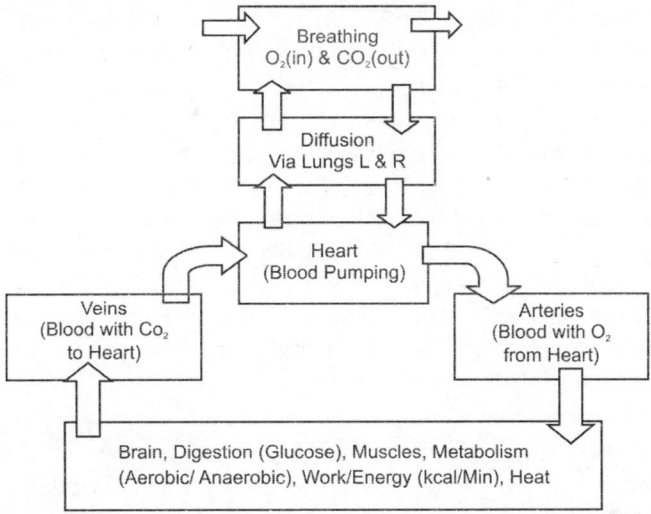

Fig. 7.2: Working of various systems to meet the requirement of energy expenditure for physical work.

The human body has a very restricted capability for ATP storage. For example, any individual with 75 kg of body weight has about one kcal of ATP-stored energy available at a time. Therefore, if a muscle relies only on its ATP storage for contraction, it would run shortage of this energy supply in a few seconds. Hence, to sustain the contractile activities of a muscle, it is a must that ATP compounds be continuously produced and refilled at the same pace at which they break down. Therefore, in human body there are three sources for supplying ATP: creatine phosphate (CP), oxidative phosphorylation (aerobic metabolism) and anaerobic glycolysis (anaerobic metabolism).

The molecules of CP contain energy which is transferred to the molecules of ADP, so that the ADP is recharged back to ATP. In this way, the CP system performs like a backup storage for ATP and gives the best means of refilling ATP in the muscle cells. The CP system has energy storage capacity that is about four times that of the ATP system; however, it is still of very limited capacity. The total energy supply from the ATP and CP systems can only support either heavy work for about 10–15 sec or moderately heavy work for a couple of minutes.

In case of need of prolonged sustainability of muscular activities, it becomes necessity that the muscle cells are able to produce ATP from other sources instead of CP. When enough oxygen is available and muscular activity is at judicious levels (i.e., moderate rates of ATP breakdown), the major portions of the ATP demand can be supplied by the process of oxidative phosphorylation. In this process, nutrients like carbohydrates and fatty acids are burned in the presence of oxygen and energy is released to form ATP for muscle work. The nutrients are obtained from the food intake, and oxygen is acquired from the air we breathe. The nutrients and oxygen are transported to the muscle cells by the blood through the circulatory system. The nutrients can also be obtained from storage in the cells. The liver and muscle cells store the carbohydrates in the form of glycogen, which is derived from glucose in the blood flow. This oxidative phosphorylation process releases energy for use by the muscles but also produces carbon dioxide as a waste derivative, which ought to be removed from the tissues by the circulatory system.

It generally requires about one to three minutes for the circulatory system to respond to increased metabolic requirements in performing physical tasks, and the skeletal muscles habitually do not have enough oxygen to carry out aerobic metabolism (oxidative phosphorylation) at the start of physical work. During this period of time, part of the energy is supplied through anaerobic glycolysis, which pertains to the production of energy through the breakdown of glucose to lactic acid in the absence of oxygen. Even though anaerobic glycolysis can produce ATP very rapidly without the presence of oxygen, it has the drawback of generating lactic acid as the by-product of this process. This lactic acid increases the acidity level in the muscle tissues and is supposed to be a major cause of muscle pain and fatigue. The lactic acid is removable if oxygen is there, hence to remove these waste products it becomes necessary that the muscle constantly consumes oxygen at a high rate even after it has stopped contraction so that its original status can be revitalized. Moreover, another shortcoming of anaerobic glycolysis is that it requires much larger quantities of glucose to produce the same amount of ATP as compared to aerobic metabolism. Therefore in case adequate supply of oxygen is available, aerobic metabolism can provide all the energy required for light or moderate muscular work. Hence under these conditions, the body is considered to be in the 'stable state'. Whereas for a very heavy work, even if adequate oxygen is available, aerobic metabolism cannot produce ATP fast enough to keep pace with the rapid rate of ATP breakdown. Thus, for very heavy work, anaerobic glycolysis serves as a supplementary source of ATP, and consequently fatigue could develop rapidly due to the accumulation of lactic acid in the muscle cells and ultimately in the blood as well. If we look at the energy conversion rate, the overall efficiency with which muscle converts chemical energy to muscular work is merely about twenty per cent (20 per cent). However metabolic heat accounts for the remaining 80 per cent of the energy released in metabolism (Edholm, 1967, 1973). Therefore, large amount of heat is produced in performing heavy work activities. Thus, it is very critical that, in hot weather, while performing the heavy activities the ambiance temperature must be controlled and brought under normal range. However, otherwise the increased level of heat production may have severe effect on the body's ability to maintain a constant body temperature.

7.4. Circulatory and Respiratory Systems

The sustainability of muscular work is only feasible when adequate amount of nutrients and oxygen is consistently available to the muscle cells and also the disposal outcome of metabolism like carbon dioxide (CO_2) is quickly removed from the body. In order to meet these requirements, the required functions are performed by the circulatory and respiratory systems. The circulatory system works as a mode of transport system of the body, as it delivers oxygen and nutrients to the tissues and removes carbon dioxide and waste products from the tissues. The respiratory system exchanges oxygen and carbon dioxide with the external environment through breathing.

7.4.1. The circulatory system

The circulatory system primarily includes blood and the cardiovascular system. This system is the main equipment that transports blood to the various parts of the body. Blood is made up of three forms of blood cells (RBC, WBC, Platelets) and plasma. Red blood cells (RBC) transport oxygen to tissues and help to remove carbon dioxide from them. White blood cells (WBC) fight with the attacking germs and defend the body against infections. Platelets help in blood clotting and stop bleeding. Plasma, in which the blood cells are suspended, contains ninety per cent of water and only ten per cent nutrient and salt solutes. Out of these three types of blood cells, RBCs have a vital importance to work physiology due to their oxygen-carrying property. Red blood cells are produced in bone marrow and carry a unique type of molecule known as the haemoglobin (Hb). A haemoglobin molecule has ability to combine with four molecules of oxygen to form oxyhemoglobin, and thus carry oxygen in the blood efficiently. Since 1 kg of blood has a volume of about 1 L and the total blood volume (in L) of an average adult is normally found approximately 8 per cent of his or her body weight (in kg). Therefore, a 70 kg adult would have a total blood volume of about 5.6 L (70 × 8/100 = 5.6), of which about 3.08 L (55 per cent) consist of plasma and 2.52 L (45 per cent) of blood cells. The capability of the blood to transport oxygen and nutrients to the tissues and take out carbon dioxide from them depends upon the blood volume. This capacity will be lesser in a person having a low blood volume or a low RBC count. Hence, if an individual has low blood volume or less RBC count or works in a polluted or poorly ventilated environment or at high altitudes where the air has a low oxygen content, will experience increased stress on the circulatory system. This is because the circulatory system has to put hard efforts to compensate for the lesser content of oxygen and nutrients in the blood to perform required functions, a circulatory system is illustrated in Figure 7.3 given below.

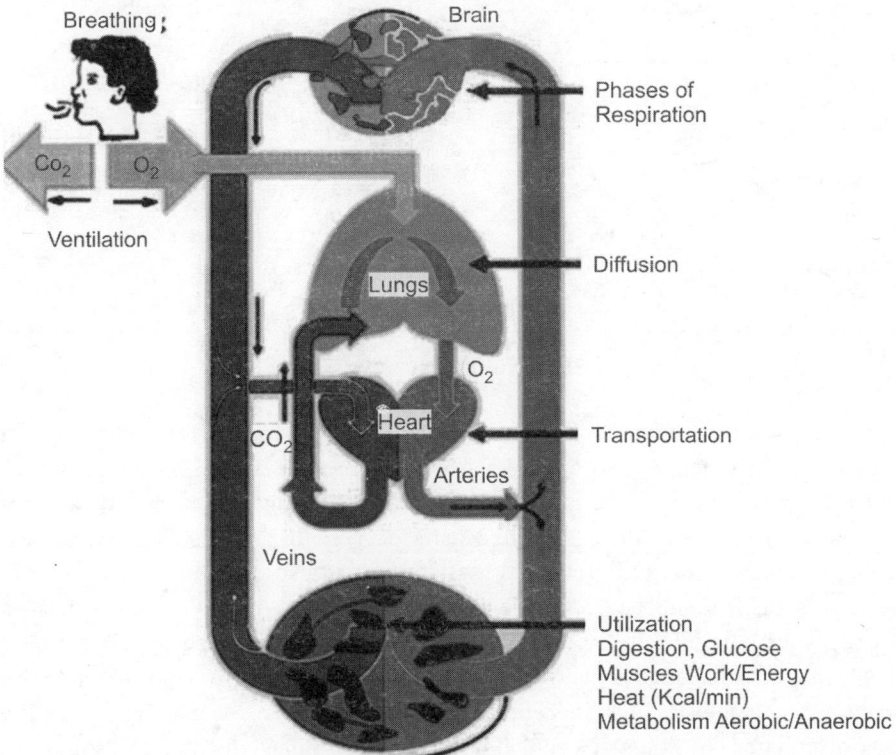

Fig. 7.3: The circulatory system.

7.4.2. The respiratory system

Respiration in humans means the exchange of gases between a living organism and its environment that occurs through lungs. It is to be noticed that the surface exposed to the environment is large than 70 square meter. The movement of air in adults occurs at the rate of 12–20 breaths per minute. During exhausting exercise, adults breathe at a rate of 35–45 per minute. The lung takes approximately ten thousand litres of air in a day and about 0.5 L per breath and the quantity increases during work, sports, or exercise, etc. A healthy person takes around 24000 breaths in a day. Human respiratory system is exhibited in Figure 7.4.

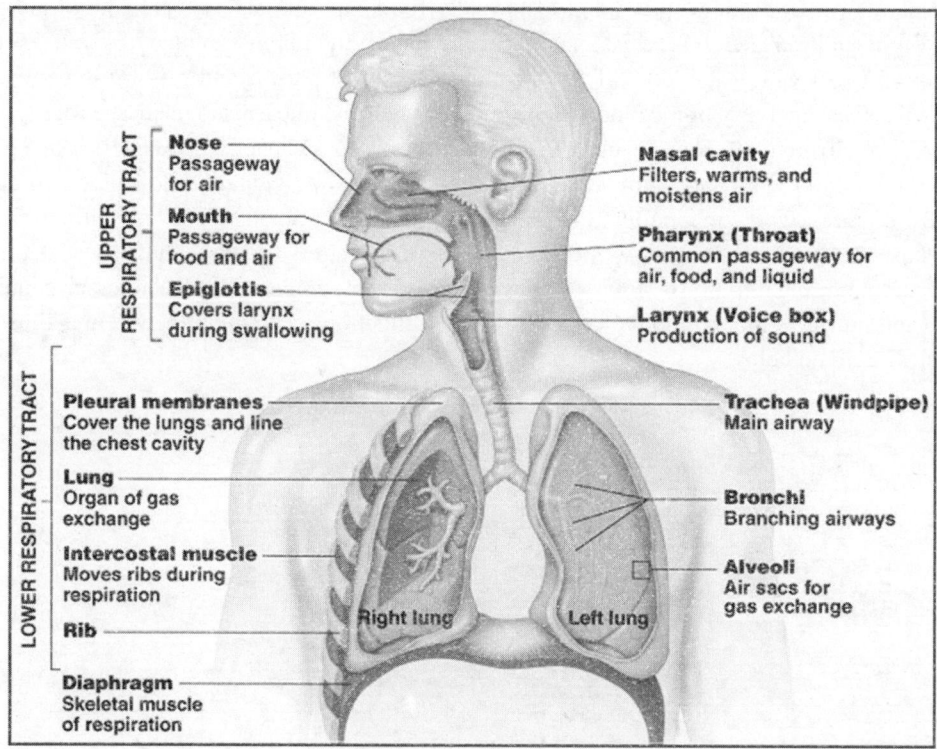

Fig. 7.4: Human respiratory system.

The upper respiratory tract starts with nose or nasal cavity which is the outermost part of the respiratory system; its function is to warm, moisten and filter the air during inhalation. Then just below the nasal cavity is pharynx (or throat) which provides passage for the air flow and it leads to trachea. Just below the pharynx, there is larynx where vocal chords are located. Larynx is also called the voice box of human. Larynx is followed by the trachea (also called as windpipe). The lower respiratory tract starts from trachea, which is lined with fine hairs, called *cilia*, which filter the air before it reaches the lungs. There are two branches at the end of the trachea, called bronchi, each leading to a lung. Each lung has a network of smaller branches leading from the bronchi into the lung tissue and ultimately to air sacs; these networks are called bronchioles. The bronchiole ends with the microscopic alveoli lined by thin moist epithelium. These are ultimate functional respiratory units in the lung where gases are exchanged. The pulmonary arteries pass on the oxygen-deficient blood to the alveoli, and the branches of pulmonary vein transport oxygen-rich blood from the alveoli to the heart. In simple words, it can be said that O_2 diffuses into the blood and CO_2 is removed out of the blood. These are also called as blood purification units; the gas exchange process is exhibited in Figure 7.5.

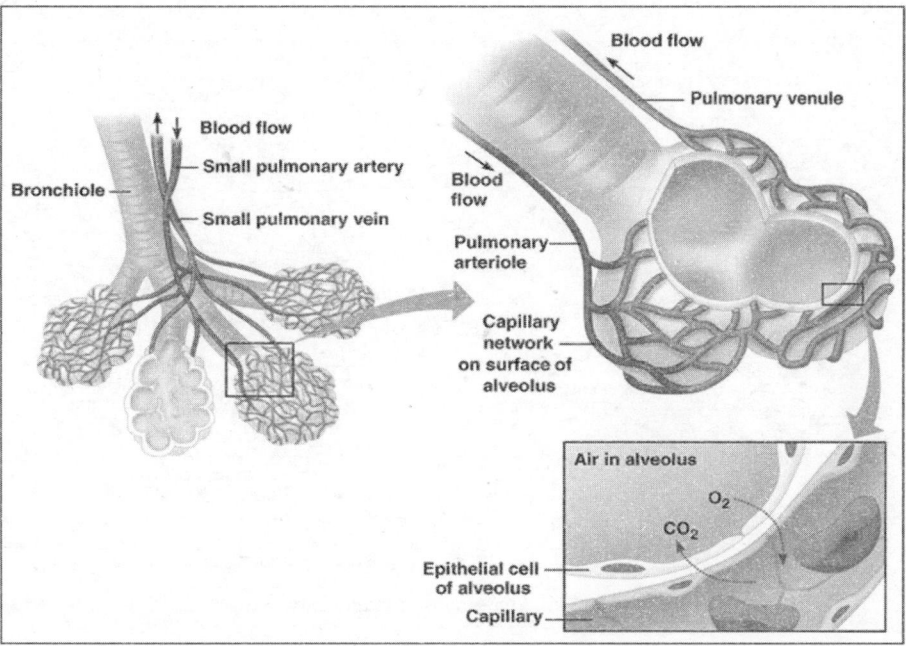

Fig. 7.5: Diffusion of oxygen into the blood in alveolis.

7.4.2.1. *Breathing cycle*

The breathing process, also called the breathing cycle, consists of two phases: inspiration, meaning air in, and expiration, meaning air out, separated by a relaxed state. In breathing cycle relaxed state means diaphragm and intercostal muscles are relaxed. Whereas in inspiration, the diaphragm contracts, muscle are pulled down, intercostal muscles contract, chest wall is elevated and volume of chest gets expanded, pressure in lungs lowers, air is inhaled. In expiration phase, the muscles relax, diaphragm resumes dome shape, intercostal muscles allow chest to lower resulting in increase of pressure in chest. The main purpose of breathing is to exchange gases with the environment. Therefore, gases diffuse according to their partial pressures. In external respiration, gases exchanged between air and blood, whereas in internal respiration, gases are exchanged with tissue fluids. Oxygen gets transported as bounded with haemoglobin in red blood cells or dissolved in blood plasma. Carbon dioxide is transported as bound to haemoglobin or dissolved in blood plasma, in the form of plasma bicarbonate.

7.4.3. **The cardiovascular system (structure and working)**

The cardiovascular system consists of heart and blood vessels; the heart is the pump that generates the flow of blood, i.e., it pushes the blood through blood vessels, and the blood vessels are channels through which blood flows. The heart is a four-chambered muscular pump located between the two lungs in the chest cavity. It is divided into right and left halves, each consisting of two chambers, an atrium and a ventricle (Figure 7.6). Between the two chambers on each side of the heart are the atrio–ventricular valves (AV valves), which force one-directional blood flow from atrium to ventricle but not from ventricle to atrium. Furthermore, the right chambers do not send blood to the left chambers, and vice versa.

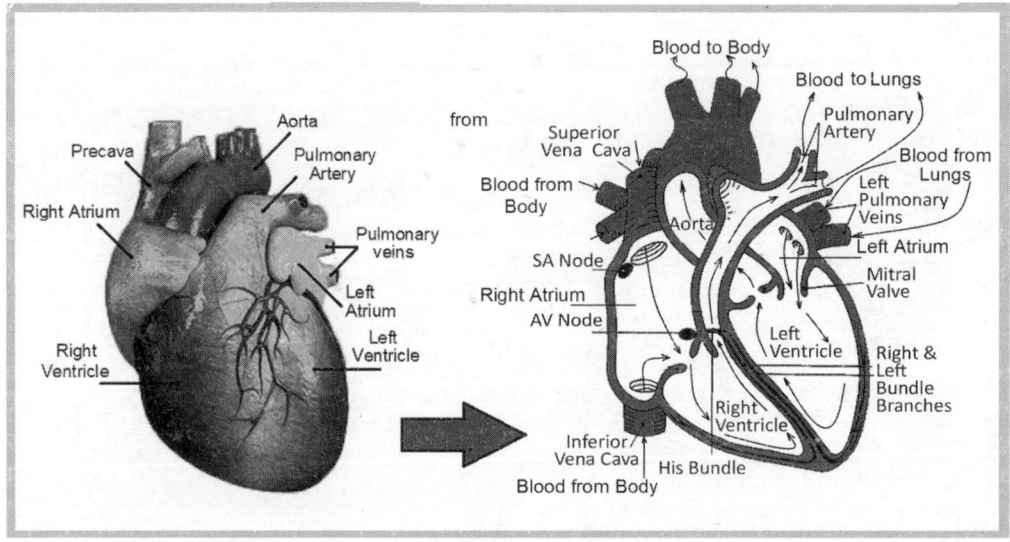

Fig. 7.6: Anatomy of human heart.

The cardiovascular system consists of two types of blood circulation circuits that both originate and end in the heart. One circuit takes away the blood from the heart through the vessels known as arteries. In the second circuit the vessels bring blood back to the heart, which are called veins. The two circulations are named as follows:

(a) Systemic circulation

(b) Pulmonary circulation

(a) Systemic circulation

It is that circulation in which fresh blood rich in oxygen and nutrients is pumped out of the left ventricle of the heart via a large artery called the aorta. From the aorta, a series of ever-branching arteries carry blood to the tissues and organs of the body. These arteries split into gradually smaller branches, and within each organ or tissue, the arteries branch into the series of vessels called the arterioles. The arterioles further split into a network of tiny thin blood vessels called capillaries permeate the tissues and organs and also through this network of capillaries that the fresh blood delivers oxygen and nutrients to the tissues. Blood also collects carbon dioxide and waste products from the tissues and carries them away on its way back to the heart. On its way back to the heart, the blood in the capillaries first merges into larger vessels called venules, and the venules are further combined into still larger vessels, veins. Ultimately, the veins from the upper half of the body are joined into a large vein, called the superior vena cava, and the veins from the lower half of the body are combined into another large vein, called the inferior vena cava. Blood returns to the right atrium of the heart via these two veins, completing a cycle of the systemic circulation.

(b) Pulmonary circulation

In pulmonary circulation, blood, rich in carbon dioxide, is pumped out of the right ventricle via the pulmonary artery, which splits into two arteries, one for each lung, Similar to the systemic circulation, the arteries branch into arterioles, which then split into capillaries. Through the bed of capillaries in the lungs, blood expels carbon dioxide and absorbs oxygen (a process called oxygenation). On its way back to the heart, the oxygenated blood in the capillaries first merges into venules and then into progressively larger veins. Finally, via the largest of these veins, that are called the pulmonary veins, the oxygenated blood leaves the lungs and returns to the left atrium of the heart, completing a cycle of the pulmonary circulation.

7.4.3.1. *Blood flow and distribution*

The heart in human body works as a pump that generates the pressure to move blood along the arteries, arterioles, capillaries, venules and veins. The heart pumps blood through its regular events of contraction and relaxation and the rate is attuned to physical workload as well as other ambiance factors such as cold/heat and humidity. Even though heart plays the critical role in producing continuous blood flow, the role of the blood vessels is much more sophisticated than that of simple lifeless plump. The blood flow encounters resistance in the blood vessels between heart and tissues, and the blood vessels can change their resistance to blood flow significantly to match the oxygen demands of various organs and tissues. The resistance to blood flow is a function of a blood vessel's radius which can changed significantly to alter the flow of blood to the muscles according to their need. Each type of blood vessel makes its own unique contribution to achieving adequate blood distribution. Because the arteries have large radius, they offer little resistance to blood flow. Their role is to serve as a pressure tank to help move the blood through the tissues. The arteries show the maximum arterial pressure during peak ventricular contraction and the minimum pressure at the end of ventricular relaxation. The maximum arterial pressure is called the systolic blood pressure, and the minimum pressure is called the diastolic blood pressure. They are recorded as systolic/diastolic; for example, 120/80 mmHg is the normal blood pressure of an adult. The difference between systolic and diastolic pressure is called as the pulse pressure.

However, in contrary to the negligible resistance offered by arteries, the radius of arterioles is small enough to provide significant resistance to blood flow. Furthermore, the radii of arterioles can be changed precisely under physiological control mechanisms. Therefore, arterioles are the major source of resistance to blood flow and are the primary site of control of blood-flow distribution.

Although capillaries have even smaller radii than arterioles, but the huge number of capillaries provides such a large area for flow that the total resistance of all the capillaries is much less than that of the arterioles. Capillaries are therefore not considered as the main source of flow resistance. However, in the capillary network, there exists another mechanism for controlling blood flow distribution like thorough fare channels; small blood vessels that provide direct links or shortcuts between arterioles and venues. These shortcuts allow blood in the arterioles to reach the venues directly without going through the capillaries and are used to move blood away from resting muscles quickly when other tissues are in more urgent need of blood supply.

The veins also contribute to the overall function of blood flow. They contain one-way valves, which allow the blood in the veins to flow only toward the heart. Furthermore, the rhythmic pumping actions of dynamic made activities can massage the veins and serve as a 'muscle pump' (also called 'secondary pump') to facilitate the blood flow along the veins back to the heart.

The amount of blood pumped out of the left ventricle per minute is called the cardiac output. It is influenced by physiological, environmental, psychological, and individual factors. The physiological demands of muscular work vary cardiac output to a great extent. At rest, the cardiac output is about 5 L per minute which may increase up to seven times depending upon the physical work load. In moderate work, the cardiac output is about three times, i.e., 15 L per minute. During heavy to very heavy work, it may increase as much as five to seven times, i.e., 25 – 35 L per minute. During work in hot and humid environmental condition cardiac output also increases, since more blood supply is required to the skin to help dissipate excess heat from the body. Cardiac output may also increase when an individual is excited or under emotional stress. Age, gender, health and fitness conditions may also influence the cardiac output of an individual under various job situations.

The cardiac output of heart can increase in two ways: either the number of beats per minute (called heart rate or HR) increases or the amount of blood per beat (called stroke volume or SV) increases. Actually, cardiac output is the product of heart rate and stroke volume as shown in the following formula:

Cardiac output (L/min) = Stroke volume (L/beat) × Heart rate (beats/minute)

Cardiac output (L/min) = SV (L/beat) × HR (beats/minute)

$$= 70 \times 72 \text{ ml/minute}$$

$$= 5040 \text{ ml/minute}$$

$$= 5 \text{ L/min (approx.)}$$

In resting condition for an adult, stroke volume is about 0.05–0.06 L/beat, for moderate work stroke volume can increase to about 0.10 L/minute. During heavy work, the increased cardiac output is accomplished largely through increased heart rate. Heart rate is one of the primary measurements of physical workload at all workload levels.

The output of the left ventricle is circulated to different body parts in proportion to their needs. The cardiac output varies with body size, for example, a large sized person requires more blood flow because of large sized organs. Hence, cardiac output per unit body surface area is somewhat constant from one individual to another and is called cardiac index.

Hence,

$$\text{Cardiac Index} = \frac{\text{Cardiac Output}}{\text{Surface Area}}$$

$$= \frac{5 \frac{\text{L}}{\text{minute}}}{1.7 \text{m } 2 \text{m}^2}$$

$$= 2.942 = 3 \text{ L/minute m}^2 \text{ (approx.)}$$

The calculation of body surface area (BSA) from height and weight, using Dubois' formula is given as under

$$\text{BSA (m}^2) = \text{Weight (kg)}^{0.425} \times \text{Height (cm)}^{0.725} \times 0.007184$$

Cardiac output is distributed as blood flow to various organs. Hence, the increased blood flow through a major vascular bed increases the entire cardiac output. A common example of such a condition is work out (exercise) in daily life, which is associated with an increase in muscle blood flow. During workout, the cardiac output may amplify up to about five times even in untrained persons and about ten-fold in trained athletes. Each tissue or organ obtains a fraction of the cardiac output. The blood-flow distribution for an adult in resting and work condition is shown in Table 7.1.

During the rest, the digestive system receives 20–25 per cent of the total cardiac output (blood flow), brain 15 per cent, kidneys 20 per cent and muscles receives around 15–20 per cent of the total cardiac output. However, in moderate work load in a hot environment at 38°C, about 45 per cent of cardiac output goes to the working muscles to meet their metabolic requirements. During very heavy work, this may increase to about 70–75 per cent, even in a moderate environment of 21°C. This is due to the reason that in hot environments more blood is distributed to the skin to dissipate the excess body heat. Whereas the portion of blood that flows to the digestive system and the kidneys strictly falls down with increased workload. An interesting aspect of blood-flow distribution is the remarkable stability of brain blood flow. The brain receives the same amount of blood under all situations, although it represents a smaller fraction of the total cardiac output in heavy work than at rest. As mentioned blood-flow distribution is made possible primarily by dilating and constricting arterioles in different organs and tissues on a selective basis.

Table 7.1: Blood flow distribution (percentage) throughout different body organs in rest and working conditions.

Sr. No.	Organs	Percentage Blood Flow Distribution at		
		Rest	*Moderate Work (at 38°C)*	*Heavy work (at 21°C)*
1	Muscles	15–20	45	70–75
2	Skin	5	40	30
3	Digestive system	20–25	6–7	3–5
4	Kidney	20	67	2–4
5	Brain	15	4–5	3–4
6	Heart	4-5	4–5	4–5

Sources: Astrand and Rodahl (1986), Eastman Kodak (1986).

7.5. Energy Expenditure and Workload

Since energy is vital for sustaining living functions, its consumption within human body is must for maintaining the basic life functions even without performing any physical activity. The bare minimum level of energy expenditure required to maintain life is known the basal metabolism. The basal metabolic rate is measured in a silent and temperature controlled environment for a person in resting condition. The person whose metabolic rate is considered is required to have certain dietary restrictions for several days and fasting for twelve hours before the measurement. The individual parameters like gender, age, and BMI influence the basal metabolic rate of a person. Human energy expenditure is measured in kilocalories (kcal). Researchers like Scbottelius and Scbottelius (1978) stated that the average basal metabolic rate of an adult is normally measured in a range of 1600–1800 kcal per day, whereas Kroemer et al. (1994) have mentioned about 1 kcal per kg of body weight per hour.

Whenever, there is any physical activity, such as low-intensity sedentary or leisurely activities, the human body requires more energy than that supplied at the basal metabolic intensity. However a variety of estimates have been made about the energy expenditures of maintaining a sedentary or non-working life. For example, it is estimated that the resting metabolism measured before the start of a working day for a resting person is about 10–15 per cent higher than basal metabolism (Kroomer et al. 1994). Luebmann (1958) and Scbottelius (1978) estimated that for basal metabolism and leisure and low-intensity daily nonworking activities the energy requirement is about 2400 kcal/day.

With the start of physical work, energy expenditure starts increasing; thereby the energy requirement of the body also exceeds the normal resting level. The body boosts its level of metabolism to meet this increased energy requirement. This increase in metabolism from the resting in the working level is called as working metabolism, or metabolic cost of work. The metabolic or energy expenditure rate during physical work is the sum of the basal metabolic rate and the working metabolic rate. Different activities require different levels of metabolic rates; e.g., a male carpenter needs an energy requirement of around 2.9–5.0 kcal/minute, and a female worker engaged in laundry work requires energy expenditure of about 3.0–4.0 kcal/minute (Durrin and Passmore, 1967). Some estimated energy expenditure rates for various types of work and daily activities are shown in Table 7.2.

Table 7.2: Estimates of energy expenditure rates for various activities.

Sr. No.	Activity	Energy Expenditure Rates (kcal/ minute)
1.	Sleeping	1.3
2.	Sitting	1.6
3.	Standing	2.3
4.	Walking at a speed of 3 km/hour	2.8
5.	Walking at a speed of 6 km/hour	5.2
6.	Carpenter-assembling	3.9
7.	Woodwork-packaging	4.1
8.	Sock room work	4.2
9.	Welding	3.4
10.	Sawing of wood	6.5
11.	Chopping of wood	8.0
12.	Athletic activities	10.0

It generally takes some time for the body to increase its rate of metabolism and meet the energy requirements of work imposed by the muscles. Normally it takes around 1–3 minutes for the circulatory and respiratory systems to regulate energy supply as per the enhanced metabolic requirements of work. During this initial warm-up period at the start of physical work, the amount of oxygen supplied to the tissues is less than the amount of oxygen needed which results as an oxygen deficiency. The time taken to acquire constant rate of metabolism to meet the energy requirements of the work imposed by the muscles is exhibited in Figure 7.7.

During this insufficient oxygen supply, anaerobic metabolism serves as the main source of energy. If the physical work is not too heavy, a steady state can be reached in which the oxidative metabolism produces sufficient energy to meet all energy requirements. The oxygen deficit incurred at the start of work must be repaid at some time either during the work, if the work is light, or during the forever period immediately after the work ceases, if the work is moderate or heavy. This is why the respiratory and circulatory systems often do not return to their normal activity levels immediately on completion of a moderate or heavy work.

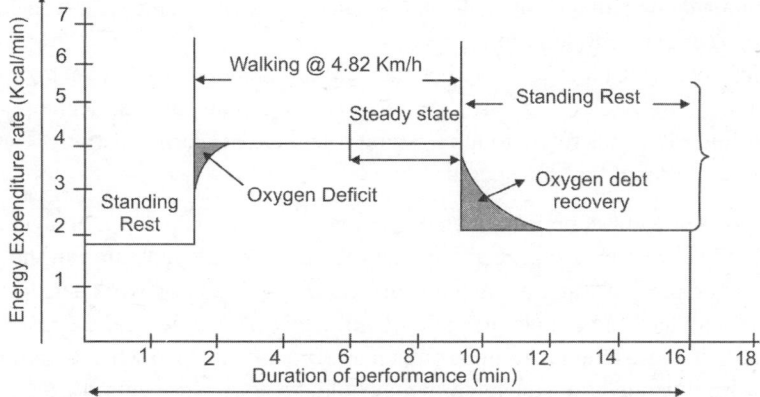

Fig. 7.7: Total energy expenditure rate in response to level of physical activity.

The physical workload can be classified as light, moderate, heavy, very heavy and extremely heavy, according to their energy expenditure requirements (Astrand and Rodahl, 1986; Kroemer et al., 1994). In case of light work, the energy expenditure rate is literally small (under 2.5 kcal/minute) and the energy demands can be easily met by the oxidative metabolism of the body. Moderate work load requires energy expenses in range of 2.5–5.0 kcal/minute, of which large portion is however met through the oxidative metabolic metabolism. Heavy work requires energy at expenditure rates between 5.0–7.5 kcal/minute. Only physically fit workers are able to carry out this type of work for a relatively long period of time with energy supplied through oxide metabolism. The oxygen deficit increased at the start of work cannot be revived until the end of work. In case of very heavy work (with energy expenditure rates between 7.5–10.0 kcal/minute) and extremely heavy work (greater than 10.0 kcal/minute), even physically fit workers cannot reach a steadystate condition during the period of work. The oxygen deficit and the lactic acid accumulation continue to increase as the work continues and make it necessary for the worker to take frequent breaks or even stop the work entirely.

7.6. Occupational Kinesiology

Kinesiology means the scientific study of the movements of body member as a function of the construction of the musculoskeletal system. Occupational kinesiology is concerned with the basic study of human movements and its limitation to work situations. It utilizes anatomy, physiology and Newtonian mechanics. It describes the laws and quantitative relationships essential for the understanding of the mechanism involved in the human performance. The general finding and applications are listed below:

1. Muscular Strength: This refers to the maximum tension per unit area that could be developed within muscle. Under normal working conditions, heavy work would generate 3.5–4.2 kg/cm^2 of tension; and light to medium work would generate 1.5–2.2 kg/cm^2 of tension. Maximum pull force of human being is 14.5 kg.

2. Muscular insufficiency: Whenever a muscle is passing over two or more joints, is shortened such that full range movements cannot be completed is called a super flexing. In such cases, a state of active insufficiency would occur and performance would be impaired. For example, i) when a person is seated too low, the knee joints are super flexed and it would become impossible to operate a pedal and ii) a flexed wrist cannot grasp a rod firmly.

3. The combined effect of tension in the muscle and rest acting on the joint surface can create conditions conducive to joint injury.
   Example: Locate push button in such a manner that include angle between upper arm and forearm being formed during push operation is between 80 and 120 degrees.
4. Postural Integrity: This refers to the postural tolerance of work situations. The following basic principle should be considered:

(a) Keep elbows down: Keeping up of an unsupported arm for a longer period would produce fatigue and severe emotional reactions with the consequent reduction in the production rate. For example, if a seat is positioned only 7.5 cm too low w.r.t. work bench will produce an angle of abduction of upper arm of approximately 45 degrees and the resulting fatigue would reduce the efficiency rating as much as 50 per cent. Also when the seat is too low during assembly operation, the left arm may hold the work piece with consequent increase in fatigue.

(b) During the lifting or holding of an article or component, bring the centre of mass as close as possible to the lumber spine. the same concept is explained in detail in chapter-8.

(c) Consider the gender difference for lifting, since the hip socket of a male person is located just below the lumber vertibra, i.e., in the same plane of centre of mass of the body. However in case of females the hip socket is located few cm further forward to the line passing through the centre of mass.

Fatigue:

Fatigue can be classified in the following three ways:

1. Physiological or muscular fatigue: It is brought by a task requiring considerable physical energy.
2. Psychological fatigue: which is an index of task 'aversion', manifested (that is seen outwardly) as a feeling of tiredness or boredom.
3. Industrial fatigue: It is defined as the one that is manifested by decrease in output after continued performance of task. The performance reduction could be due to: (a) Improper task design, (b) Non-compatibility with work environment and (c) Morale or psychological problem, indicating a desire a change in routine.

Applications of method study seek for proper task design with the reduction unwanted manual efforts. However, it has been noticed that main cause for industrial fatigue in many cases is due to non-compatibility of the worker with the work environment. These usually develop a state of 'tension', which may manifest itself in different forms of physiological and psychological effects. Hence, the major efforts in the field of ergonomics, is mostly based on the reduction of non-compatibility and subsequent 'tension'.

Physical effort tasks:

This is referring to the task, where the efforts are the main inputs to the performance. These are considered in the following three categories.

1. Full body dynamic work: Here the utilization of large muscular groups (2/3–3/4) is involved during the performance of task; e.g., walking, climbing, carrying, lifting, full body pushing or pulling.
2. Localized muscle work: Here fewer muscle groups are used while performing the task. We consider the weight involved, speed of operation and of reaches during the performance.

It should be noted that even though certain work for job may said to be 'light', it may become highly demanding, if proper precautions and considerations are not taken; e.g., bench work, assembly work, and motion of one or both arms.

3. Static muscular work: In this case, the muscles exert force, but no work or mechanical work takes place. It requires forceful muscular contraction; example: holding power sander.

Bio-energetics:

Bio-energetics is concerned with the various level of energy consumption and utilization of human being at work. Human being get energy due to the combination of oxygen with the nutrients (in the form of glycogen) contained in the body. The energy is expressed in kcal. It has been noted that only 22 per cent of energy from the food is available for work and the rest dissipates as heat.

In general, a normal healthy person gets energy from food in a range of 4500 – 4800 kcal per day. The energy requirement for the basal metabolism is about 500 kcal per day. The energy required to keep the body active varies from 1500 to 2000 kcal per day. Therefore, the energy available to perform physical work is 2500 kcal per day. This is fairly on the higher side. However, it has been estimated that for actual work on the functional job, the required energy level is 1500–2000 kcal per day. The maximum energy expenditure rate, at which a person can work continuously, is 5 kcal per minute and 4 kcal per minute, for male and female workers, respectively. However, maximum average heart rate is observed between 115 and 120 beats per minute. Healthy people generally have reserve energy of 25 kcal. Therefore, as long as one is working at the rate of energy expenditure of 5 kcal per minute, one need not take from the reserve energy. For performing any work with higher rate, one has to take from the reserve energy only. If the reserve energy is over, then the individual would feel exhausted and requires rest. During the rest period, one can recover back the reserve energy at a rate of 3–4 kcal per minute. However, it has been noted that the recovery period depends on the working rate and the period of continuous working. Hence, in order to calculate the required rest period, following formula may be used.

$$T = \frac{W \times (E - S)}{(E - 1.5)}$$

where, T = rest period required in min, over the working time

E = average energy expenditure in kcal/min during the work period

S = standard level of energy expenditure (5 kcal/min)

W = continuous working time in min.

Example:
A person is working at an energy expenditure rate of 7.5 kcal/minute. What would be the continuous period of work for him? What would be the period of rest needed after this continuous period of work?

Solution:
Energy expenditure rate (E) = 7.5 kcal/minute

Normal level of energy expenditure (S) = 5 kcal/minute

Excess rate of energy expenditure (E-S) = 7.5 - 5 = 2.5 kcal/minute

The excess energy requirements will be met from the reserve energy 25 kcal.

Hence, the continuous working time possible for him = $\dfrac{25}{(E-S)} = \dfrac{25}{2.5} = 10$ minute.

If this is taken as continuous work period for him, hence continuous working time (w) = 10 minute. The rest period required over the working time (t)

$$T = \frac{W \times (E - S)}{(E - 1.5)}$$

$$T = \frac{10 \times (7.5 - 5)}{(7.5 - 1.5)}$$

$$T = \frac{25}{6} = 4\frac{1}{6} \text{ minutes.}$$

The working time (W) = 10 minute

The rest period required = ($W + T$) = $10 + 4\dfrac{1}{6}$ = 15 minute (approx..).

7.7. Conclusion

Physical work is possible only when there is enough energy to support muscular contractions. In this chapter, we saw how the cardiovascular and respiratory systems work together to meet the energy requirements of work and how these requirements can be measured quantitatively and considered in the analysis of physical work. Poorly designed workstations and manual material handling may cause both physical and psychological stress, but they are not the only causes of stress in life and work. Other factors such as noise and vibration, as well as time pressure and anxiety, may cause stress as well. These stressors are the topic of the next chapter.

Unsolved Problems

1. Two workers, one male and a female, are packing hand tools in carton boxes and their energy consumption rate is 6.0 kcal/minute. Another worker whose energy expenditure is 7.0 kcal/minute is loading the same carton boxes. What would be the continuous period of work for all the workers? Calculate the period of rest needed for both of them after their continuous periods of work.

Multiple Choice Questions

1. How many types of muscles structures does human body have?
   (a) Four
   (b) Five
   (c) Three
   (d) None of the above

2. Which of the following systems controls the contraction of smooth muscles?
   (a) Autonomic nervous system
   (b) Central nervous system
   (c) Conscious
   (d) None of these

3. The skeletal muscles are the main muscles in the human body. What do these account for?
   (a) 40% of body weight
   (b) 65% of body weight
   (c) 80% of body weight
   (d) None of these

4. Contraction of skeletal muscles is controlled by
   (a) ANS
   (b) CNS
   (c) Conscious
   (d) All of the above

5. The energy required for muscular contraction and various other physiological functions of the body itself, is available in the form of
   (a) ATP and CP
   (b) ATP only
   (c) CP only
   (d) ADP

6. For light to moderate activity in the availability of sufficient amount of oxygen, energy requirement is fulfilled by.
   (a) Anaerobic glycolysis
   (b) Aerobic glycolysis
   (c) Oxidative phosphorylation
   (d) All of the above

7. Red blood cells (RBC) have a vital importance to work physiology due to their property for
   (a) Carrying carbon dioxide
   (b) Carrying oxygen
   (c) Defend the body against infection
   (d) None of above

8. An individual having low blood volume or less RBC count or works in a polluted or poorly ventilated environment or at high altitudes where the air has a low oxygen content, will experience:
   (a) Higher fatigue
   (b) Increased stress on the circulatory system
   (c) Breathlessness
   (d) All of these

9. The movement of air in adults occurs at the rate of 12–20 breaths per minute but during exhausting exercise it can increase up to a rate of
   (a) 24–35 per minute
   (b) 35–45 per minute
   (c) > 55 per minute
   (d) None of the above

10. The cardiac output is same as that of
   (a) Heart rate only
   (b) Product of Heart rate and stroke volume
   (c) Stroke volume only
   (d) Any one of the above

11. During the rest, the digestive system receives approximately
   (a) 10–15% of the total cardiac output
   (b) 15–20% of the total cardiac output
   (c) 20% of the total cardiac output
   (d) 20–25% of the total cardiac output

12. A healthy person normally has reserve energy of approximately
   (a) 25 kcal
   (b) 15 kcal
   (c) 35 kcal
   (d) None of the above

Answers

1. (c)   **2.** (a)   **3.** (a)   **4.** (c)   **5.** (a)   **6.** (b)   **7.** (b)   **8.** (b)   **9.** (b)   **10.** (b)
11. (d)   **12.** (a)

Biomechanics of Manual Lifting Tasks

8.1. Introduction to Musculoskeletal System

The musculoskeletal system is composed of bones, muscles and connective tissues, which include ligaments, tendons, fascia and cartilage. Bone can also be considered as a connective tissue. The main function of the musculoskeletal system is to support and protect the body and body parts, to maintain posture and produce body movements, and to generate heat and maintain body temperature.

8.2. Functions of the Muscular System

Muscles are basically composed of many bundles of tough fibres bound together in bunches (Figure 8.1). Muscles are further connected to bones by tendons. Tendons are tough, fibrous connective tissues that attach muscles to bones and transmit the forces exerted by the muscles to the attached bones. Ligaments are also dense fibrous tissues, but their function is to connect the articular extremities of bones and help to stabilize the articulations of bones at joints. Bones are connected to each other by ligaments. Muscles, when stimulated by a nerve to act, contract and produce body heat. Some involuntary functions of the muscular system include; muscles to help you breathe (lungs), make your heart beat and help move food through the digestive system. Some voluntary functions of the muscular system are such as playing piano, running, playing video games and throwing a ball. About 40–45 per cent of body weight is muscle weight. The 206 bones of the skeletal framework are covered by nearly 650 muscles.

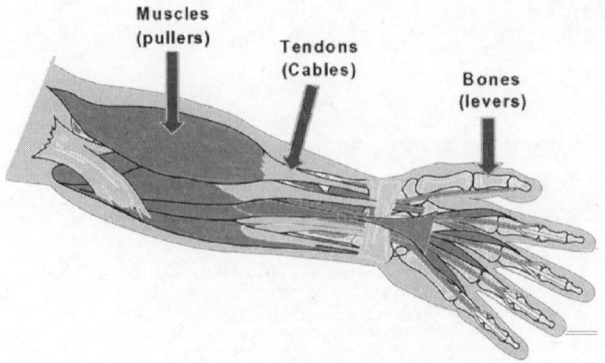

Fig. 8.1: The composition of muscles.

8.3. Bones and Connective Tissues

The rigid skeletal structure of human body composed of bones plays the major supportive and protective roles in the body. The skeleton establishes the body framework that holds all body parts together. Some bones protect internal organs, such as the skull that covers and protects the brain, and the rib cage that shields the lung and heart from the outside. Some bones, such as long bones of the upper and the lower extremities, work with the attached muscles to support body movements and activities.

8.4. Bone Joints

Two or more bones are linked with each other at joints, which can be classified into three types as follows:

 (i) Synovial joints: These are the most prominent joints, in which no tissue exists between the highly lubricated joint surfaces.
 (ii) Fibrous joints: These joints include the bones connecting the bones such as the bones of the skull connected through fibrous tissues.
(iii) Cartilaginous joints: These are such joints like bridging vertebral bones and intervertebral discs.

Joints can also be classified on the basis of type of allowed degree of movement; these are enumerated as follows:

 (a) Non-mobility joints: Non-mobility joints are those which do not support movement, such as seams in the skull of an adult.
 (b) Hinge joints: A hinge joint is that which permits motion in only one plane, such as elbow joint.
 (c) Pivot joints: A pivot joint is that which allows two degrees of freedom in movement, such as wrist joint.
 (d) Ball-and-socket joints: A ball-and-socket joint has three degrees of freedom; examples are the hip and shoulder joints.

It is significant to understand that bones can fracture when they are exposed to excess or repetitive loading in the form of bending forces, torsion forces, or combined forces. Therefore, the amount of load, the number of repetitions, and the frequency of loading are the three most important factors that can cause bone fracture. It is also true that a bone is itself capable of repairing small fractures, provided an adequate recovery time is given, which is generally referred as rest. Thus, the repetitive rate of manual exertions or the recovery period after exertions can become significant factors (Chaffin et al., 1999). Connective tissue may also be damaged after excessive or repetitive use. For example, heavy loads may increase tension in tendons and cause tendon pain. Excessive use of tendons may also cause inflammation of tendons.

8.4.1. Work-related musculoskeletal disorders (MSDs)

MSDs are cumulative, generally developed slowly in response to prolonged, repeated activities that affect the soft tissues of the body. Usually rapid, repetitive movement, sustained, constrained or awkward postures or forceful movement are involved. Prolonged repetitive movements can cause the tendons and tendon sheaths to become inflamed as the supply of the fluid that lubricates the tendons is exhausted. As the result, severe muscle strain of the forearm, upper arm, shoulders, neck and back develops either alone or combination with pain in the hands. These disorders are generally incremental and may take months or years to appear. Likely, if the symptoms are ignored, it may take long time for the injury to be repaired (Oxenburgh, 1991). MSDs are injuries or disorders of the muscles, tendons, joints, spinal discs, nerves, ligaments or cartilage. MSDs develop as a result of repeated exposure to ergonomic risk factors. Work-related MSDs are those disorders to which the work environment and the performance of work contribute significantly. Another familiar and related term is cumulative trauma disorders (CTDs). Common examples of MSDs include herniated spinal discs, low back pain and some MSDs related to upper distal extremities such as carpel tunnel syndrome (CTS), tendinitis and tenosynovitis.

There are alternative approaches, based on simplified methods to find ergonomic exposures. Quantification of ergonomic exposures, based on comprehensive information on the frequency and duration of particular postures and movements, is now common. During the last decades, numerous ergonomic risk assessment methods have been developed for assessing exposure to risk factors for MSDs. Many of them assess the risk of the various regions of the body such as low back, upper back, neck, shoulder, arms, legs, knees, thighs, ankles and wrists. Originally, ergonomic risk factors include the workstations, tools, equipment, work methods, work environment, worker's personal characteristics, metabolic demands, physical stress and emotional stress. Professionals from mechanical engineering, industrial engineering, occupational hygiene, occupational medicine, occupational therapy, kinesiology, psychology and many other fields provide unique insights into the relationship between worker/workplace and MSDs. Understanding ergonomic risk factors are essential because there is indication that ergonomic risk factors are deeply related to musculoskeletal disorders of the upper extremities and the low back.

In performing physical work, excessive loading can cause musculoskeletal problems such as bone fracture and muscle fatigue. To determine whether a load is excessive for a body segment, we need to quantify the magnitude of physical stress imposed on the body segment in performing the task. How do we obtain these quantitative estimates? Biomechanical modelling provides an important method for answering this question.

8.5. Biomechanics

Biomechanics is an interdisciplinary study of human movements with respect to range, strength and speed. It basically comprises physics, mathematics, physiology, anatomy and several other sciences. Among other aspects, study of biomechanics is useful in sports and manual industrial works for improving our training methods and preventing injuries.

8.6. Biomechanical Models

Biomechanical models are mathematical models of mechanical properties of the human body. In biomechanical modelling, the musculoskeletal system is analysed as a system of mechanical links, and the bones and the muscles act as a series of levers. Biomechanical models allow one to predict the stress levels on specific musculoskeletal components quantitatively with established methods of physics and mechanical engineering and thus can serve as an analytical tool to help job designers identify and avoid hazardous job situations.

8.7. Lever Systems

Muscles and bones act as a set of levers. Any weight to be lifted at a distance from joint will cause external torque. In response to the same, forces are developed by the muscles–tendon complex and produce torque (rotation) around the joint, which is called internal torque. A typical type of lever is shown in Figure 8.2. The external torque and the intended direction of motion determine the internal torque. If internal torque becomes equal to external torque, the state of equilibrium is achieved. If internal torque is larger than the external torque, then state of trunk extension will occur. When a body or body segment is not in motion, it is described as in static equilibrium. For an object to be in static equilibrium, two conditions must be met: The sum of all external forces acting on an object in static equilibrium must be equal to zero and sum of all external moments acting on the object must be equal to zero. These two conditions play an essential role in biomechanical modelling.

In biomechanics of lifting following abbreviations are used:

 W = Weight of the object (lbs or kg)
MW = Moment arm of the weight, W (inch or cm)
BW = Body weight is above L5/S1 disc
MB = Moment arm of body weight
AF = Applied force (lbs or kg)
MA = Moment arm of the applied force, AF (inch or cm)
MF = Muscle force (lbs or kg)
MM = Moment arm of the muscle force, MF (inch or cm)
CF = Compressive force (lbs or kg)
SF = Shear force (lbs or kg)

Let us try to calculate the force required to lift a weight of 45 kg in lever system as shown in Figure 8.2.

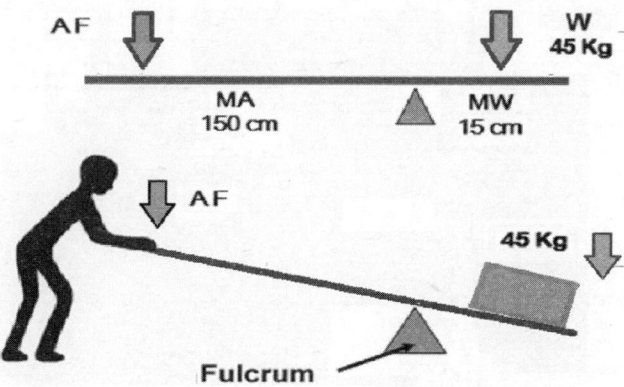

Fig. 8.2: A typical type of lever system.

The given weight can be lifted by applying required force, which can be calculated by using equation of equilibrium as follows:

Moment arm (MA) of the applied force, AF (inch) = Moment arm of the weight, W (inch)

$$MA \times AF = MW \times W$$

$$150 \times AF = 15 \times 45$$

$$AF = \frac{MW \times W}{MA}$$

$$AF = \frac{15 \times 45}{150}$$

$$AF = 4.5 \text{ kg}$$

The following is a description of a planer, static model of isolated body segment based on planer models (also called 2-D models), these are often used to analyze symmetric body postures with forces acting in a single plane. Static models assume that a person is in a static position with no movement of the body or body segments. Although the model is elementary, it illustrates the methods of biomechanical modelling. Complex 3-D, whole-body models can be developed as expansion of elementary models.

8.8 Single-Segment Planer Static Model

A single-segment model was given by Chaffin et al. (1999) which analyses an isolated body segment with laws of mechanics to identify the physical stress on the joints and muscles involved. As an illustration, suppose a person is holding load of 20 lbs mass with each hands and his forearms are horizontal. The load is balanced in each hand. Let us calculate the muscles force (MF) required to lift the weight when the distance between the load and the elbow is 14 inches. The calculations for one hand, forearm, and elbow are performed, the required muscle force (MF) is found to be 140 lbs which is enumerated as follows.

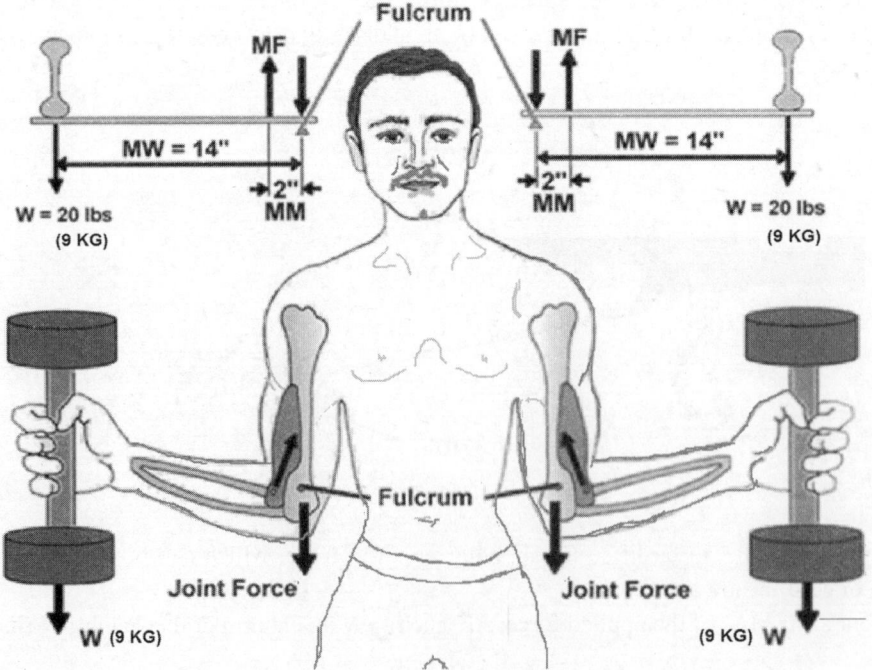

Fig. 8.3: A person lifting weight on forearm and biceps.

$$MF \times MM = MW \times W$$

$$MF \times 2 = 14 \times 20$$

$$MF = \frac{14 \times 20}{2}$$

Therefore, MF = 140 lbs

8.9. Low–Back Problems

Lower back pain (LBP) is ranked first as a cause of disability and inability to work, and is expected to affect up to 90 per cent of the world's population at some point in their lifetime (Graham et al., 2007). Prospective studies demonstrate that low back problems do not display a six-week spontaneous recovery pattern, as was once believed. The condition is regularly seen to worsen over time, becoming a chronic disorder, influenced by both physical and psychosocial factor. In industry as well LBP is the costliest and prevalent work-related musculoskeletal disorder. According to the estimates of the National Council on Compensation Insurance, low-back pain cases account for approximately one-third of all workers' compensation payment. When indirect costs are included, the total costs estimates range from about $27–$56 billion in United States (Pope et al., 1991). About 60 per cent of overexertion injuries reported each year in the United States are related to lifting (NIOSH, 1981). Further it is estimated that low-back pain may affect as much as 50–70 per cent of the general population due to occupational and other unknown factors (Anderson, 1981; Waters et al., 1993). Manual material handling involving lifting, bending, and twisting motions of the torso is a major cause of work-related low-back pain and disorders, both in the occurrence rate and degree of severity. However, low-back problems are not restricted to these situations. Low-back pain is also common in sedentary work environments requiring a prolonged static sitting posture. Thus, manual handling and seated work become two of the primary job situations in which the biomechanics of back should be analyzed. Wherever manual material handling or lifting is involved, i.e., low back bio-mechanics is applicable. Therefore there is need to understood about the human spine.

8.9.1. Human spine

Human spine comprises 33 vertebrae joined together by multiple ligaments and intervening cartridges, their functions are briefly mentioned in the previous Section 8.4. For the convenience of description the vertebrae are divided into four regions which correspond roughly to the changes in the shape of the spine. These regions are: topmost seven cervical (neck), and then twelve thoracic, and five lumber, followed by five fused sacral and four fused coccygeal vertebrae, known as sacrum. Human spine is supported by various muscles, e.g., the lumber region is supported by the erector spinal muscles. The various muscles that provide maximum support during the lifting tasks are shown in Figure 8.4. Therefore, from the viewpoint of lifting load, the orientation of the lumber and sacral vertebrae is very important, since these vertebrae and their respective discs and muscles bear most of the load in manual lifting tasks. In daily life human being is engaged in different tasks and consequently different spinal segments and respective disks are exposed to different kinds of loads such as compression, tension, shear and torsion. The mechanism of disk compression with respect to time is shown in Figure 8.5. Due to repeated exposure to different segmental loads, disk degeneration occurs and ultimately results in the distortion of disk and rupture of the casting, which leads to distortion of nerve root. The mechanism of disk degeneration and narrowing down of nerve root is shown in Figure 8.6.

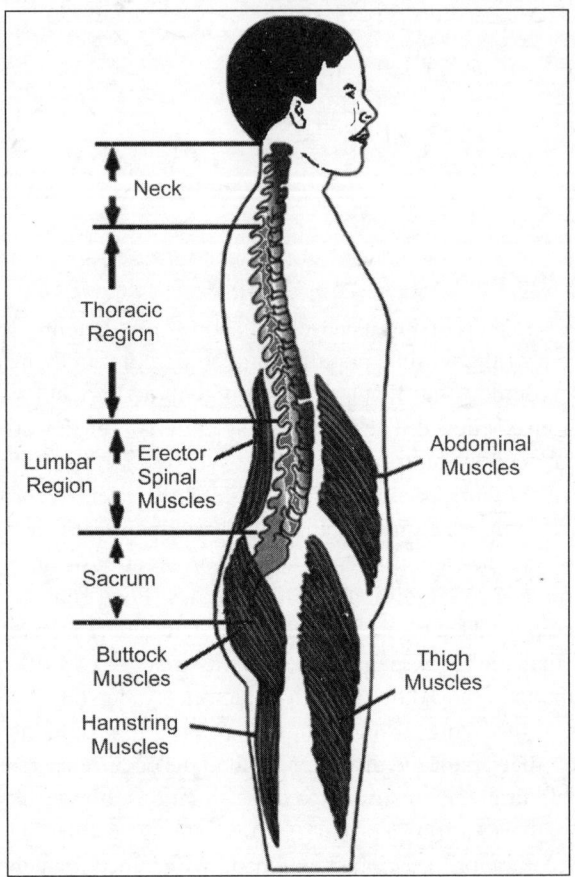

Fig. 8.4: Human spine and muscles supporting the lower spine.

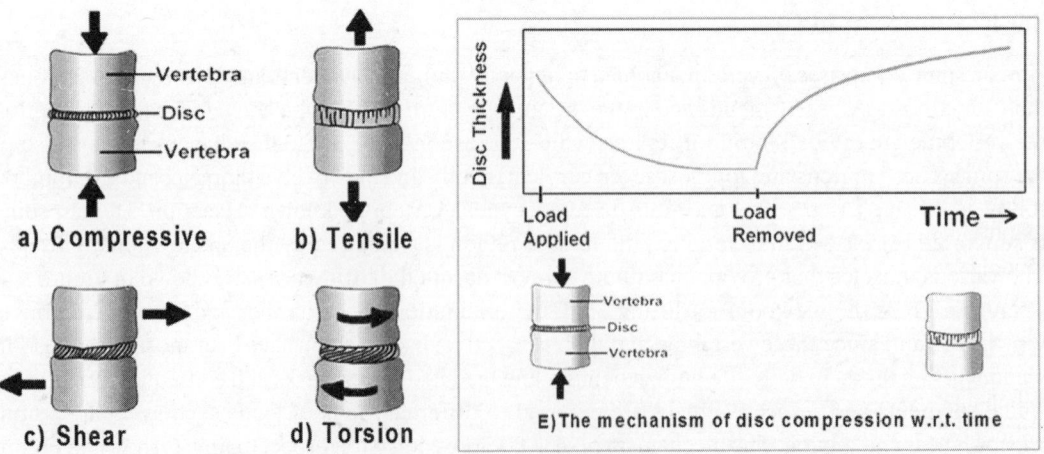

Fig. 8.5: Loads (A–D) on spinal motion segments and E) the mechanism
of disc compression with respect to time.

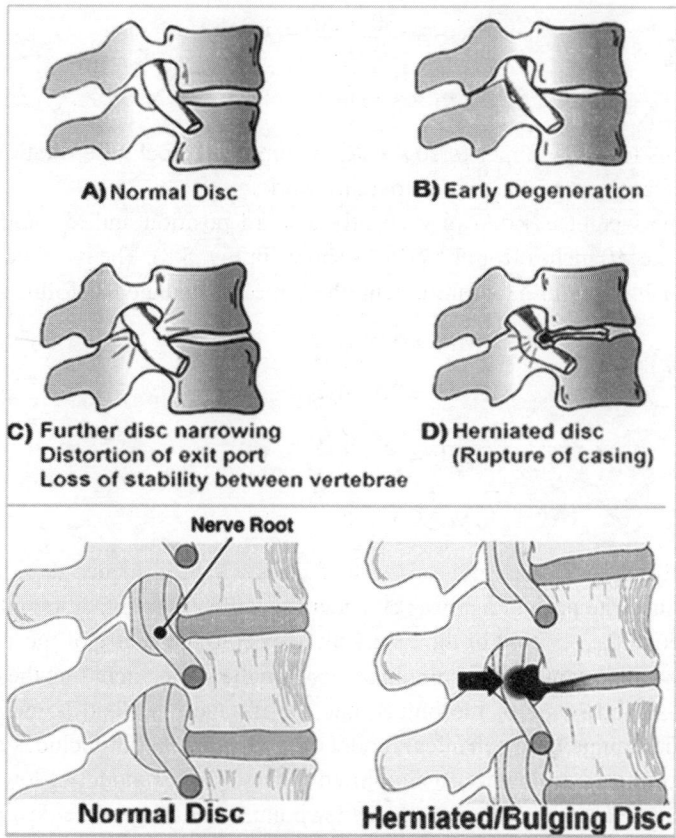

Fig. 8.6: Mechanism of disc degeneration.

8.9.2. Low-back biomechanics of lifting

The lower back is perhaps the most vulnerable link of musculoskeletal system in material handling because it is most distant from the load handled by the hands. Both the load and the weight of the upper torso create significant stress on the body structures at the low back, especially at the disc between the fifth lumber and the first sacral vertebrae (called the L5/S1 disc).

A more accurate determination of the reactive forces and the moments at L5/S1 disc requires the use of multi-segment model. It also requires the consideration of abdominal pressure, created by the diaphragm and abdominal wall muscles (Morris et al., 1961). However, a simplified single-segment model can be used to obtain a quick estimate of the stress at the low-back (Chaffin et al., 1999).

8.9.3. Low back as a lever

When a person lifts a load of 30 lbs in forward bending (flexion) posture at a horizontal distance of 20 inches from L5/S1 as shown in Figure 8.7, the required muscle force can be calculated as:

$$MF \times MM = MW \times W$$

$$MF \times 2 = 30 \times 20$$

$$MF = \frac{30 \times 20}{2} = 30 \times 10$$

$$MF = 300 \text{ lbs}$$

In addition to this load, the upper torso creates a combined clockwise rotational moment. Disc compression at this level can be hazardous to many workers.

Now, if the same weight is lifted in vertically upward position and at relatively lesser (half) horizontal distance i.e. 10 inches from L5/S1 (As shown in Fig. 8.8). The required muscle force will be reduced to one half of the initial requirement, the same is calculated as follows:

$$MF \times MM = MW \times W$$

$$MF \times 2 = 30 \times 10$$

$$MF = \frac{30 \times 10}{2} = 15 \times 10$$

$$MF = 150 \text{ lbs}$$

In executing a lifting task, there are numerous factors which influence the load stress beard or tolerated by the spine. The present analysis considers specifically two factors only, first the weight to be lifted and second the position of the load with respect to the centre of the spine. A number of other factors are also important in determining the load on the spine, including the degree of twisting of the torso, the size and shape of the object, and the distance the load is moved. Developing a comprehensive and accurate biomechanical model of the low back that includes all these factors is beyond the scope of this text; however a simplified biomechanical model of low back is shown in Figure 8.9; from practical ergonomics analysis viewpoint, the lifting guide developed by National Institute for Occupational Safety and Health (NIOSH) is of huge importance.

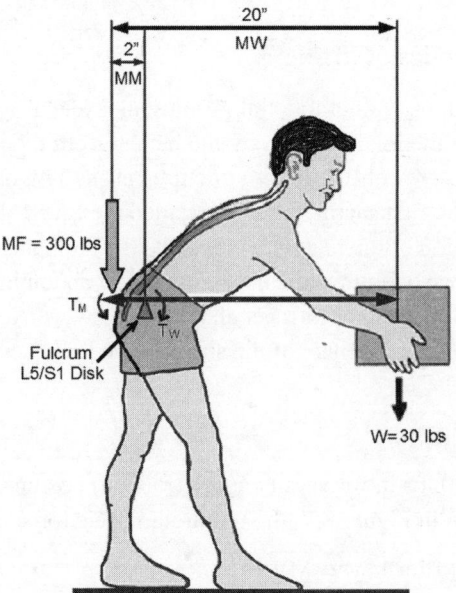

Fig. 8.7: The person lifting a weight at distance of 20 inches from L5/S1.

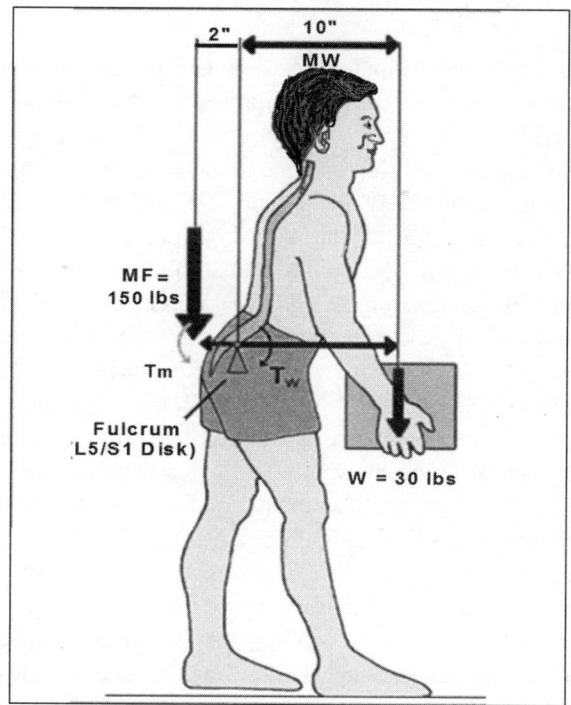

Fig. 8.8: The person lifting a weight at distance of 10 inches from L5/S1.

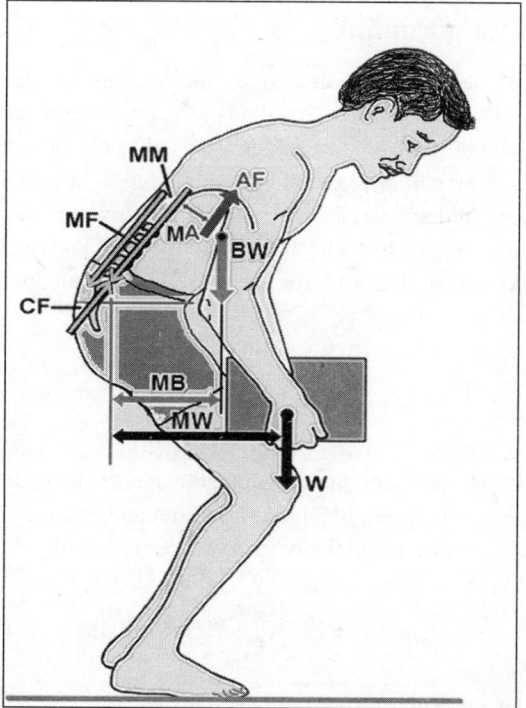

Fig. 8.9: A simplified biomechanical model of low back.

8.9.4. Low back pain

Low back pain (LBP) and injuries attributed to manual lifting activities continue as one of the leading occupational health and safety issues facing preventive medicine. Despite efforts at control, including programs directed at both workers and jobs, work-related back injuries still account for a significant proportion of human suffering and economic cost. More than three decades ago, the National Institute for Occupational Safety and Health (NIOSH) recognized the growing problem of work-related back injuries and published the Work Practices Guide for Manual Lifting (NIOSH WPG, 1981). Afterwards in 1985, NIOSH organized an ad-hoc committee of experts who reviewed the current literature on lifting, including the NIOSH WPG (1981). The literature review was summarized in a document entitled Scientific Support Documentation for the Revised 1991 NIOSH Lifting Equation: Technical Contract Reports, May 8, 1991, which is available from the National Technical Information Service (NTIS No. PB - 91 - 226 - 274). The equation was publicly presented in 1991 by NIOSH staff at a national conference in Ann Arbor, Michigan, entitled 'A National Strategy for Occupational Musculoskeletal Injury Prevention – Implementation Issues and Research Needs'. Subsequently, NIOSH staff developed the documentation for the equation and played a prominent role in recommending methods for interpreting the results of the lifting equation.

The NIOSH lifting equation is only one tool in a comprehensive effort to prevent work-related low back pain and disability. Moreover, lifting is only one of the causes of work-related low back pain and disability. Other causes which have been hypothesized or established as risk factors include whole body vibration, static postures, prolonged sitting, and direct trauma to the back. Psychosocial factors, appropriate medical treatment, and job demands (past and present) also may be particularly important in influencing the transition of acute low back pain to chronic disabling pain.

8.10. The Revised Lifting Equation

The revised lifting equation uses six variables which includes horizontal location from mid-point between the ankles, vertical location, vertical travel distance, asymmetric angle, lifting frequency and coupling along with the load constant. The variables are used to calculate their respective multiples, and these multipliers are used to calculate the recommended weight limits (RWL), which further is utilized to calculate the lifting index (LI). The current section provides the technical information for using the revised lifting equation to evaluate a variety of two-handed manual lifting tasks. Definitions, restrictions/limitations, and data requirements for the revised lifting equation are also provided, which are described in the following sub-sections.

8.10.1. Recommended weight limit (RWL)

The RWL is the principal product of the revised NIOSH lifting equation. For a specific set of task conditions, the RWL is defined as the weight of the load that nearly all healthy workers could perform over a substantial period of time (e.g., up to 8 hours) without an increased risk of developing lifting-related LBP. By healthy workers, we mean the workers who are free of adverse health conditions that would increase their risk of musculoskeletal injury. The RWL is defined by the following equation:

$$RWL = LC \times HM \times VM \times DM \times AM \times FM \times CM$$

where, LC = Load Constant
HM = Horizontal Multiplier
VM = Vertical Multiplier

    DM = Distance Multiplier
    AM = Asymmetric Multiplier
    CM = Coupling Multiplier
    FM = Frequency Multiplier

Load Constant (LC) = 51 lbs (23 kg) for US population; however for Indian population LC can be further reduced to a fraction of 75–80 per cent or so. If the weight of the object is more than 51 lbs (23 kg), the lifting task will be unsafe to some workers even under ideal lifting conditions.

8.10.2. Lifting index (LI)

The term lifting index (LI) provides a relative estimate of the level of physical stress associated with a particular manual lifting task. The estimate of the level of physical stress is defined by the relationship of the weight of the load lifted and the recommended weight limit. The LI is defined by the following equation:

$$LI = \frac{\text{Weight of the Object Lifted}}{\text{Recommended Weight Limit}}$$

8.10.3. Terminology and data definitions

The following are the brief definitions useful in applying the revised NIOSH lifting equation.

(a) Lifting task – It is defined as the act of manually grasping an object of definable size and mass with two hands, and vertically moving the object without mechanical assistance.

(b) Load weight (L) – Weight of the object to be lifted in pounds or kilograms, it also includes the weight of container/box.

(c) Horizontal location (H) – Distance of the hands away from mid-point between the ankles, in inches or centimetres. It is measured at origin and destination of a lift.

(d) Vertical location (V) – Distance of the hands above the floor, in inches or centimetres. It is measured at origin and destination of a lift.

(e) Vertical travel distance (D) – Absolute value of the difference between the vertical heights at the destination and origin of the lift, in inches or centimetres.

(f) Asymmetry angle (A) – Angular measure of how far the object is displaced from the front (mid-sagittal plane) of the worker's body at the beginning or ending of the lift, in degrees. It is measured at origin and destination of a lift. The asymmetry angle is defined by the location of the load relative to the worker's mid-sagittal plane, as defined by the neutral body posture, rather than the position of the feet or the extent of body twist.

(g) Neutral body position – It is the position of the body when the hands are directly in front of the body and there is minimal twisting at the legs, torso or shoulders.

(h) Lifting frequency (F) – Average number of lifts per minute over a period of 15 minutes.

(i) Lifting duration – It is specified by the distribution of work-time and recovery-time or work pattern. Duration is classified as either short (1 hour), moderate (1–2 hours) or long (2–8 hours), depending on the work pattern.

(j) Coupling classification – It the classification of the quality of the hand-to-object coupling (e.g., handle, cut-out or grip). Coupling quality is classified as good, fair or poor.

(k) Significant control – It is defined as a condition requiring precision placement of the load at the destination of the lift. This is usually the case when; (1) the worker has to re-grasp the load near the destination of the lift, or (2) the worker has to momentarily hold the object at the destination or (3) the worker has to carefully position or guide the load at the destination.

8.11. Lifting Task Limitations

The lifting equation is a tool for assessing the physical stress of two-handed manual lifting tasks. As with any tool, its application is limited to those conditions for which it was designed. The following are the assumptions and limitations for implementation of the revised NIOSH lifting equation:

1. It is based on the assumption that manual handling activities other than lifting are minimal and do not require significant energy expenditure, especially when repetitive lifting tasks are performed. Examples of non-lifting tasks include holding, pushing, pulling, carrying, walking and climbing. However, applicable if carrying is limited to one or two steps and holding should not exceed a few seconds.

2. It does not include the unpredicted conditions, such as unexpectedly heavy loads, slips or falls. It also does not apply for lifting/lowering for over 8 hours.

3. Moreover, if the environment is unfavourable (e.g., temperatures or humidity significantly outside the range of 19–26 °C or 35–50 per cent respectively), independent metabolic assessments would be needed to gauge the effects of these variables on heart rate and energy consumption.

4. The equation is not applicable to assess the tasks like one-handed lifting, lifting while seated or kneeling, or lifting in a constrained or restricted work space. The equation also does not apply to lifting unstable loads. The equation does not apply to lifting of wheel barrows, shovelling, or high-speed lifting. For such task conditions, independent and task specific biomechanical, metabolic, and psychophysical assessments may be needed.

5. The revised lifting equation assumes that the worker/floor surface coupling provides at least a 0.4 (preferably 0.5) coefficient of static friction between the shoe sole and the working surface.

6. The revised lifting equation assumes that lifting and lowering tasks have the same level of risk for low back injuries (i.e. that lifting a box from the floor to a table is as hazardous as lowering the same box from a table to the floor). This assumption may not be true if the worker actually drops the box rather than lowering it all the way to the destination.

8.11.1. The equation and its function

The revised lifting equation for calculating the Recommended Weight Limit (RWL) is based on a multiplication model that put on a penalty in the form of multiplier for each task variable. The weightings (penalties) are the coefficients that serve to decrease the load constant, which represents the maximum recommended load weight to be lifted under ideal conditions. As already described that the RWL is defined by the following equation:

$$RWL = LC \times HM \times VM \times DM \times AM \times FM \times CM$$

The term task variables refers to the measurable task descriptors (i.e., H, V, D, A, F and C); whereas, the term multipliers refers to the reduction coefficients in the equation (i.e., HM, VM, DM, AM, FM and CM). These multipliers are in-fact penalties imposed on LC, which reduces its

(LC's) value and gives recommended weight limit (RWL) to make the lift safe. Each multiplier should be computed from the appropriate formula as shown in Table 8.1.

Table 8.1: The formula for calculation of multipliers for each of the six variables of NIOSH lifting equation (1991).

Parameter	Abbreviation	Metric (units)	U.S. System
Load Constant	LC	23 kg	51 LB
Horizontal Multiplier	HM	(25/H)	(10/H)
Vertical Multiplier	VM	$(1-0.003 |V-75|)$	$(1-0.0075 |V-30|)$
Distance Multiplier	DM	0.82 + (4.5/D)	0.82 + (1.8/D)
Asymmetric Multiplier	AM	(1-0.0032A)	(1-0.0032A)
Frequency Multiplier	FM	From Table 8.2	From Table 8.2
Coupling Multiplier	CM	From Table 8.3	From Table 8.3

Source: Revised NIOSH Lifting Equation (1991).

Table 8.2: Coupling multiplier.

Coupling	V < 30"	V > 30"
Good	1.00	1.00
Fair	0.95	1.00
Poor	0.90	0.90

Source: Revised NIOSH Lifting Equation (1991).

Table 8.3: Frequency multiplier.

Frequency	Vertical Location of Lift					
	≤ 8 Hours		≤ 2 Hours		≤ 1 Hours	
Lifts/Minutes	V < 30"	V ≥ 30"	V < 30"	V ≥ 30"	V < 30"	V ≥ 30"
0.2	0.85	0.85	0.95	0.95	1	1
0.5	0.81	0.81	0.92	0.92	0.97	0.97
1	0.75	0.75	0.88	0.88	0.94	0.94
2	0.65	0.65	0.84	0.84	0.91	0.91
3	0.55	0.55	0.79	0.79	0.88	0.88
4	0.45	0.45	0.72	0.72	0.84	0.84
5	0.35	0.35	0.60	0.60	0.80	0.80
6	0.27	0.27	0.50	0.50	0.75	0.75
7	0.22	0.22	0.42	0.42	0.70	0.70
8	0.18	0.18	0.35	0.35	0.60	0.60
9	0.00	0.15	0.30	0.30	0.55	0.55

Contd.

Contd.

Frequency	Vertical Location of Lift					
	≤ 8 Hours		≤ 2 Hours		≤ 1 Hours	
Lifts/Minutes	V < 30"	V ≥ 30"	V < 30"	V ≥ 30"	V < 30"	V ≥ 30"
10	0.00	0.13	0.26	0.26	0.45	0.45
11	0.00	0.00	0.00	0.23	0.41	0.41
12	0.00	0.00	0.00	0.21	0.37	0.37
13	0.00	0.00	0.00	0.00	0.00	0.34
14	0.00	0.00	0.00	0.00	0.00	0.31
15	0.00	0.00	0.00	0.00	0.00	0.28

Source: Revised NIOSH Lifting Equation (1991).

8.11.2. The lifting index (LI)

The lifting index (LI) provides a relative estimate of the physical stress associated with a manual lifting job.

$$LI = \frac{\text{Load Weight}}{\text{Recommended Weight Limit}}$$

where, Load Weight (L) = weight of the object lifted (lbs or kg).

8.11.3. Use of RWL and LI to Design the Lifting Task

The ergonomic design can be used to guide lifting index (LI) and recommended weight limit (RWL) index in several ways:

1. The job related problems are specifically identified by using the specific multipliers. The relative contribution of each task variable is indicated by the relative magnitude of each multiplier (e.g., horizontal, vertical, frequency, etc.).
2. The RWL can be utilized to guide the re-design of present manual lifting tasks or to design new manual lifting jobs. For instance, if the lifting task variables are preset, then the maximum weight of the load could be selected so as not to exceed the RWL; if the weight to be lifted is permanent, i.e., cannot be altered, then the task variables could be optimized so as not to exceed the RWL.
3. The LI can be applied to estimate the relative level of physical stress for a particular task. The greater the value of LI, the smaller the fraction of staff capable of safely sustaining the extent of activity. Thus, two or a lot of job designs can be compared.
4. The LI can be utilized to prioritize the ergonomic re-design of jobs, for example, a series of suspected unsafe jobs could be ranked according to the LI and on the basis of LI rank a control strategy can be developed. For instance jobs with lifting indices above 1.0 or higher would benefit the most from redesign. Job should be deigned to keep the LI ≤1, lower the LI safer is the job.
   If LI > 1, job is unsafe to some workers, and also a job with and LI = 4 is not twice as unsafe as a job with an LI = 2.

8.11.3.1. *Procedural steps of calculation of lifting index*

The procedure for analyzing a lifting task is shown in Figure 8.10. Following are the steps for calculation of lifting index and then analyze the jobs.

1. Determine the task variables as follows:

   (a) Horizontal location (H)
   (b) Vertical location (V)
   (c) Vertical travel distance (D)
   (d) Asymmetry angle (A)
   (e) Coupling classification (C)
   (f) Lifting frequency (F)

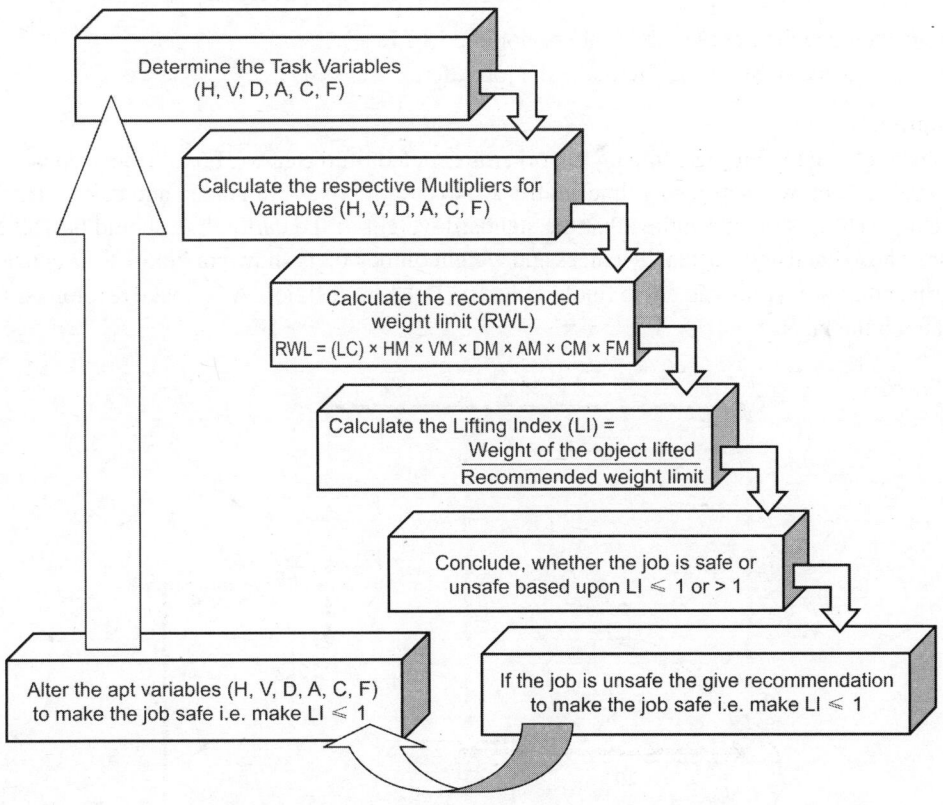

Fig. 8.10: Procedural steps for analysis of lifting task.

2. Determine the respective multipliers variables as follows:

   (a) Calculate the Horizontal Multiplier (HM)

   HM = (10/H)

   H: Horizontal distance of hands in inches.

   (b) Calculate the Vertical Multiplier (VM)

   VM = (1 - .0075 |V - 30|)

   V: Vertical location of hands in inches.

   (c) Calculate the Distance Multiplier (DM)

       DM = .82 + (1.8/D)

       D: Travel distance in inches.

   (d) Calculate the Asymmetric Multiplier

       AM = 1 - (.0032 × A)

       A: Asymmetric angle in degrees.

   (e) Find out the Coupling Multiplier (CM) and Frequency Multiplier (FM) from tables 8.2 and 8.3 respectively.

3. Calculate the RWL = 51 × HM × VM × DM × AM × CM × FM

4. Calculate the Lifting Index (LI)

$$LI = \frac{\text{Weight of the Object Lifted}}{\text{Recommended Weight Limit}}$$

5. Conclude whether the job is safe or not i.e. LI ≤1.

6. Alter the variables or LC to make the job safer.

Example 8.1:

In a hand tool manufacturing company a worker is deputed lift the carton boxes from a conveyor and load on a truck at two steps away. In each truck 240 carton boxes are loaded and daily 2 trucks are dispatched. The customer requires that the standard weight of the carton box should be 15 kg. The details of horizontal and vertical distances and weight of the box is shown in Figure 8.11 considering fair grip and zero asymmetric angle find out the RWL, LI for task job. Also make recommendations based upon the facts.

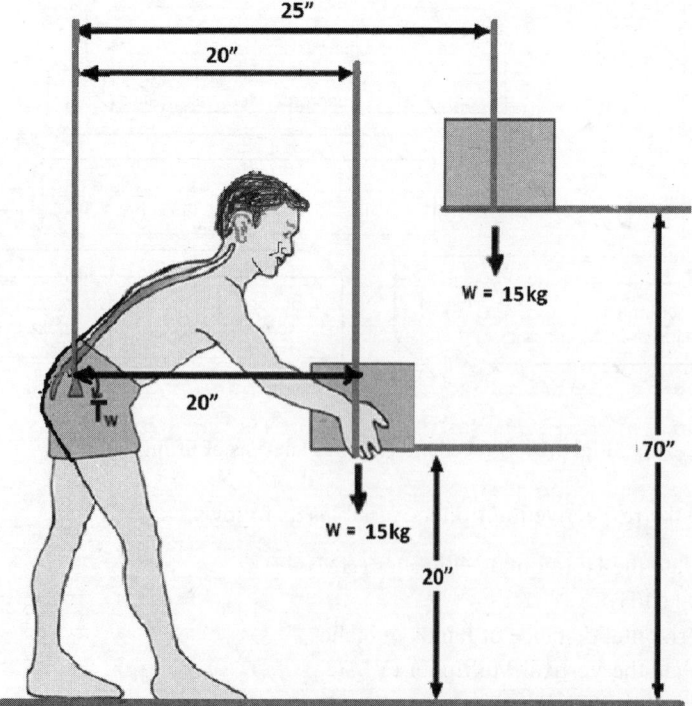

Fig. 8.11: A worker lifting load in hand tool industry.

The variables of the task are enumerated as follows:

Variable	H	V	D	A	C	F
Origin	20	20	50	0	Good	1
Destination	25	70	50	0	Fair	1

Calculations of multipliers at origin

$HM_{origin} = 10/20 = 0.50$

$VM_{origin} = (1 - 0.0075 \times |20 - 30|)$

$VM_{origin} = 0.925$

$DM_{origin} = 0.82 + (1.8/50)$

$DM_{origin} = 0.86$

$AM_{origin} = 1 - (.0032 \times 0)$

$AM_{origin} = 1$

CM from Table 8.2 = 1

FM from Table 8.3 = 0.75

Calculations of multipliers at destination

$HM_{dest} = 10/25 = 0.40$

$VM_{dest} = (1 - 0.0075 \times |70 - 30|)$

$VM_{dest} = \mathbf{0.70}$

$DM_{dest} = 0.82 + (1.8/\mathbf{50})$

$DM_{dest} = 0.86$

$AM_{dest} = 1 - (.0032 \times 0)$

$AM_{dest} = 1$

CM from Table 8.2 = 1

FM from Table 8.3 = 0.75

Variable	LC (lb/kg)	HM	VM	DM	AM	CM	FM	RWL (kg)
Origin	51 (23)	0.50	0.93	0.86	1	1	0.75	6.88
Destination	51 (23)	0.40	0.70	0.86	1	1	0.75	4.14

$$LI = \frac{15}{4.14} = 3.62$$

Since LI > 1, therefore the job is unsafe.

Now, let us redesign the task so that the horizontal distance (H) at origin and destination is reduced to 10 inches whereas the vertical distance is made 30 inches at origin and reduced from 70 inches to 40 inches. This will also reduce the distance traveled 'D' from 40 to 10 inches. At the same time we depute two workers simultaneously so that the frequency of lift is reduced from 1.0/minute to 0.5 /minute. Thereby the penalty to the load constant will be reduced, consequently, RWL will be very near to the weight lifted, hence the lifting index is 0.95, i.e.,< 1 and the job is safe. The lifting variables and corresponding multipliers are shown as follows:

Variable	H	V	D	A	C	F
Origin	10	30	10	0	Good	0.5
Destination	10	40	10	0	Good	0.5

Variable	LC	HM	VM	DM	AM	CM	FM	RWL
Origin	51(23)	1	1	1	1	1	0.81	18.63
Destination	51 (23)	1	0.93	1	1	1	0.81	17.23

$$LI = \frac{15}{17.23} = 0.87 \text{ i.e.} < 1$$

Since Li is less than 1.0, hence, job is safe for majority of the workers.

Example 8.2

In a casting industry, a worker is deputed to load casted components on shot blasting machine hanger. The worker hangs nearly 240 components in 4 hours. The standard weight of the component is 14 kg. The details of horizontal distance and vertical distance are as under:

$$H_D, H_O = 16$$
$$V_O = 12$$
$$V_D = 44$$

Consider good grip and 30° asymmetric angle find out RWL, LI for task job, also make recommendations based upon the facts.

Solution:

From the given data, lifting variables are as under.

Variable	H	V	D	A	C	F
Origin	16	12	32	30	Good	1
Destination	16	44	32	30	Good	1

Calculations of multipliers at Origin-

$$HM_O = 10/H$$
$$= 10/16 = 0.625$$
$$VM_O = (1 - 0.0075 \times |V - 30|)$$
$$= (1 - 0.0075 \times |12 - 30|) = 0.865$$
$$DM_O = 0.82 + 1.8/D$$
$$= 0.82 + 1.8/32 = 0.876$$
$$AM_O = 1 - (0.0032 \times A)$$
$$= 1 - (0.0032 \times 30) = 0.904$$

CM (from table 8.2) = 1

FM (from table 8.3) = 0.75

Calculations of multipliers at Destination:

$HM_D = 10/H$

$\quad = 10/16 = 0.625$

$VM_D = (1 - 0.0075 \times |V - 30|)$

$\quad = (1 - 0.0075 \times |44 - 30|) = 0.895$

$DM_D = 0.82 + 1.8/D$

$\quad = 0.82 + 1.8/32 = 0.876$

$AM_D = 1 - (0.0032 \times A)$

$\quad = 1 - (0.0032 \times 30) = 0.904$

CM (from table) = 1

FM (from table) = 0.75

Variable	LC	HM	VM	DM	AM	CM	FM	RWL
Origin	23	0.625	0.865	0.876	0.904	1	0.75	7.385
Destination	23	0.625	0.895	0.876	0.904	1	0.75	7.64

$LI = WL/RWL$

$\quad = 14/7.385 = 1.895$

Since LI > 1, therefore job is unsafe.

Redesign 1:

Now, let us redesign the task so that H reduces from 16 inches to 10 inches and 'v' may be made 30 inches and 40 inches at origin and destinations respectively. At the same time we depute two workers so that the frequency is reduced from 1/min to 0.5/min. Thereby the penalty to the load constant (LC) will be reduced, consequently, RWL will be very near to weight lifted, hence the lifting index is approximately near to 1 and the task will be comparatively safe.

The lifting variables and corresponding multipliers are shown as under

Variable	H	V	D	A	C	F
Origin	10	30	10	30	Good	0.5
Destination	10	40	10	30	Good	0.5

Variable	LC	HM	VM	DM	AM	CM	FM	RWL
Origin	23	1	1	1	0.904	1	0.81	15.594
Destination	23	1	0.925	1	0.904	1	0.81	14.42

Now, LI = 14/14.42 = 0.94, which is < 1, hence the lifting job is safe for majority of the workers.

Unsolved Problems

Students are advised to analyze the job and give recommendations

Problem 1: In a casting industry, a worker is deputed to prepare moulds using automatic ramming machine. The worker performs the same task for 6 hours and prepares nearly 360 moulds. The standard weight of the moulding box is supposed to be 25 kg. The details of horizontal distance and vertical distance are as under:

$$H = 15$$
$$V_O = 15$$
$$V_D = 34$$

Consider fair grip and 30° asymmetric angle find out RWL, LI for task job. Also make recommendations based upon the facts.

Example 2: In a casting industry, a worker is deputed to prepare moulds using automatic ramming machine. The worker performs the same task for 6 hours and prepares nearly 360 moulds. The standard weight of the moulding box is supposed to be 25 kg. The details of horizontal distance and vertical distance are as under:

$$H = 12$$
$$V_O = 6$$
$$V_D = 37$$

Consider fair grip and 45° asymmetric angle find out RWL, LI for task job. Also make recommendations based upon the facts.

Example 3: In a casting industry, a worker is deputed to load the final casting bags in the trolley. The worker performs the same task for 4 hours and loads nearly 960 bags in the trolley. The approximate weight of the bag is supposed to be 20 kg. The details of horizontal distance and vertical distance are as under:

$$H = 10$$
$$V_O = 5$$
$$V_D = 45$$

Consider good grip and 0° asymmetric angle find out RWL, LI for task job. Also make recommendations based upon the facts

Multiple Choice Questions

1. The composition of human musculoskeletal system includes:
   (a) Muscles and Tendons
   (b) Tendons and Bones
   (c) Muscle, Tendons and Bones
   (d) Muscles and Bones
2. The total 206 bones of the skeletal framework are covered by nearly
   (a) 700 muscles
   (b) 650 muscles
   (c) 206 muscles
   (d) None of these

3. Which of the following is a type of bone joint?
   (a) Synovial joint   (b) Fibrous joint
   (c) Cartilaginous joint   (d) All of these

4. The term MSDs stands for
   (a) Muscle skeletal disorders   (b) Musculoskeletal disorders
   (c) Muscular statute disorders   (d) None of these

5. WMSDs stands for
   (a) Work related musculoskeletal disorders   (b) Work repetitive musculoskeletal disorders
   (c) Work related muscle disorders   (d) None of these

6. The CTD refers as
   (a) Cumulative trauma disorders   (b) Cumulation trauma disorders
   (c) Continue trauma disorders   (d) None of these

7. An interdisciplinary study of human movements with respect to. range, strength and speed is called
   (a) Biomechanics   (b) Ergonomics
   (c) Work measurement   (d) None of these

8. In biomechanics of lifting the term AF stands for
   (a) Arm force   (b) Applied force
   (c) Aggregate force   (d) None of these

9. For an object to be in static equilibrium, the external forces acting on object must be
   (a) Uniform   (b) Non uniform
   (c) Zero   (d) Constant

10. Which of the following is a part of biomechanics?
    (a) Weight of object   (b) Arm moment
    (c) Muscle force   (d) Applied force
    (e) All of these

11. Moment arm is calculated as
    (a) $MA = MW \times W/AF$ (b) $MA = BW \times W/AF$
    (c) $MA = MW \times WF/AF$ (d) $MA = MW \times WA/AF$

12. The 3-D whole models can be developed as an expansion of
    (a) Material models   (b) Elementary models
    (c) Material and Elementary models   (d) None of these

13. The term NIOSH stands for
    (a) National institute of occupational safety and health
    (b) National institute of occupational safer health
    (c) National initiative of occupational safety and health
    (d) None of these

14. The low back pain can be due to
    (a) Working in forward bending posture   (b) Lifting of heavy weight
    (c) Repetitive lifting of weight   (d) All of these

15. NIOSH equation was publically presented in
    (a) 1989   (b) 1991
    (c) 1990   (d) 1988

16. The term RWL stands for
    (a) Recommended weight limit   (b) Required weight level
    (c) Required weight limit   (d) None of these

17. The number of variables in revised lifting equation are
    (a) 7   (b) 6
    (c) 5   (d) 8

18. LI stands for
    (a) Lifting index
    (b) Lifting indices
    (c) Lower index
    (d) None of these
19. The value of load constant (LC) is
    (a) 23 lbs
    (b) 51 lbs
    (c) 21 lbs
    (d) 61 lbs
20. Absolute value of the difference between the vertical heights at the destination and origin of the lift is
    (a) Vertical location
    (b) Vertical travel distance
    (c) Horizontal location
    (d) Neutral body position
21. Revised NIOSH lifting equation is applicable on lifting, and carrying of loads up to steps?
    (a) three
    (b) six
    (c) four
    (d) two
22. How many parameters are considered to calculate RWL in Revised NIOSH lifting equation?
    (a) three
    (b) six
    (c) four
    (d) five
23. What value of lifting index is the indicator of risk of low back injury?
    (a) >2
    (b) ≥ 3
    (c) ≤1
    (d) >1
24. What is the normal value of horizontal location in SI units?
    (a) >10 inches
    (b) ≤25 inches
    (c) 10 inches
    (d) None of the above
25. what is the normal value of vertical location in SI units?
    (a) >30 inches
    (b) < 30 inches
    (c) 30 inches
    (d) None of the above
26. The value of distance traveled multiplier in metric system is found as
    (a) 10/H
    (b) 0.82 + (4.5/D)
    (c) 0.82 + (1.8/D)
    (d) None of these
27. Which one is the correct value of load constant in metric conversion?
    (a) 51 kg
    (b) 23 kg
    (c) 24 kg
    (d) 23.4 kg
28. What is the formula for horizontal multiplier in metric conversion?
    (a) (24/H)
    (b) (23/H)
    (c) (25/H)
    (d) (26/H)

Answers

1. (c)	**2.** (b)	**3.** (d)	**4.** (b)	**5.** (a)	**6.** (a)	**7.** (a)	**8.** (b)	**9.** (c)	**10.** (e)
11. (a)	**12.** (b)	**13.** (a)	**14.** (d)	**15.** (b)	**16.** (a)	**17.** (b)	**18.** (a)	**19.** (b)	**20.** (b)
21. (d)	**22.** (b)	**23.** (d)	**24.** (c)	**25.** (c)	**26.** (b)	**27.** (b)	**28.** (c)		

Risk Assessment for Distal Upper Extremities (DUE) Disorders

9.1. Introduction

In general, the upper limb or distal upper extremity (DUE) includes the muscle forming the rounded contour of the shoulder to the hand which is called deltoid region. Therefore, DUEs are the limbs like armpit, upper and lower arm, elbow, wrist and hand. These are perceived to be more associated with nerve injury. The risk to these DUE disorders can be assessed using some qualitative and quantities methods. Moore and Garg (1995) proposed a method 'The Strain Index' to analyze jobs for risk of distal upper extremity disorders and to distinguish between jobs which are associated with distal upper extremity disorders from those which are not. However, Drinkaus et al. (2005) proposed that the Strain Index (SI) may be modified to estimate the risk of distal upper extremity injury. The SI was developed and validated using single task jobs. It was proposed that the SI would be more useful; however, if it could be extended to estimate the risk of a job with more than one task. Two methods were mentioned: the first is maximum task and the second is CARD (Cumulative Assessment of Risk to the Distal Upper Extremity) approach. Stephens et Al. (2009) evaluated the SI for various worksites with job titles including assemblers, painters and office workers (which performed very different tasks between the different worksites even with the same job titles), and some unique job titles to specific worksites such as hook fabricator and deck hand. These authors used different SI computation methods and resulted in significantly different SI scores along with different risk level classifications calculated by the different computation methods. Vishal et al. (1999) developed a cumulative trauma disorders (CTD) risk assessment model for predicting injury incidence rates. The model was found best suited for job tasks with cycle times greater than 4 sec. Moore and Garg (1998) evaluated the effectiveness of a corporate ergonomics programme that used a participatory approach to solving problems related to musculoskeletal hazards. The corporation experienced a significant decrease in the percentage of recordable disorders related to musculoskeletal risk factors, a marked decrease in the lost-time incidence rate, and a marked decrease in total and per capita annual workers' compensation costs. Garg et al. (2007) presented three different studies showing that the SI is capable of identifying jobs with no distal upper extremity morbidity as 'safe' and jobs with distal upper morbidity as 'hazardous'. Evidence was provided for the SI's generalizibility and predictive validity, Ergonomics today (2002). There are various types of DUE disorders, such as Carpal tunnel syndrome, Cubital tunnel syndrome, Thoracic outlet syndrome, Raynaud's syndrome (white finger),

Rotator cuff syndrome, DeQuervain's disease, Tendinitis, Tenosynovitis, Trigger finger and Ganglion cyst; these are briefly described in the following sections.

9.2. Carpal Tunnel Syndrome (CTS)

Carpal tunnel is the flexor tendons and median nerve passes through eight bones in the wrist wrapped by ligaments that forms a rigid tunnel about the size of a dime (a silver coin of US of worth ten cent of a dollar). Due to repetitive motion, the tendons constantly move back and forth through the sheath. This decreases the synovial fluid and increases the friction between tendons and the sheath. As a result the area becomes inflamed and median nerve is pinched. The main causes of CTS are repetitive motion and poor wrist posture. The mechanism of carpal tunnel syndrome is exhibited in Figure 9.1. The symptoms of CTS includes pain, numbness and tingling of the hands.

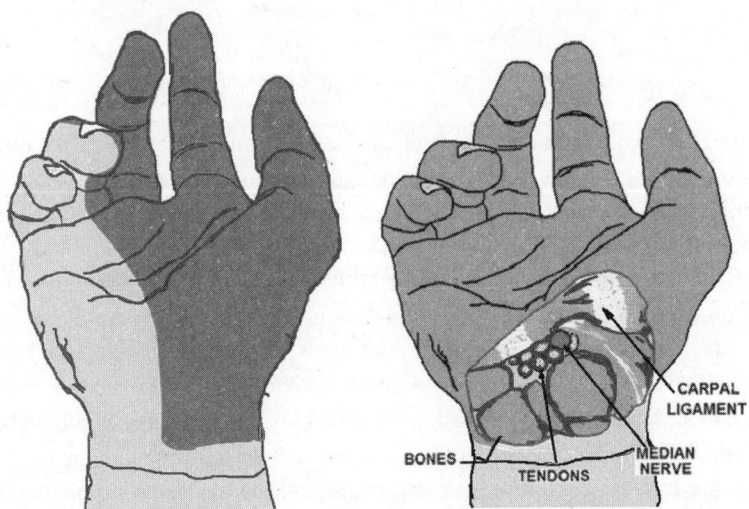

Fig. 9.1: Mechanism of carpal tunnel syndrome.

9.3. Cubital Tunnel Syndrome

It is caused by resting the elbows on hard surfaces such as unpadded tables or armrests. The ulnar nerve, which feeds the ring and little fingers, can be damaged from pressure near the elbows. The mechanism for cubital tunnel syndromes is exhibited in Figure 9.2. The symptoms for this disorder include pain in the ring and little fingers, tingling and numbness in these areas.

9.4. Thoracic Outlet Syndrome

It is caused by frequent reaching above shoulder level, by carrying heavy objects or poor posture involving a forward head tilt. A neurovascular bundle, called the brachial plexus, which passes between the collar bone and the top rib, can happen to be impaired from pressure associated with movements that causes these two bones to be positioned close together. The mechanism for thoracic outlet syndrome is shown in Figure 9.3. The symptoms for this disorder include the arms falling asleep, weakened pulse, numbness in the fingers.

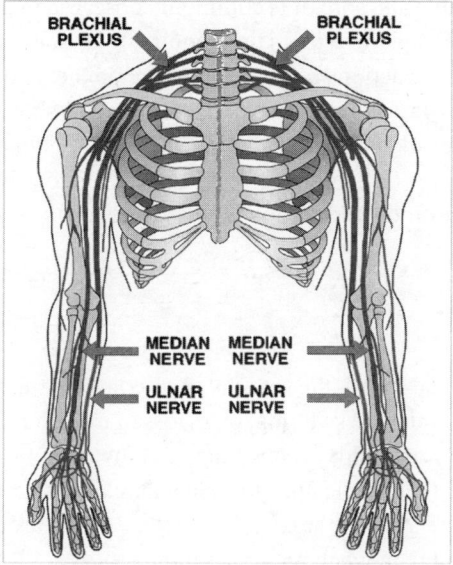

Fig. 9.2: Mechanism for the cubital tunnel syndromes.

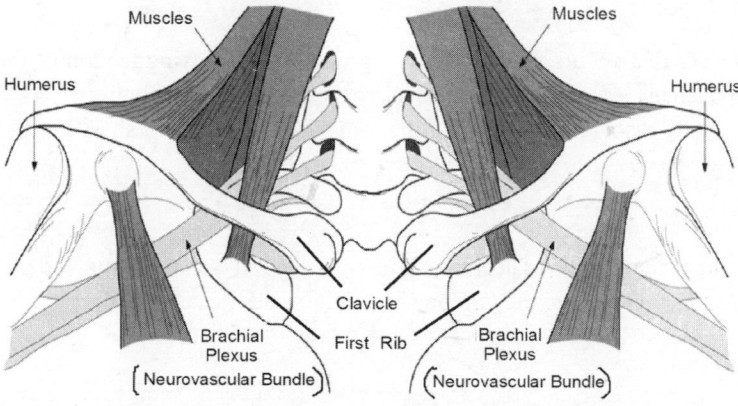

Fig. 9.3: Mechanism for the thoracic outlet syndrome.

9.5. Rotator Cuff Syndrome

It is a disorder involving swelling and pain of tendons comprising the rotator cuff muscle group such as subscapularis, supraspinatus, infraspinatus and teres minor. The common symptoms of this syndrome are pain when one bends the arm and rotates it outwards against resistance; pain on the outside of the shoulder possibly radiating down into the arm; pain in the shoulder, which is worse at night; stiffness in the shoulder joint.

9.6. Tendinitis

It is basically related to tendons that join muscles to the bones. When these tendons become irritated or swollen, it is called tendinitis. Hence, it occurs when the tendon cables are continually stressed, causing them to become irritable and sore. It is commonly observed in the wrist, elbow and shoulder. It often arises due to excessive repetitive tasks that require, for instance, the shoulder to be elevated, overexerted and overused. These actions put extra pressure on the tendon, and generally cause pain in joint areas, such as the wrists, elbows, knees, hips, heels and shoulders. It is of two types, one lateral epicondylitis and other medial epicondylitis; these are commonly known as tennis elbow and golfer's elbow, respectively. The common symptoms of tendonitis are, aching sensation at the joints, discomfort with certain movements, tenderness to touch, swelling, pain radiating down to back of hand.

9.7. Tenosynovitis

It is swelling of the sheath that covers the tendon from constant rubbing against the tendon. The common symptoms of this disorder are swelling, pain, loss of motion and loss of strength. The wrists, hands, and feet are commonly affected, because the tendons are long across those joints. But, the condition may occur with any tendon sheath, symptoms may include any of these; difficulty moving a joint, swelling in joint of the affected area, pain and tenderness around a joint, especially in the hand, wrist, foot, or ankle, and pain when moving a joint.

9.8. Trigger Finger

It is a tendon disorder that occurs when there is a groove in the flexing tendon of the finger. If the tendon becomes locked in the sheath, attempts to move the finger cause snapping or jerking movements. This disorder is usually associated with using the tools that have hard or sharp edges in handles.

9.9. Ganglion Cyst

It is usually found at the hand and wrist as shown in Figure 9.4. It is like a bump under the skin caused by an accumulation of fluid within the tendon sheath. It causes occasional pain.

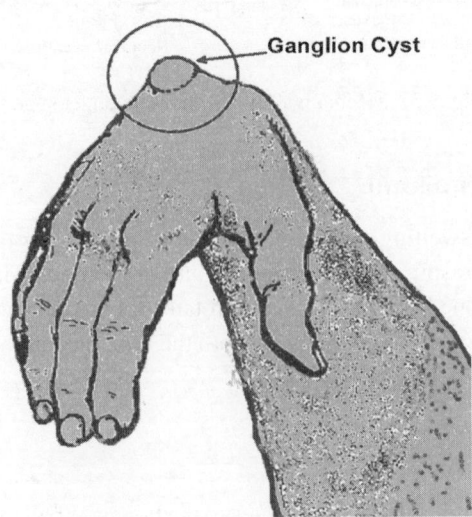

Fig. 9.4: Ganglion cyst at the wrist.

9.10. Cumulative Trauma Disorder (CTDs)

Cumulative Trauma Disorders (CTDs) are the result of repeated exposure to ergonomic stress factors over a period of time. CTDs are not caused by a single event like fall or struck by an object, although the pain and suffering is the same. There can be numerous causes of CTD; however, the few main causes include forceful exertion like stapling of thick layers of documents together, sustained/awkward body posture like sitting with forward bending over a chair and compression on hard and sharp edges like; resting the arm against the edge of the desk while using mouse.

9.11. Distal Upper Extremities Disorder Risk Factors

Development of a DUE disorder depends upon four factors and results in absorbed dose or strain and other exposure or stress level. The flow-diagram for the development of DUE is shown in Figure 9.5. The fours factors include, job risk factors, individual factors, psychological factors and work organizational factors. These can be described as follows:

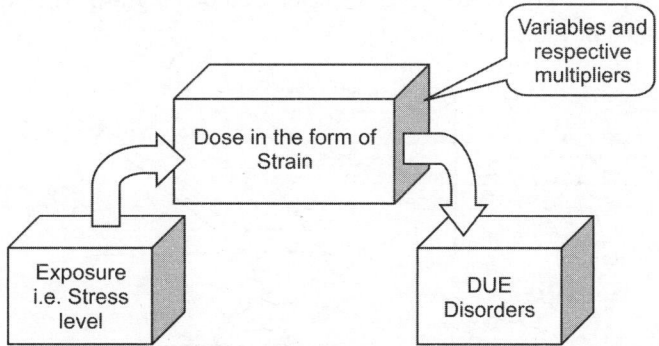

Fig. 9.5: Flow diagram for development of DUE disorder.

1. *Ergonomic Factors or Job Physical Factors:* These include risk factors related to the physical job, like forceful exertions, stress induced due to the job, concentration of force required to perform a task, high repetition, awkward posture, duration of exertion, duration of recovery, duration of exposure, speed of work, static muscular work, pinch grasp, vibrating tools, cold temperature, poorly fitting gloves, etc.
2. *Individual or Personal Risk Factors:* These include individual health fitness, physique, work capacity, level of endurance to exertions, height weight, age, awareness, work experience, etc.
3. *Psychosocial Risk Factors:* These include perceived stress level, mental status, motivation and will power, etc.
4. *Work Organization Risk Factors:* These include work place design, work hours per day or week, rest pauses, competition among peer group, training and awareness at organizational level, etc.

9.12. Wrist Posture in Daily Life Activities

In industries, workers perform fine manipulative work (speed and precision, e.g., holding a pen) with wrist flexion and radial deviation. Some of the activities involve slower and stronger movements and may require wrist extension and ulnar deviation like tightening the screw with manual screw driver under the dashboard of a car, etc. In our daily life, we often encounter with many such activities in which we make our wrist adopt different postures. For example, in simple typing job wrist is normally in extension and radial deviation. However for using computer mouse, our wrist experience extension, radial and ulnar deviations. Therefore, maximum voluntary range of motion in different postures is exhibited in Figure 9.6. However, the resting position of wrist is considered in 12° wrist extension. An extreme flexion or extension can cause muscle insufficiency, as the flexed wrist cannot grasp tool firmly and an extended wrist affects finger movement. Female wrist strength is 17.4 lbs in neutral and 11.6 lbs in 30° flexion, which is relatively lower than the male. If we consider the grip in daily life, it is of three types: power, hook and oblique. In power grip the position of thumb is directly oblique and opposes the fingers, both fingers and thumb are wrapped around the object. The hook grip position includes a flat hand (palm) and curled fingers, whereas thumb is not used to grasp an object. In oblique grip the thumb is extended to stabilize the grasp. However in pinch grasp we require less strength as compared to a grip. The various types of grips are exhibited in Figure 9.7. The grip strength is gender variable, the females have relatively lower grip strength as compared to male counterparts. A power grip is five times stronger than pinch grasp. The grip strength of male and female is compared in Table 9.1.

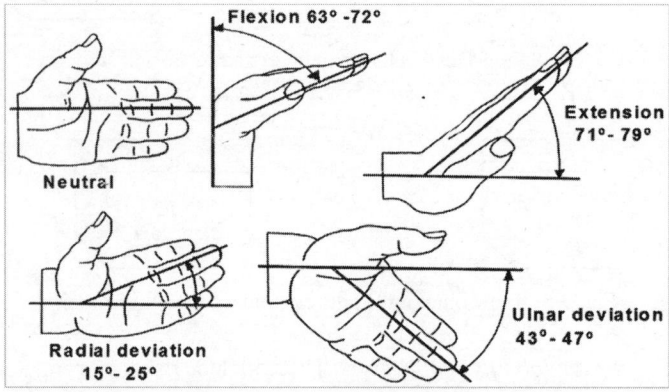

Fig. 9.6: Various wrist postures

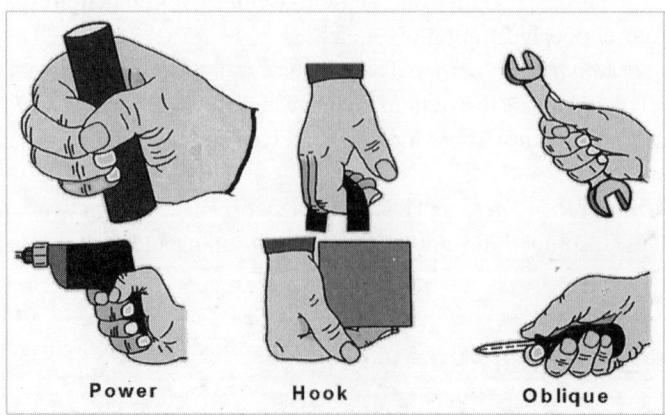

Fig. 9.7: Various types of grips in daily life.

Table 9.1: Various types of grip strength.

Grip Type	Grip Strength (lbs)	
	Males	*Females*
Two point pinch	17	11
Three point pinch	23	16
Lateral pinch	24	16
Oblique grip	65	38
Power grip	100	59
Hook	100	59
** Power grip ≈ 5 times pinch grip		

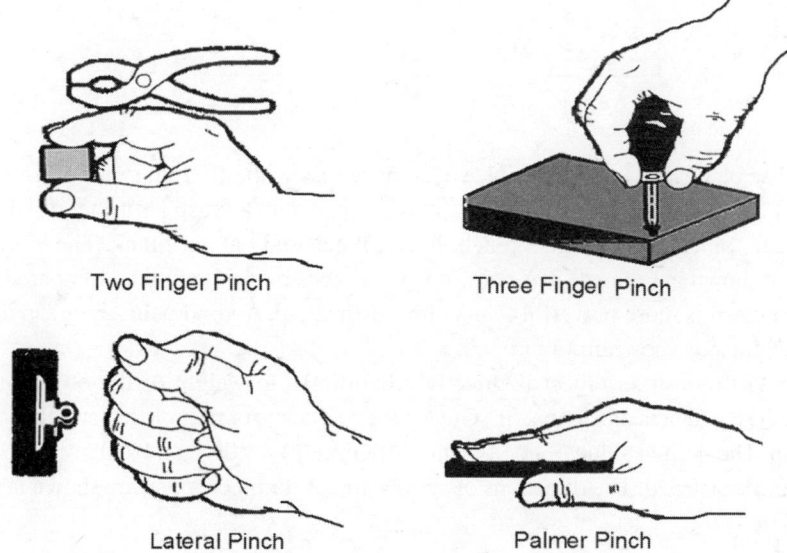

Two Finger Pinch   Three Finger Pinch

Lateral Pinch   Palmer Pinch

Fig. 9.8: Various types of pinch grasp.

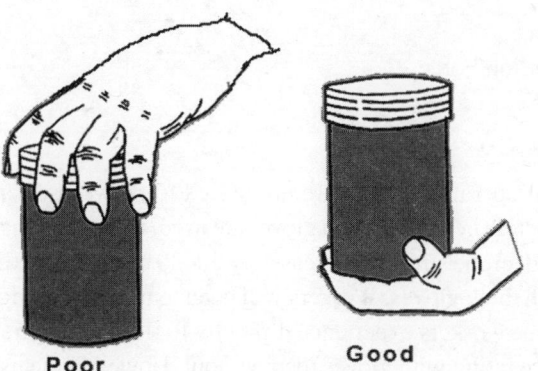

Poor   **Good**

Fig. 9.9: Palm-up grasp.

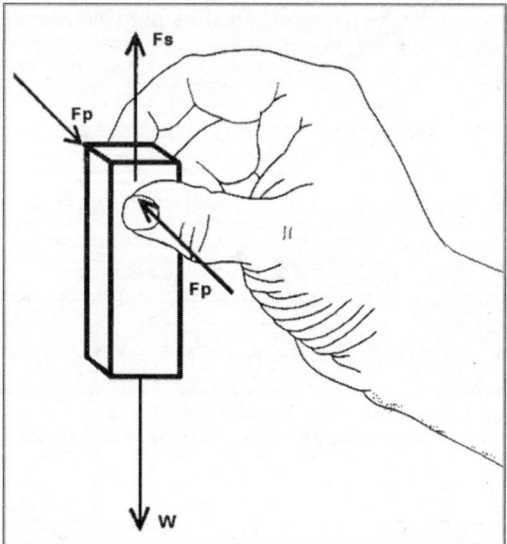

Fig. 9.10: Directions of object weight and pinch force.

Beside the use of various types of grips, we come across with different types of grasps also. The grasp can be of two types: pinch grasp and palmer grasp. A pinch grasp is further divided into three types: two fingers pinch, three fingers pinch, lateral pinch and palmer pinch. The various types of pinch grasps are shown in Figure 9.8. A palm-up grasp needs more strength as compared to the pinch grasp. Palm-up grasp is more powerful than palm-down grasp. A good palm-up grasp in contrast to a poor palm-up grasp is shown in Figure 9.9.

Force required to grasp an object is directly proportional to weight of the object and inversely proportional to the coefficient of friction, 'µ'. Greater the coefficient of friction, lesser is the force required to make a grasp. The normal value of coefficient of friction (µ) for dry hand is 0.5, which increases to 2 when hands are moistened. The directions of various forces for pinch grasp are shown in Figure 9.10.

Where, $FS = \mu\, Fn$

$W = \mu\, (2 \times Fp)$

$W = 0.5\, (2 \times 2)$

$W = 2$ lbs

Fp = Pinch Force

FS = Static Friction Force.

Use of gloves and grip effectiveness

In most of the industrial operations, gloves are required to be worn often e.g. during hot forging, pouring of molten metal etc. Therefore, when gloves are used, the grip strength decreases by 15–20 per cent, whereas pulling and torque strength increases by 20–30 per cent. Tactile and sensory feedback is reduced particularly with thick gloves. Workers will need to utilize more force to hold or manipulate the tools. At the same time workers experience difficulty to flex the fingers if they are apart. Overall, there is a lower grip force rating with gloves than without. However, the uses of gloves prevent tools from rotating or sliding out of hand.

9.13. Introduction to Strain Index Method

It has already been mentioned in the introduction section of this chapter that, the Strain Index (SI) was developed by Moore and Garg (1995) and it is primarily a method to 'analyze jobs for risk of distal upper extremity disorders'. There are six task variables used to determine the SI score. These variables are both qualitative and quantitative. Each of these variables attempts to quantify the amount of physical and physiological strain experienced by the muscle–tendon units of the distal upper extremity due to physical activity or the stress associated with the task. Three out of the six task variables (duration of exertion, number of exertions per minute, and duration of task per day) are quantitative and rest are qualitative. The procedural steps for the analysis using SI is shown in Figure 9.11 given below.

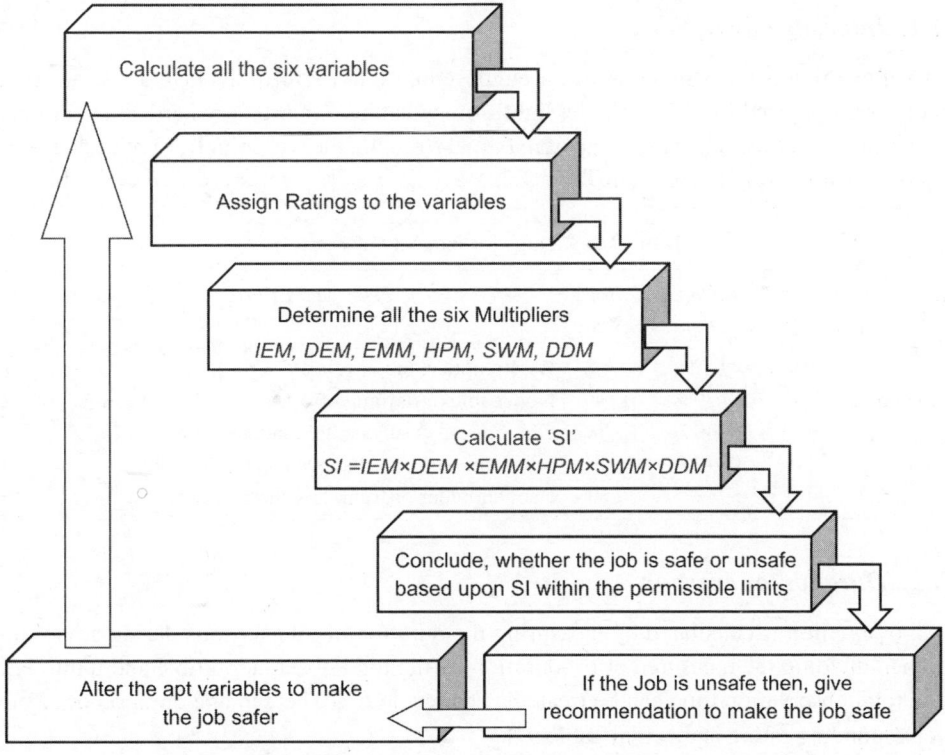

Fig. 9.11: Overall average SI of casting, forging and car assembly line.

9.13.1. Variables of Strain Index

There are six task variables used to determine the SI score. These variables are both qualitative and quantitative. Each of these variables attempts to quantify the amount of physical and physiological strain experienced by the muscle – tendon units of the distal upper extremity due to physical activity or the stress associated with the task. Each of the six task variables is assigned a rating of 1–5 (Tables 9.2–9.8). Three out of the following six task variables (duration of exertion, number of exertions per minute and duration of task per day) are quantitative.

1. *Intensity of exertion:* the force required for a single performance of the task.
2. *Duration of exertion:* the proportion of the exertion cycle. The average length of the exertion divided by the cycle time multiplied by 100 gives the duration per cent.

3. **Efforts per minute:** is synonymous with frequency of exertions per minute.
4. **Hand/wrist posture:** relates the anatomical posture of the hand.
5. **Speed of work:** estimates the perceived pace of the task and accounts for the stresses associated with dynamics of work.
6. **Duration of task per day:** is a measure of how much time of the workday is allocated to performing that task.

Each task needs to be observed and six aspects of each task were recorded and compared with quantitative and qualitative measures provided by the SI authors. Ratings corresponded with each measure.

9.13.1.1. *Intensity of exertion*

Intensity of exertion is an estimate of the strength required to perform the task one time. There are five rating criterion used for this variable: light, somewhat hard, hard, very hard and near maximal. These criteria can be linked to percentage of maximal strength, the ten-point Borg scale and perceived effort and intensity rating is shown in Table 9.2.

Table 9.2: Ratings for intensity of exertion.

Rating Criterion	%Maximum Strength	Borg Scale	Perceived Effort	Rating
Light	<10%	< 2	Barely noticeable or relaxed effort.	1
Somewhat hard	10–29%	4	Noticeable or definite effort.	2
Hard	30–49%	6	Obvious effort; unchanged facial expression.	3
Very hard	50–79%	8	Substantial effort; changes facial expression.	4
Near maximal	>79%	10	Uses shoulder or trunk to generate force.	5

9.13.1.2. *Duration of exertion*

Duration of exertion is calculated by measuring the duration of all exertions during an observation period, then dividing the measured effort duration by the total observation time and multiplying by 100. The total observation time can be treated as the cycle time. Percentage duration of efforts and their corresponding rating is given in Table 9.3.

$$\% \, Duration \, of \, Exertion = \frac{Effort \, Duration}{Cycle \, Time} \times 100$$

Table 9.3: Ratings for percentage duration of exertion.

% Duration of Effort	Rating	% Duration of Effort	Rating
<10	1	50–65	3.5
10–19	1.5	65–79	4
20–29	2	80–90	4.5
30–39	2.5	> 90	5
40–49	3		

9.13.1.3. *Efforts per minute*

Efforts per minute are determined by counting the number of exertions that occur during an observation period, then dividing the number of exertions per cycle by the duration of the observation period, measured in minutes. The efforts per minutes can vary from 4 to 24 and above. The rating for various ranges of efforts per minutes is shown in Table 9.4.

$$Efforts\ per\ Minute = \frac{Number\ of\ Exertions\ per\ Cycle}{Cycle\ time\ (minute)} \times 100$$

Table 9.4: Rating for efforts per minute.

Efforts/Minute	Ratings	Efforts/Minute	Ratings
<4	1	15–17	3.5
4–6	1.5	18–19	4
6–8	2	20–24	4.5
9–11	2.5	≥24	5
12–14	3		

9.13.1.4. *Hand/wrist posture*

Hand/wrist posture is an estimate of the hand or wrist position relative to neutral. The rating for various wrist positions is given in Table 9.5.

Table 9.5: Rating for various positions of hand/wrist.

Rating Criterion	Wrist Extension (degrees)	Wrist Flexion (degrees)	Ulnar Deviation (degrees)	Perceived Posture	Ratings
Very good	0–10	0–5	0–10	Perfectly neutral	1
Good	11–25	6–15	11–15	Near neutral	2
Fair	26–40	16–30	16–20	Non-neutral	3
Bad	41–55	31–50	21–25	Marked deviation	4
Very bad	>55	>50	>25	Near extreme	5

9.13.1.5. *Speed of work*

Speed of work is an estimate of how fast the worker is working. Rating for speed of work varies from 1 to 5, which is shown in Table 9.6.

Table 9.6: Ratings for different categories of speed of work.

Rating Criterion	Perceived Speed	Rating
Very slow	Extremely relaxed pace	1
Slow	Taking one's own time	2
Fair	Normal speed of motion	3
Fast	Rushed, but able to keep up	4
Very fast	Rushed and barely or unable to keep up	5

9.13.1.6. *Duration of task*

Duration of task per day is a measure of the total time that a task is performed per day. It is either measured or obtained from plant personnel. Various categories of duration of task per day and their corresponding ratings are shown in Table 9.7.

Table 9.7: Ratings for various categories of duration of task per day.

Duration of Task per Day (hours)	Ratings
<1	1
1 to 2	2
2 to 4	3
4 to 8	4
≥8	5

Table 9.8: Combined rating for all six variables.

Intensity of Exertion	Duration of Exertion	Efforts per Minute	Hand/Wrist Posture	Speed of Work	Duration per Day (hours)	Rating Values
Light	<10	<4	Very good	Very slow	0–1	1
Somewhat	10–29	4–8	Good	Slow	1–2	2
Hard	30–49	9–14	Fair	Fair	2–4	3
Hard	50–79	15–19	Bad	Fast	4–8	4
Very hard	≥80	≥20	Very bad	Very fast	>8	5
Near maximal						

Once all the six variables are determined, the next is to obtain their respective multipliers, that are obtained by using the subsequent table 9.9.

9.14. Case Study

The present study represent a case on comparison of a small scale forging unit without automation, a semi-automated casting unit and a large scale automotive car assembly industry in northern India. Samples of 50 workers engaged in various repetitive tasks (one sample from firm each) from small scale casting and forging units was selected. Another sample was selected from large scale automotive plant. In small scale casting and forging units, data were collected in two parts first through personal interviews and second through video analysis, whereas, due to lack of approachability to the workers in large scale automotive plant, data were collected through video analysis only. A comprehensive questionnaire was developed for calculating the self-reported Strain Index (SI). Each question was asked in local language and the strain parameters were explained to the workers.

9.14.1. Materials and methods

In the present case study jobs at various sections in each industry were identified. Videos of each job were recorded and observed. The focus for videography was the upper extremities of the worker. Videography was done repeatedly to ensure that the full cycle of the worker was covered so that the focus area was clear. Each job was divided into work elements or tasks. These tasks were defined as sub-activities of a job that could be done before transferring the job to the next workstation. Usually these tasks involved the same motion and intensity of efforts in the repetitive cycles. There are no rules to judge movements as either high or low in repetition. However, some researchers classify a job as 'high repetitive' if the time to complete such a job was less than 30 sec or 'low repetitive' if the time to complete the job was more than 30 sec. Although no one could really predict that at what point WMSDs may develop or workers performing repetitive tasks are at risk for WMSDs. There can be any cycle time. The next step was to calculate the cycle time by taking three observations of full cycle at the factory shop floor and taking the average. In each cycle two time attributes were measured with a stopwatch. The first timed attribute was the total time of the cycle and second was effort duration. Effort duration is the time during which the operator is exerting a significant force for that task. This time of exertion is used to determine the 'Duration of Effort' variable of strain index.

9.14.2. Analysis of video

After the videography each of the casting and forging units as well as the assembly lines of automotive plant, the videos were analyzed by a team of three members. The three members were selected to reduce the biasness during the video analysis, the three quantitative variables; cycle time, number of efforts/ minute and duration of exertion were determined. Afterwards the ratings for two variables; (No. of efforts/minute and duration of exertion) was done. The three qualitative variables, intensity of exertion, hand/wrist movement and speed of work, were judged for their ratings. Since, as per the survey, a shift for the worker is >10 hours, the average duration of task per day was taken as 10 hours and a rating of 1.75 was given for Duration per Day Multiplier (DDM). Rating for the wrong or difficult posture was incorporated into the hand wrist rating itself.

9.14.3. Determine the multipliers

Each variable's multiplier is determined by its rating, these multipliers are in-fact penalties w.r.t. the rating value of variables. The relationship between the SI rating scores and the SI multipliers are illustrated in Table 9.9.

Table 9.9: Value of multipliers w.r.t. rating scores variables.

Rating Values	Intensity of Exertion Multiplier (IEM)	Duration of Exertion Multiplier (DEM)	Efforts per Minute Multiplier (EMM)	Hands/Wrist Posture Multiplier (HPM)	Speed of Work Multiplier (SWM)	Duration per Day Multiplier (DDM)
1	1	0.5	0.5	1	1	0.25
1.5	2	0.75	0.75	1	1	0.375
2	3	1	1	1	1	0.5
2.5	4.5	1.25	1.25	1.25	1	0.625
3	6	1.5	1.5	1.5	1	0.75
3.5	7.5	1.75	1.75	1.75	1.25	0.875
4	9	2	2	2	1.5	1
4.5	11	2.5	2.5	2.5	1.75	1.25
5	13	3	3	3	2	1.5

9.14.4. Calculations of SI index

On the basis of the ratings obtained during the video analysis, all the six multipliers were obtained using table 9.9. The multipliers are: intensity of exertion (IEM), duration of exertion (DEM), efforts per minute (EMM), hands/wrist posture (HPM), speed of work (SWM), and duration per day multiplier (DDM). On the basis of these multipliers, the SI Index was calculated as:

$$Strain\ Index\ (SI) = IEM \times DEM \times EMM \times HPM \times SWM \times DDM$$

9.14.5. Results and discussions

The SI scores obtained through video analysis of tasks and self-reported SI scores at different sections are shown in Tables 9.10–9.12. In the forging unit SI scores calculated from video analysis ranged from 6.56 to 26.25 (average = 15.35, SD = 7.74) whereas the SI score obtained through subjective responses ranged from 5.25 to 23. 62 (average = 14.72, SD = 7.42). In the casting unit the SI ranged from 3.52 to 17.35 (average = 7.52, SD = 4.62) whereas the self-reported SI ranged from 4.25 to 19.25 (average = 8.78, SD = 5.03). The average SI score of the automated car assembly unit was 8.5 (SD = 4.72). The comparison of average SI score of the three plants is shown in Figure 9.12. The data reveals that the automation in casting has significantly reduced SI score, and thus the risk of DUE disorders has been reduced except the tasks like moulding and grinding of medium components of automobile or valves. However workers were suffering from back ache, neck stiffness, whereas DUE disorders were found less often in the tasks under consideration. The job of forger in the hammer section was near the normal SI value (7.0) in standard method as well as in self-reported subjective response (Ergonomics Today @ ergoweb August 20, 2002). The job of rope puller was found to be highly stressful to the upper limbs with the highest SI values both in the video analysis as well as in the self-reporting response analysis. The workers performing other tasks were also prone to high strain except in broaching and grinding of small size components. The majority of the workers are found with high strain index. It is very clear from the overall SI scores that workers of small scale forging industry are exposed to great distal upper extremity strain compared to those in the semi-automated small casting unit. It is therefore observed that there is a strong need for implementing the job rotation for jobs with a high SI to lower the possibility of musculoskeletal disorders of upper limbs with provision of regular rest pauses. The study therefore recommends that time standards for workers performance levels should be prepared by incorporating the job rotation as well as the regular rest pauses. It is also recommended that low cost automation and gravity chutes must be introduced for material handling at small distances. As far as the self-reporting SI values are concerned, these were found to be lower (except hot material handling task) compared to the SI values found by the video analysis, which can be attributed to the following reasons:

1. The workers participating in the study have been working in the same job for at least 5 years and thus they have become accustomed to working at the observed level of intensity and speed of work. Therefore, they perceived the job as less strenuous.
2. The majority of the workers engaged in casting and forging firms were habitual smokers and tobacco consumers during their working schedule; this could be one of the reasons that the workers do not feel much strain (Singh et al., 2008).
3. The majority of the workers are illiterate or less educated; therefore they are not aware of musculoskeletal disorders caused by strenuous repetitive jobs.

Although the SI values of the forging unit found by both methods differed by -0.44 - 7.00, there was a significant correlation (0.5547 at 99 per cent confidence level) for the forging section and 0.5305 at 99 per cent confidence level for the machining section between the SI score obtained by both the methods. The highest difference between average video analysis SI score and self-reported SI score for hot material handling (from furnace to forger) was at the forging section (Table 9.10). The reason for high self-reported SI scores for hot material handling task could be that the workers may feel greater strain due to high ambient temperature and insufficient ventilation.

Table 9.10: Strain Index score associated with various processes in forging unit.

Sr. No.	Section	Job Description	Strain Index (SI) Video Analysis	Strain Index (SI) Self-reported
Forging Section				
1.	Small Components	Rope puller	26.25	15.75
		Forger	9.84	5.25
		Hot material handler	14.76	23.63
2.	Medium Components	Rope puller	26.25	18.38
		Forger	8.20	7.88
		Hot material handler	14.76	23.63
3.	Large Components	Rope puller	21.00	18.38
		Forger	6.56	9.18
		Hot material handler	19.68	23.63
Machining Section				
1.	Punching section	Large component 1	27.56	15.75
		Very small Components 2	8.20	9.84
		Medium Components 3	13.13	15.75
		Small Components 4	9.84	9.84
2.	Broaching section	Small Components1	6.56	7.87
		Medium Components 2	9.18	9.18
		Large Components 3	10.5	10.5
3.	Grinding section	Belt grinding small components 1	6.56	7.87
		Duplex grinding Medium Components 2	26.25	31.5
		Duplex grinding Large components 3	26.25	7.00
		Wheel grinding 4	15.75	23.62
		Average (SI)	15.35	14.72
		Std. Deviation (SI)	7.74	7.42

Table 9.11: Strain Index score associated with various processes in automated casting unit.

Sr. No.	Job Description	Strain Index (SI) Video Analysis	Strain Index (SI) Self-reported
1.	Cleaning and matching of moulding boxes	3.52	4.25
2.	Clamp fitting at moulding boxes (1)	4.69	5.60
3.	Clamp fitting at moulding boxes(2)	4.10	5.20
4.	Large M/C moulding man(1)	3.94	3.65
5.	Large M/C moulding man(2)	4.50	4.56
6.	M/C moulder (medium box) machine control	7.20	8.10
7.	M/C moulder (medium box) with shovelling	13.12	15.90
8.	M/C moulder (small box) machine control	4.50	6.89
9.	M/C moulder (small box) with shovelling	6.56	10.25
10.	Manual moulding	7.03	8.15
11.	Grinding (small components)	13.75	13.67
12.	Grinding (medium components)	17.35	19.25
	Average (SI)	7.52	8.78
	Std. deviation (SI)	4.62	5.03

Table 9.12: Average SI score associated with various processes in automated car assembly unit.

Assembly Line	Average SI Score	SD of SI Score
Trim-1	7.87	2.21
Trim-2	8.44	4.44
Trim-3	6.71	4.07
Chassis 1	10.99	6.97
Chassis 2	8.4	5.3
Final Assembly	8.59	5.34
Grand Average	8.5	4.72

9.14.6. Conclusions

The jobs with high strain index should be rotated after a regular interval of time with the low SI score jobs. For the non-rotatable jobs, regular rest pauses must be provided for jobs at various sections of forging units. The automation in small scale casting unit has resulted in lower SI score. Therefore, low-cost automation should be implemented for material handling and lifting. In order to minimize the rate of DUE disorders, performance standards should be calculated by incorporating rest pauses.

Multiple Choice Questions

1. The term DUE refers to
   - (a) Distal Upper Extremities
   - (b) Dual Upper Extremities
   - (c) Damaged Upper Extremities
   - (d) None of these

2. DUEs are the limbs like
   - (a) Shoulder
   - (b) Upper and lower arm
   - (c) Elbow
   - (d) Wrist and hand
   - (e) All of these

3. To analyze jobs for risk of distal upper extremity disorders and to distinguish between jobs which are associated with DUE from those which are not, the method is called as
   - (a) Lifting index
   - (b) Strain index
   - (c) Stress index
   - (d) Both strain and stress index

4. The tern CTD stands for
   - (a) Cumulative trauma disorders
   - (b) Casual trauma disorders
   - (c) Cumulative trauma diseases
   - (d) Critical trauma disorders

5. The cumulative trauma disorders risk assessment model is best suited for job tasks with cycle times
   - (a) 2 sec
   - (b) 3 sec
   - (c) 4 sec
   - (d) >4 sec

6. Which of the following is a type of distal upper extremity disorders?
   - (a) Carpel tunnel syndrome
   - (b) Cubital tunnel syndrome
   - (c) Thoracic outlet syndrome
   - (d) Raynaud's syndrome
   - (e) All of these

7. CTS stands for
   - (a) Carpal tunnel syndrome
   - (b) Cumulative trauma syndrome
   - (c) Cumulative tunnel syndrome
   - (d) None of these

8. Tendons that join muscles to the bones become irritated or swollen then the disorder is
   - (a) Tendinitis
   - (b) Cubital Tunnel Syndrome
   - (c) Thoracic Outlet Syndrome
   - (d) All of these

9. The neutral strength of female wrist is
   - (a) 17.4 lbs
   - (b) 27.4 lbs
   - (c) 16.4 lbs
   - (d) 15.4 lbs

10. If coefficient of friction is increased then the force required to make a grasp
    - (a) Increases
    - (b) Decreases
    - (c) May increase or decrease
    - (d) All of these

11. The normal value of coefficient of friction (μ) for dry hand is
    - (a) 0.5
    - (b) 0.3
    - (c) 0.6
    - (d) 0.4

12. The formula of coefficient of friction (μ) is
    - (a) $\mu = Fn/FS$
    - (b) $\mu = Fn/F$
    - (c) $\mu = Fn/S$
    - (d) $\mu = Fn/P$

13. The strain index was developed by
    - (a) Griffin
    - (b) Moore
    - (c) Garg
    - (d) Moore and Garg

14. The multipliers if strain index are
    - (a) IEM
    - (b) DEM
    - (c) EMM
    - (d) All of these

15. The equation of strain index is
    (a) SI = IEM × DEM × EMM × SWM × DDM
    (b) SI = IEM × DEM × EMM × HPM × SWM × DDM
    (c) SI = IEM × DEM × EMM × HPM × SWM
    (d) SI = IEM × DEM × EMM × HPM × DDM

Answers

1. (a)   **2.** (e)   **3.** (b)   **4.** (a)   **5.** (d)   **6.** (e)   **7.** (a)   **8.** (a)   **9.** (a)   **10.** (b)

11. (a)   **12.** (a)   **13.** (d)   **14.** (d)   **15.** (b)

Work Related MSDs Risk and Work Postures Assessment

10.1. Introduction

The reasons behind injuries in a variety of work-related circumstances have been possible to observe since last few decades. For example, consider the case of physically demanding jobs such as casting process in steel industries where a number of tasks involve heavy lifting, carrying, pulling, pushing and moving of equipment in awkward postures that may result into musculoskeletal disorders (MSDs). So, the primary cause of musculoskeletal disorders (MSDs) has been found as manual material-handling tasks.

Likewise in the service industry, IT professionals devote six to eight hours on video display terminals (VDT) and consequently they undergo some MSD complaints. Employees of any organization, who are physically engaged in the task, are unknowingly exposed to the bad work posture. In hospitals the professionals such as dentists, ophthalmologists and nurses are exposed to frequent bending posture while attending patients. Similarly in manufacturing industry also the shop floor workers are working in bad postures, chronic effect of ergonomically bad posture could result in MSD complaints like low back ache, neck pain, wrist stiffness, etc. have been reported by different authors like; Singh et. al (2010). These further could result in low back injuries, cervical problems and carpals tunnel syndromes, etc. Therefore, it becomes very essential that as a preventive measure, the work posture of employees/ workers must be investigated using suitable tools. Although, there are a number of ergonomic tools available for the MSD risk assessment out of which RULA (Rapid Upper Limb Assessment) and REBA (Rapid Entire Body Assessment) are the most prevailing for work posture assessment.

The objective of this chapter is to make the reader understand about the procedural steps of using the ergonomic tools to analyze the working postures of an employee or workers. Hence the cases of workers engaged in various processes of small scale casting industry have been taken and discussed. The present chapter is confined to the used two assessments tools, i.e., Rapid Upper Limb Assessment (RULA) and Rapid Entire Body Assessment (REBA) that recommend the various body postural changes to be made while working.

In 1993, McAtamney and Corlett develop an ergonomics tool known as RULA which is used to assess the risk of musculoskeletal disorders caused in sedentary tasks having high demand of upper body. It is a tool for quick assessment of neck posture, trunk and upper limbs along with muscle function and the external loads experienced by the body with the use of any special equipment. The tool also indicate the level of intervention required to reduce the risks of injury due to physical loading on the operator in terms of an action list by using a coding system.

10.2. Assessment of Work Postures Using RULA Method

The main focused of RULA method is put on assessment of MSD risk for upper limbs (body parts); arms, neck and trunk. There are four categories of MSD risk based upon rating score, these include; as negligible, low, medium and high. Some urgent actions are required to reduce the risk level as in case of medium and high risk category. The flow chart of RULA is exhibited in Figure 10.1, which provides an appropriate procedure for assessing the posture. The very first limb to be assessed is arm which includes upper arm, lower arm and wrist. Let us see how the movement of limbs is scored in different positions.

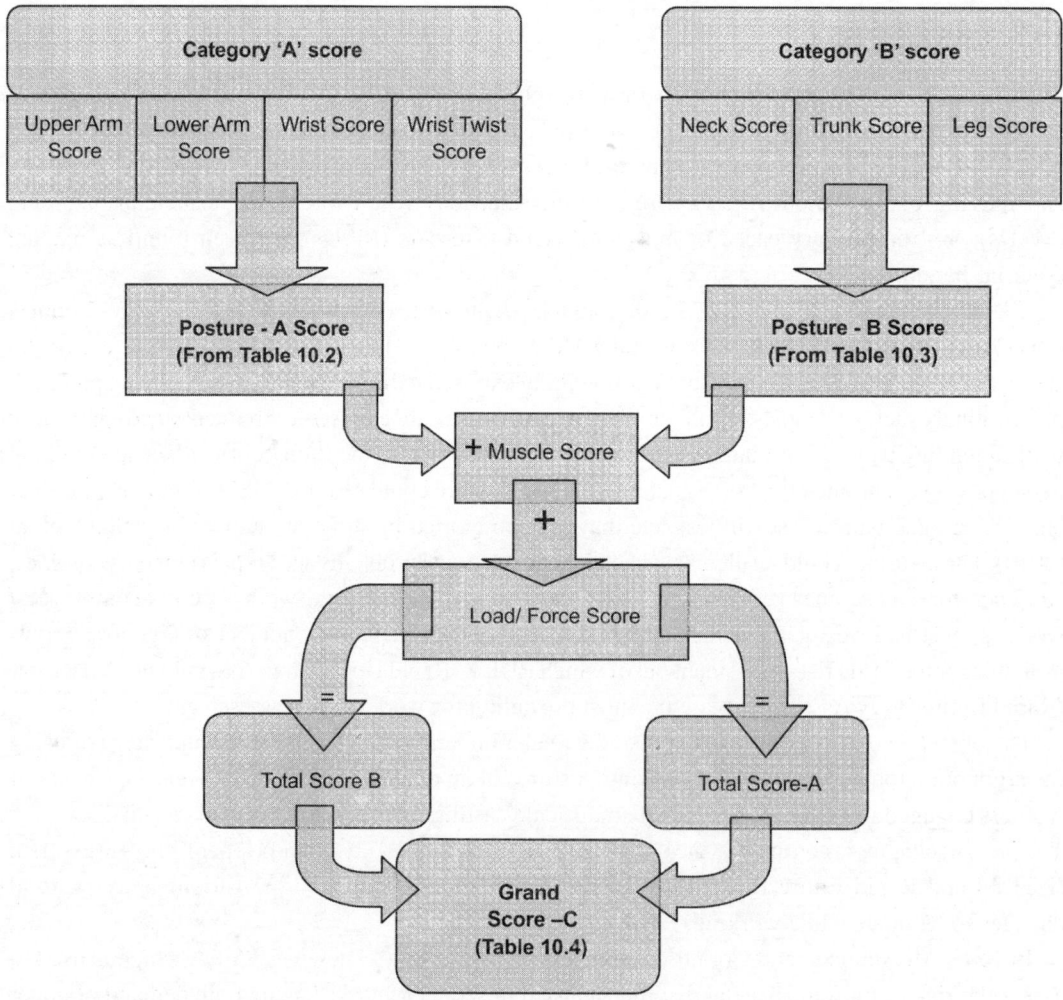

Fig. 10.1: Flow chart for calculation of RULA score.

10.2.1. Upper arm score

Upper arm can be in extension or flexion position; there are five positions as shown in Figure 10.2, If shoulder is raised or upper arm is abducted the basic rating need to be enhanced by +1. In case a subject is leaning or the weight of the arm is supported, the basic rating may be reduced by -1.

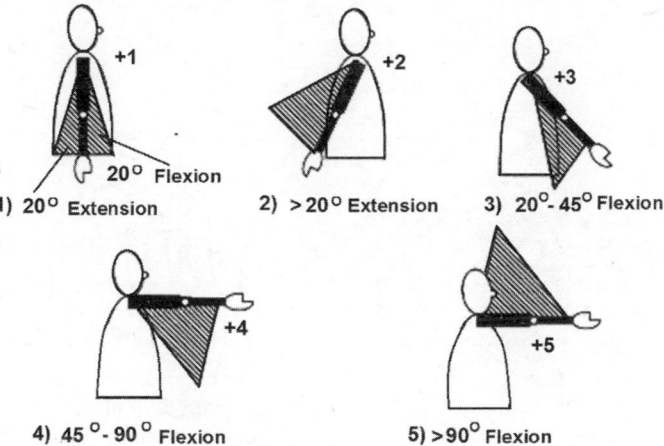

Fig. 10.2: Movements of upper arm and allocation of score*.

Similarly rating for the positions of lower arm is shown in Fig. 10.3. Rating is increased by +1 if working across the midline of the body or out of the side.

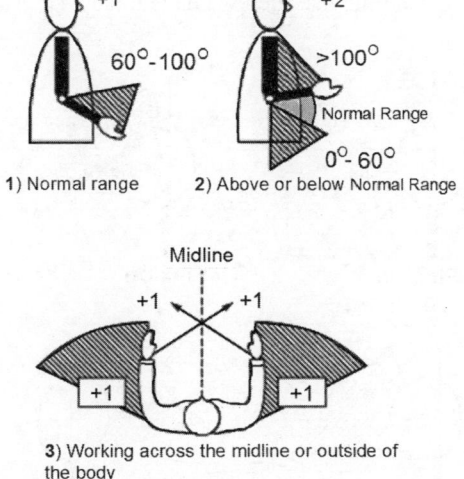

Fig. 10.3: Movements of lower arm and allocation of score*.
Source: McAtamney, L. and Corlett, E. N. (1993).

Wrist positions are shown in Figure 10.4. If wrist is bent away from the midline, rating score is increased by adding + 1. In case wrist is twisted mainly in mid-range of twist or near end of the twisting range, rating is increased by +1 or +2, respectively.

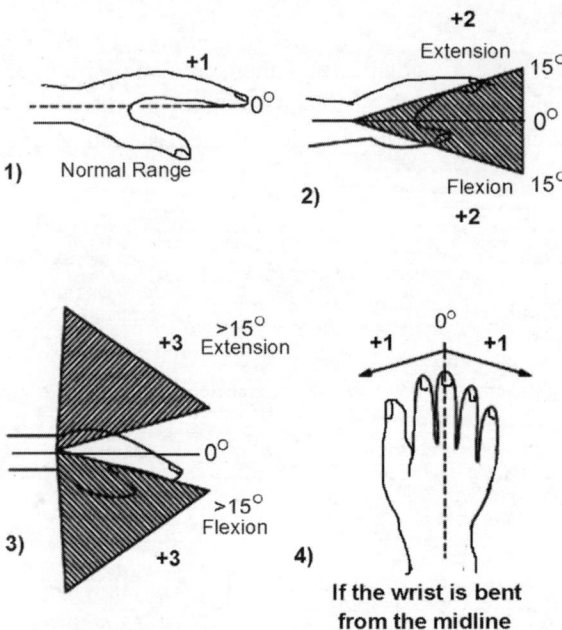

Fig. 10.4: Movements of wrist and allocation of score*.

The range of neck positions (angles) with respect to vertical line and their respective ratings are shown in Figure 10.5; it can have flexion (forward bending) or extension (backward bending) position. In case neck is either twisting or side way bending, in addition to the basic rating we have to add +1. If neck is twisting and side way bending also then we have to add +2.

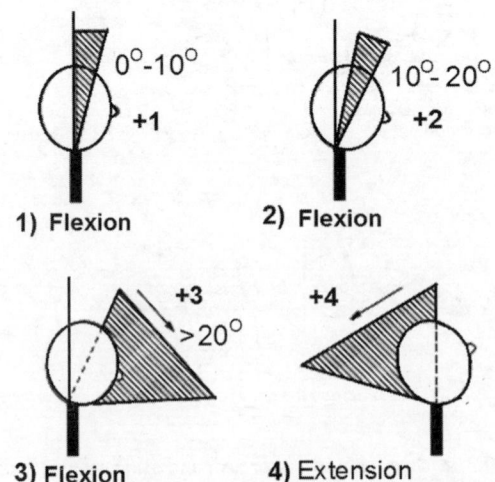

Fig. 10.5: Movements of neck and allocation of score*.

The trunk in various positions (angles) along with their respective rating is exhibited in Figure 10.6. If trunk is twisting or bending sideway, the basic rating is enhanced by adding +1.

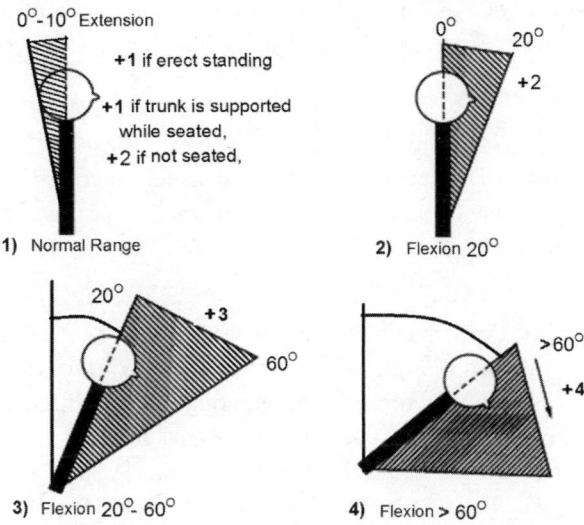

Fig. 10.6: Movements of trunk and allocation of score*.

Source: McAtamney, L. and Corlett, E. N. (1993).

As far as the positions of legs are concerned, RULA considers only two cases, first; leg and feet are well supported in an evenly balanced posture with rating 1. Second case; legs and feet are not well supported or not in an evenly balanced posture, therefore rating is 2. The basic score for upper limbs can be calculated using Table 10.2, whereas score for neck trunk and legs is obtained from Table 10.3.

Once the basic score is obtained, the muscle score is to be in added to it. If the posture is mainly static e.g. held for longer than 1 minute, or posture is repeated more than 4 times per minute, raise the basic score by +1. Similarly, load/ force score is added to the basic scores. The criteria for selection of load force score is shown in Table 10.1.

Table 10.1: Various load/force score for RULA*.

Score	Description of Load
0	No resistance or less than 5 lb (2kg) intermittent load force
1	5–20 lb (2–10 kg) intermittent load force
2	5–20 lb (2–10 kg) static or repeated load force
3	More than 20 lb (10 kg) static or repeated loads or forces. Shock or forces with rapid build-up

Source: McAtamney, L. and Corlett, E. N. (1993).

Action Level

The following action levels are available for recommendation based upon overall RULA score.

- Action level 1: a score of 1 or 2 indicates that posture is acceptable if it is not maintained or repeated for long periods
- Action level 2: a score of 3 or 4 indicates that further investigation is needed and changes may be required

- Action level 3: a score of 5 or 6 indicates that investigation and changes are required soon
- Action level 4: a score of 7 indicates that investigation and changes are required immediately

10.2.2. Procedural steps for calculation of RULA score

In order to explain the procedural steps of RULA method, let us consider an illustration of two workers 'A' and 'B' both are engaged in grinding job in small scale casting industry as shown in Figure 10.7. The procedure followed and calculation of score at each stage for worker 'A' and 'B' is given in Figures 10.8 and 10.9, respectively.

1. The very first step is the calculation of arm and wrist score, hence for the upper arm select the appropriate position from Figure 10.2 in case of both of the workers A and B upper arm is in normal position; therefore upper arm rating is 1 for both the workers. Similarly rating for lower arm is selected from Figure 10.3; it is again 1 for both of them. Whereas the wrist score obtained from Figure 10.4, is 3 for worker A and 1 for worker B.
2. The second step is calculation of score – A from Table 10.2. Hence the posture score is 2 for worker A and 1 for worker B, after adding the muscle (+1) and load/force (+1) score, the overall scores – A from Table. 10.2 for worker A and B are 4 and 3, respectively. The moves in Table A for selection of score are shown with arrowheads and selected score is distinguished by dark and light asterisks for worker A and B, respectively.
3. The third step is selection of score for neck, trunk and leg. The position of neck is referred from Figure 10.5; it is 3 for worker A and 2 for worker B. Similarly rating for trunk position is obtained from Fig. 10.6, it is 3 for worker A and 2 for the worker B. Since both of the workers are in sitting posture, we have selected rating score 1 for legs.

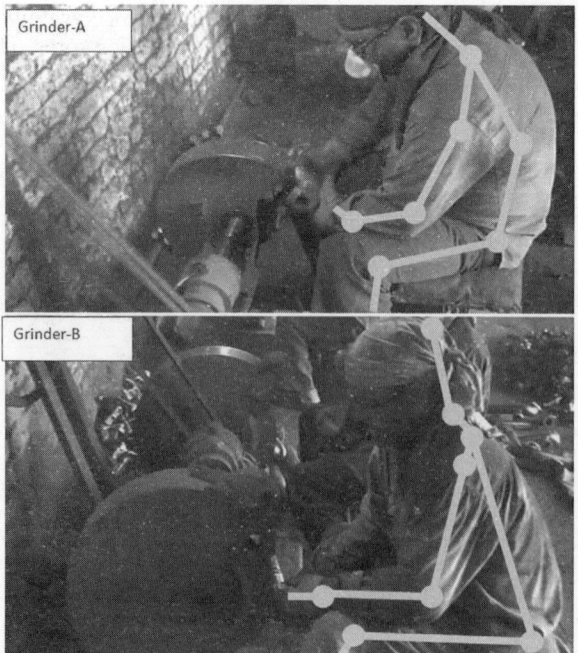

Fig. 10.7: Worker A and B are engaged in grinding jobs.

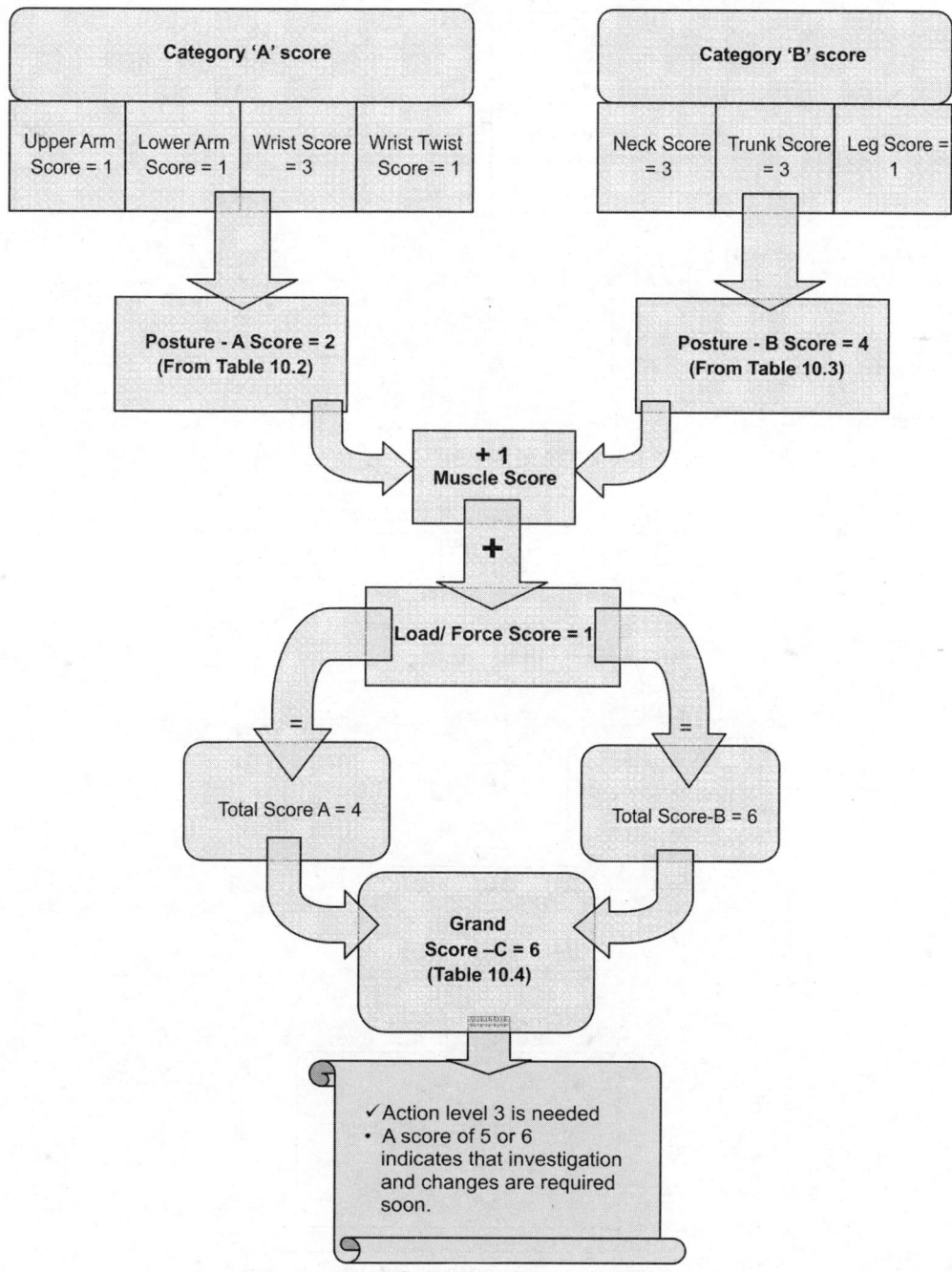

Fig. 10.8: Assessment of RULA score for worker A.

4. Fourth step involves the calculation of score – B from Table 10.3; hence it comes out to be 4 for the worker A and 2 for the worker B. After adding the muscle (+1) and for load/force (+1) score the overall neck, trunk and leg score for worker A and B becomes 6 and 4, respectively.

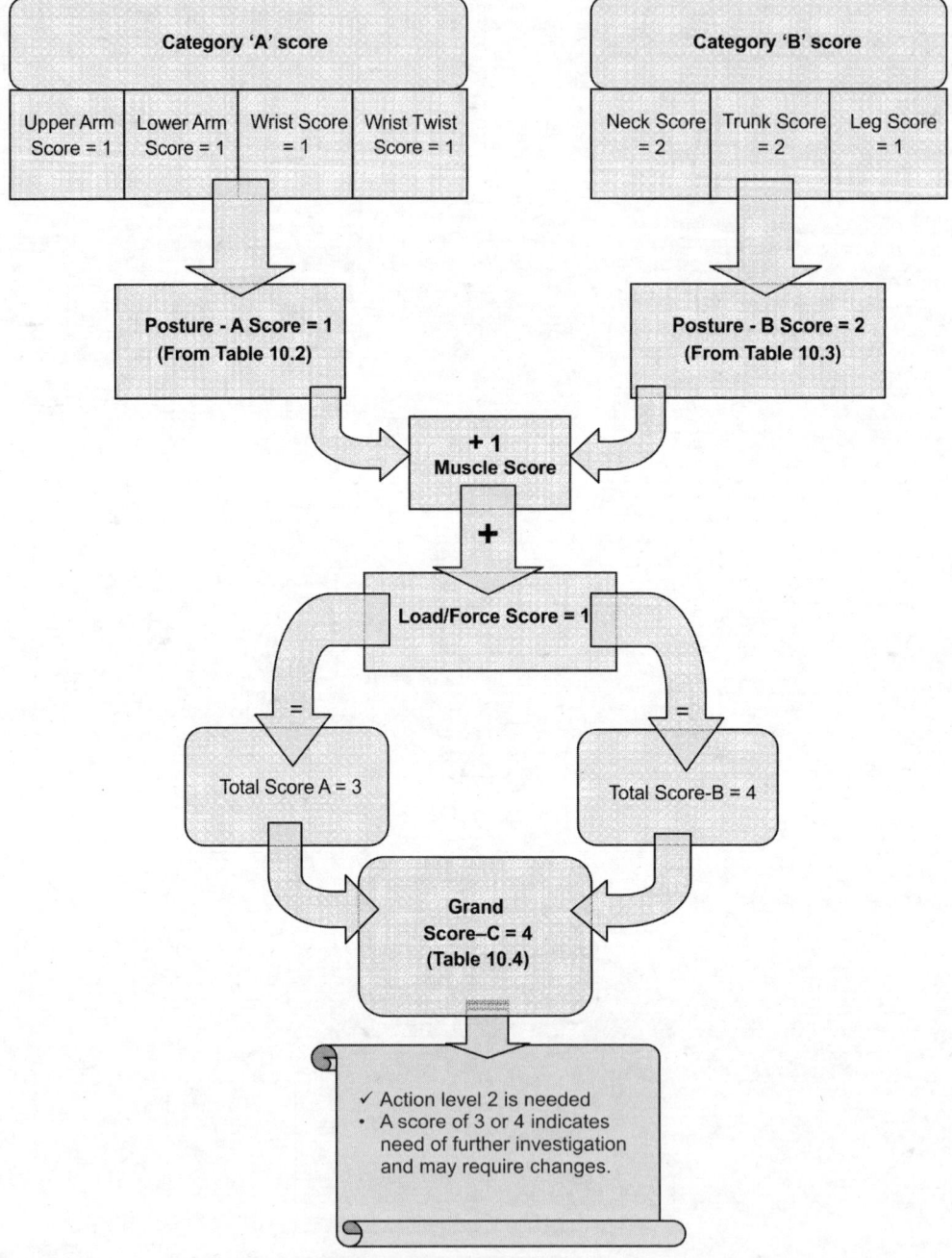

Fig. 10.9: Assessment of RULA score for worker B.

5. Fifth step is to calculate the grand score – C; from Table 10.4. Hence we got a grand score of 6 for worker A and 4 for the worker-B.

6. The last step is the selection of action level and recommendation with respect to the grand score –C obtained from Table. 10.4. Hence we select action level of 3 for worker-A and action level 2 for the worker-B. The action levels are as follows:

Worker-A: Action level 3- a score of 5 or 6 indicates that investigation and changes are required soon.

Worker-B: Action level 2- a score of 3 or 4 indicates that further investigation is needed and changes may be required.

Once we have suggested the appropriate action level, it is also the duty of the posture investigator to identify the limbs which are contributing towards the higher grand score and recommend the correct posture of the respective limbs. In present case of worker-A, wrist, neck and trunk are the main contributory factors towards higher grand score. Worker A is working in relatively more forward bending posture than the worker B. For worker A; the wrist is in more flexion, neck and trunk are also forward bending, and the same needs to be corrected very soon. Hence, the recommendations are made to correct the position of these body parts.

Table 10.2: Upper limb posture score table-A of RULA*.

Upper Arm	Lower Arm	Wrist							
		1★		2		3★		4	
		Wrist Twist		Wrist Twist		Wrist Twist		Wrist Twist	
		1	2	1	2	1★	2	1	2
1 W_A and W_B	1	1☆	2	2	2	2★	3	3	3
W_B = 1	2	2	2	2	2	3	3	3	3
	3	2	3	2	3	3	3	4	4
2	1	2	2	2	3	3	3	4	4
	2	2	2	2	3	3	3	4	4
	3	2	3	3	3	3	4	4	5
3	1	2	3	3	3	4	4	5	6
	2	3	4	4	4	4	4	5	5
	3	4	4	4	4	4	5	5	5
4	1	4	4	4	4	4	5	5	5
	2	4	4	4	4	4	5	5	5
	3	4	4	4	5	5	5	6	6
5	1	5	5	5	5	5	6	6	7
	2	5	6	6	6	6	6	7	7
	3	6	6	6	7	7	7	7	8
6	1	7	7	7	7	7	8	8	9
	2	8	8	8	8	8	9	9	9
	3	9	9	9	9	9	9	9	9

W_A = 2

Table 10.3: Neck, trunk, and legs posture score table-B of RULA*.

Neck Posture Score	Trunk Posture Score											
	1		2☆		3★		4		5		6	
	Legs		Legs★		Legs		Legs		★Legs		Legs	
	1	2	1☆	2	1★	2	1	2	1	2	1	2
1	1	3	2	3	3	4	5	5	6	6	7	7
2 W_B ☆	2	3	2☆	3	4	5	5	5	6	7	7	7
3 W_A ★	3	3	3	4	4★	5	5	6	6	7	7	7
4	5	5	5	6	6	7	7	7	7	7	8	8
5	7	7	7	7	7	8	8	8	8	8	8	8
6	8	8	8	8	8	8	8	9	9	9	9	9

$W_B = 2$

$W_A = 4$

Table 10.4: Grand score table–C of RULA*.

Score-C		Score-B (Neck, Trunk, Leg)						
		1	2	3	4☆	5	6★	7+
Score-A (Upper Limb)	1	1	2	3	3	4	5	5
	2	2	2	3	4	4	5	5
	3☆	3	3	3	4☆	4	5	6
	4★	3	3	3	4	5	6★	6
	5	4	4	4	5	6	7	7
	6	4	4	5	6	6	7	7
	7	5	5	6	6	7	7	7
	8+	5	5	6	7	7	7	7

$W_B = 4$

$W_A = 6$

Source: McAtamney, L. and Corlett, E. N. (1993).

10.2.3. Case study

In India, more than 5 million employees are working in various occupations of casting industries involving manual material handling tasks in awkward postures which result into musculoskeletal disorders among the workers. A small scale casting unit situated in Punjab region was taken for the case study. The present study is focussed to analyze postures of workers engaged in various occupations of small scale casting industry and need of effective changes have been assessed in the postures during their work by using RULA (Rapid Upper Limb Assessment). As analyzing the different sections in case organization it has been observed that workers engaged in pattern making, mould making, core making, moulding, sand preparation, and molten metal pouring, grinding, trimming etc., were suffering from hazardous body postures, which was recorded in a video film. Approximate 100 snapshots were cropped from both left hand (LH) and right hand (RH) side of the workers, out of these 88 (LH) and 92 (RH) were selected to analyze the postures of different body parts by filling the scores in RULA score sheet (Figure 10.1).

10.2.4. Results and discussion

The results shown in Table 10.5 reveals that about 60.22 per cent of left hand (LH) side postures and 69.58 per cent of right hand side postures were found at high risk of MSDs as reflected in RULA scores. Therefore there is a need to investigate the working postures and make the changes very soon. Similarly the percentage of medium to low risk was 25 & 13.6 per cent and 18.47 & 11.95 per cent for left and right hand side postures respectively.

Table 10.5: Distribution of risk level and their corresponding action levels for RULA.

RULA Score	Risk Level	Action	No. of Workers		%age Workers	
			L	R	L	R
1–2	Negligible	Acceptable	-	-	-	-
3–4	Low	Investigate further	12	11	13.60	11.95
5–6	Medium	Investigate further and change soon	22	17	25.00	18.47
7	High	Investigate and change immediately	53	64	60.22	69.56
Total			88	92		

Table 10.6: Distribution of RULA score process-wise.

Process	Scores							
	1–2		3–4		5–6		≥7	
	L	R	L	R	L	R	L	R
Mould making	-	-	-	-	9	6	13	15
Mould lifting	-	-	8	8	3	3	17	19
Pattern making	-	-	-	-	4	2	2	5
Molten metal lifting to mould box	-	-	-	-	2	1	4	5
Molten metal pouring	-	-	4	3	4	5	17	20

In this casting unit around 24 per cent workers were found at high risk in mould making process which arise the need of eager examination of working postures. Further analysis is required for 38.49 per cent workers that lie in medium risk level category. If we talk about mould lifting activity 30.76 per cent under high level of risk, 15.38 per cent in medium level of risk and 69.56 per cent of the workers were found in low level of risk category. The suitable preventive actions were suggested to apply in the specific occupation consequently.

However, some required actions are to be made in pattern making occupation where 6.08 per cent workers are at high risk, 14.97 per cent workers are in medium level of risk category.

On the other hand, 7.93 per cent workers found to work in awkward postures at high level of and 7.48 per cent at medium level of risk category in molten metal lifting/ carrying to mould box. About 32.68 per cent of workers were at high level of risk of MSDs in metal poring activity as they do not work in suitable manner which arise the need of some remedial actions ergonomically such as automatic devices and techniques.

10.2.5. Conclusion

The present case study concluded that, the workers engaged in various activities in casting industry were working at high rate of risk of musculoskeletal disorders (MSDs) due to awkward postures. The level of risk need to be eliminated or reduced by means of ergonomic interventions, low cost automation and mechanical aids. The proper training of metal pouring by means of ergonomics techniques should be introduced.

10.3. Work Posture Assessment Using Rapid Entire Body Assessment Tool (REBA)

Hignett, S. and McAtamney, L. in 2000 developed REBA (Rapid Entire Body Assessment), an analysis tool for rapid and easy observation of postural whole body activities. REBA provides a score system for muscle activities by dividing the body into segments coded separately with respect to movement planes. REBA reflects the importance of coupling during handling of loads and immediate action is shown to be taken by means of an action level. The five categories of REBA scores are: negligible, low, medium, high and very high.

10.3.1. Procedural steps of REBA for employee assessment

The body is divided into two segments, first includes; neck, trunk and legs, then second consists of arms and wrists. Each segment is to be rated individually with reference to movement planes. Therefore Table 10.8 (for category - A score) is used to calculate posture score considering the neck, trunk and legs of a subject, and Table 10.9 (for category - B score) is used to determine a score from postures of upper limbs; arm and wrist. There are thirteen steps in REBA for assessment of posture of a person engaged in an activity are described as follows:

Step 1: Neck score
First of all, we have to select neck score for different position as shown in Fig. 10.10, and score can be selects as per following criterion.

- If neck is in normal range of forward bending position i.e. flexion angle ≤ 20°, the score is 1.
- If neck is in forward bending position i.e. flexion angle > 20°, then score is 2.
- If there is any extension, then again score is 2.
- If neck is twisted or sideway bent, then add +1 to the basic score.

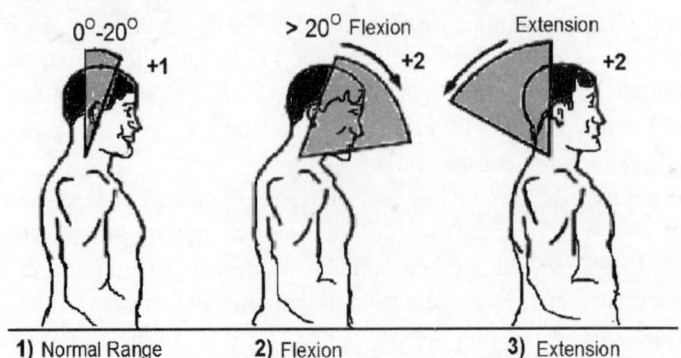

Fig. 10.10: Score with respect to variations in posture of neck.

Step 2: Trunk score

Second step is to select score for trunk position as shown in Fig 10.11, the rating score is selected as per the following criteria.

- If trunk is in vertically straight (normal) position, then score is +1.
- If trunk is extension position, then score is +2.
- If trunk is in forward bending between 0 and 20°, then score is 2.
- If trunk is in forward bending between 20 and 60° then score is 3.
- If movement of trunk is more the 60°, then score is 4.
- In case of twisting or side bending of trunk, +1 is added to the basic score. If trunk is twisted and side bent simultaneously, then +2 is added to the basic score.

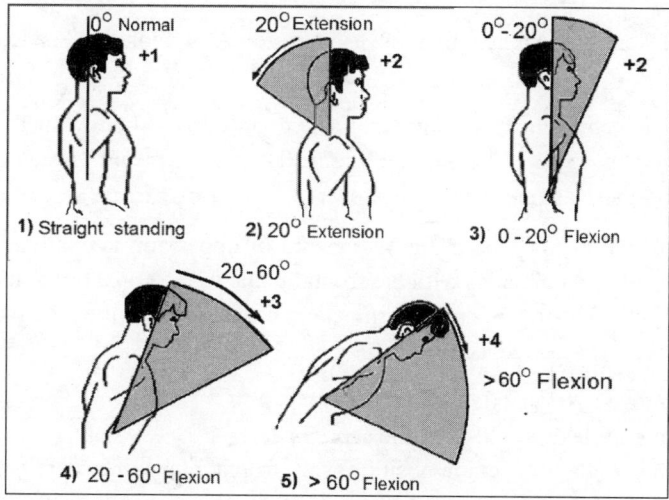

Fig. 10.11: Score with respect to variation in posture of trunk.

Step 3: Legs score

Third step is to obtain score for legs in different positions as shown in Figure 10.12. The score for legs is selected as follows:

- If subject is standing with straight legs, score is 1.
- If subject is standing on one leg and second leg is relaxed, score is 2.
- If legs are bent from knees between 30 and 60°, add + 1, or if legs are bent from knees >60°, add +2 to the basic score.

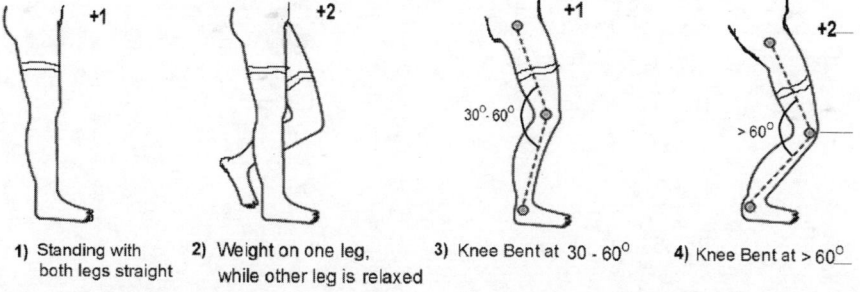

Fig. 10.12: Score with respect to variation in posture of legs.

Step 4:

Once the score for neck, trunk and legs is determined, the next step is to determine category-A score considering these scores using Table 10.8.

Step 5:

Once the scores for neck, trunk and legs is determined, next step is to add load/force score if subject is applying any force or carrying any load, as follows:

- If subject is carrying load <11 lbs, score is 0
- If subject is carrying between 11 and 22 lbs, score is 1
- If subject is carrying >22 lbs, score is + 2
- If there are shocks or rapid build-up force, only add +1 to the basic score

Add the score of steps 5 and 4, and then the total score - A of Table 10.8 is determined.

Step 6:

The overall score obtained from steps 4 and 5 is to be used to mark the row in Table 10.10.

Once the category 'A' score is determined, the next steps are meant for category 'B' score i.e. arms and writs score, these are describes as follows.

Step 7: Identify position of upper arm, the movement of upper arm is exhibited in Figure 10.13. When the center line of arm coincides with the frontal plane of the body, or the movement of arm is only 20° flexion or 20° extension i.e. upper arm is in normal position, then the score is 1, for rest of the cases the score will be selected as follows:

- If the arms move >20° in extension, the score is 2
- If the arms are in flexion >20–45°, the score is 2
- If the arms are in 45–90° flexion position w.r.t. frontal plane, the score is 3
- If the arms are at 90–180° flexion position w.r.t. frontal plane, the score is 4.

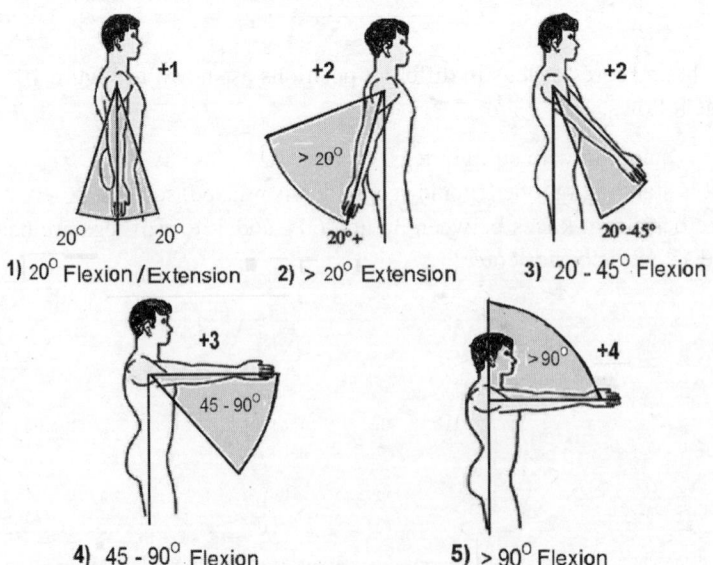

Fig. 10.13: Score with respect to variation in posture of upper arm.

- If the shoulders are raised or if upper arm is abducted, then add +1 in the upper arm basic score. If arm is supported or person is learning, then subtract –1 from the upper arm basic score

Step 8: Location of lower arm position

The movement of lower arm is exhibited in Figure 10.14, the score for the lower arm is selected as follows:

- If lower arm is flexed between 60 – 100° with respect to vertical line of frontal plane, the score is 1
- If lower arm flexed <60° or >100° with respect to vertical line, then the score is 2

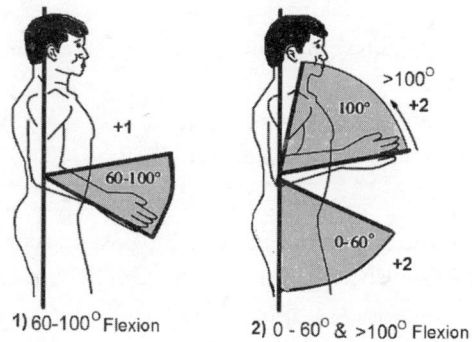

Fig. 10.14: Score with respect to variation in posture of lower arm.

Step 9: Identify the wrist position

The wrist position is shown in Figure 10.15. The score for wrist position is selected as follows:

- If the wrist is in flexion or extension up to 15°, the score is 1
- If the wrist is moving up (extension) and down (flexion) more than 15°, the score is 2
- If the wrist is bending from the middle or twisting, add +1 to the basic wrist score

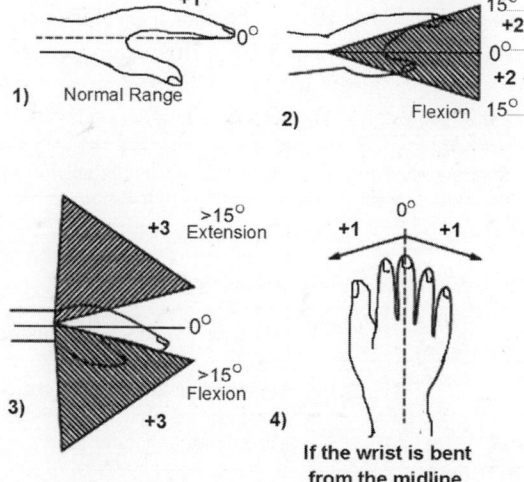

Fig. 10.15: Score for various wrist positions.

Step 10:
Once arms (upper and lower) and wrist score is determined, next step is to determine overall posture score-B from Table 10.9 by using the scores obtained in steps 7–9.

Step 11: Coupling score
Once the posture score-B is obtained from Table 10.9, the next step is to add the coupling score as follows:
- Well-fitting handles and mid rang grip rating is considered good, and coupling score is 0.
- Acceptable but not ideal hand hold or coupling acceptable with another body part then grip rating is considered fair, and coupling score is +1.
- Hand hold not acceptable however possible then grip rating is considered poor, and coupling score is +2.
- Awkward, unsafe grip, no handles; coupling is unacceptable using other parts of the body, then coupling score is +3.

Once the category score-B, obtained by adding values of scores obtained in steps 10 and 11, the same will be to be used to obtain the overall score -C, from Table 10.10.

Step 12:
In this step the overall REBA score is determined using the posture score - A and score -B obtained from Tables 10.8 and 10.9, respectively, the overall score- C is determined from Table 10.10.

Step 13: Activity score
Activity score is added to the score obtained from Table as follows:
- If one or more body parts are held for longer than 1 minute (static), then add +1 to the score - C obtained from step 12.
- If repeated small range action (more than 4 times per minute), add +1 to the score - C obtained from step 12.
- Action cause; rapid large range change in posture or unstable base, then add +1 to the score - C obtained from step 12.

The overall final REBA score is determined by adding activity score obtained from step 13 to the category score - C obtained from Table 10.10. The action level is decided based upon the overall REBA score, which is shown in Table 10.7. The REBA flow chart is exhibited in Figure 10.16.

Table 10.7: Distribution of scores and their corresponding action levels according to REBA*.

REBA Score	Risk Level	Action Required
1	Negligible risk	No change required
2–3	Low risk	Change may be needed
4–7	Medium risk	Further investigation, change soon
8–10	High risk	Investigate and implement change
11+	Very high risk	Implement change

*Source: McAtamney, L. (2000).

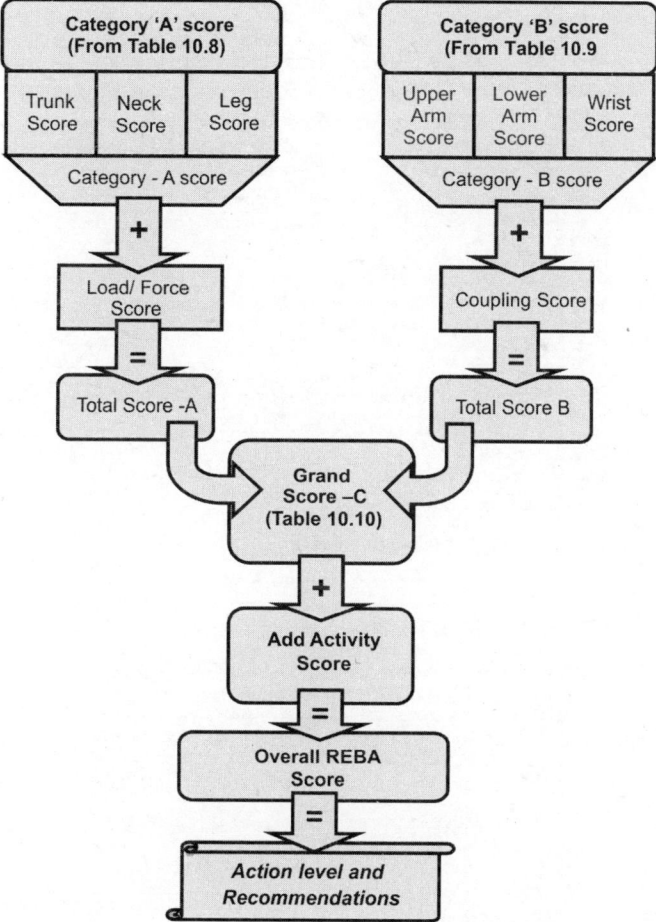

Fig. 10.16: Flow chart for calculation of REBA scores.

10.3.2. Case analysis

Let us try to assess and compare the work posture of two workers (W_A and W_B) engaged in pouring task in a small scale casting industry. The working postures of worker A and B is shown in Figures 10.17 and 10.18, respectively.

Step 1: (To locate neck position, refer to Figure 10.10)
Referring to Figure 10.10, it is ascertained that, the neck of worker A (W_A) is in extension. Therefore, his neck score is 2. The neck of worker B (W_B) is bent forward ≤20°; therefore, neck score is 1.

Step 2: (To locate the trunk position, refer to Figure 10.11)
The worker (W_A) has bent his trunk in forward direction ≥60°, therefore his trunk score is 4; in addition, the trunk is also twisted; therefore, we add +1, thus the total trunk score of W_A = 4 + 1 = 5
   On the other hand worker (W_B) has bent his trunk ≤20°, therefore his trunk score is the score is 2.

Step 3: (To locate legs position, refer to Figure 10.12)
Since both the workers (W_A and W_B) are standing on both the legs bent at their knees >60°; therefore, their legs score is 1 + 2 = 3

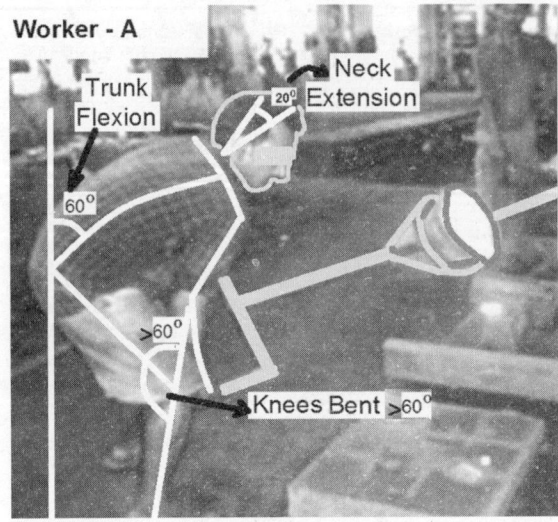

Fig. 10.17: Worker (W_A) in a bending posture at pouring task.

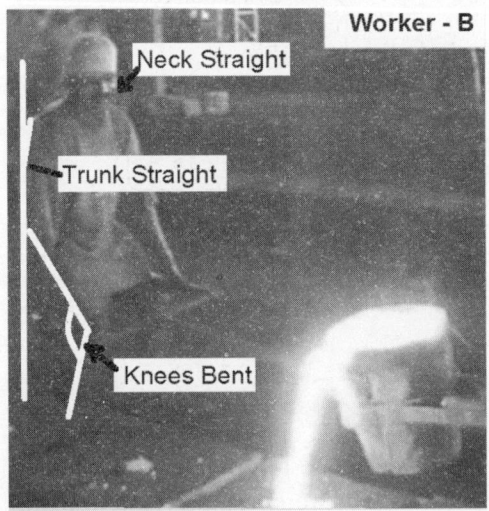

Fig. 10.18: Worker (W_B) in relatively better posture at pouring task.

Step 4: (Look-up final posture score of Table 10.8, A)

For determining the posture score for neck, trunk and legs, we have to use Table 10.8 (Score- A). Since for worker W_A neck position score is 2, we select column 2. Next we have to select column for legs position out of four columns. Since the legs score is 3, we move downward in column 3. Using the trunk position score, i.e., 5, we move horizontally. Horizontal row of trunk position and vertical column for neck and legs position intersect each other at box with score 8. Similarly posture score for W_B is obtained. Therefore total posture score from Table A for both the workers is obtained as follow:

$$W_A = 8, \ W_B = 4$$

Table 10.8: Neck, trunk and leg score Table-A of REBA*.

Score-A		Neck												
		1				*2*				*3*			
	Legs	1	2	3	4	1	2	3	4	1	2	3	4
	1	1	2	3	4	1	2	3	4	3	3	5	6
Trunk Posture Score	2	2	3	4	5	3	4	5	6	4	5	6	7
	3	2	4	5	6	1	5	6	7	5	6	7	8
	4	3	5	6	7	5	6	7	8	6	7	8	9
	5	4	0	7	8	0	7	8	9	7	8	9	9

$W_B = 4$

$W_A = 8$

Source: McAtamney, L. (2000).

Step 5: (Adding force/load score in posture score-A)

Both the workers are carrying load >22 lbs so, the force/load score is +2 for both of them.

Adding the posture score from Table 10.8 with force/load score, we get score-A

$$W_A = 8 + 2 = 10, W_B = 4 + 2 = 6$$

Step 6: (Final score-A of Table 10.8)

Score-A of both the workers will be used to select required row from Table 10.10 for score-C

Once the category-A score is obtained, next step is to determine the scores of arms (upper and lower) and writs i.e. category-B score, using table 10.9.

Step 7: (Location of upper arm position)

Both the workers have their upper arms position in between 20 and 45°. So, both of they score is 2

$$W_A = 2, W_B = 2$$

Step 8: (Location of lower arm position)

Both the workers are have their lower arm between 0 and 60°; therefore, lower arm score for both of them is 2.

$$W_A = 2, W_B = 2$$

Step 9: (Locate wrist position)

Both the workers have their wrist 15° flexion or extension; hence wrist score for both of them = 1

$$W_A = 1, W_2 = 1$$

Step 10:

For determining the overall posture score arms and wrist, we have to use Table 10.9 (score- B). Since for both workers *(W_A and W_B) upper arm position is 2, therefore we select row 2, similarly the lower arm position for both workers is 2, therefore we move down ward in columns 2. Next each lower arm column has three positions for wrist and since both the workers have wrist position 1. Therefore total posture score from Table 10.9 for both the workers is obtained as follow:

$$W_A = 2, W_B = 2$$

Table 10.9: Upper arm, lower arm and wrist score Table-B of REBA*.

Score- B	Lower Arm					
		1			*2*	
Upper Arm Score	**Wrist**	**1**	**2**	**3**	**1**	**2**	**3**
	1	1	2	2	1	2	3
	2	1	2	3	2	3	4
	3	3	4	5	4	5	5
	4	4	5	5	5	6	7
	5	6	7	8	7	8	8
	6	6	7	8	8	9	0

$W_A = 2$
$W_B = 2$

**Source:* McAtamney, L. (2000).

Step 11: (Adding coupling score in posture score for arm and wrist)

Since, there are well-fitting handles and mid-range power grip, coupling score is 0 for both W_A and W_B
Adding the posture score from Table 10.9 with coupling score we get score-B

$$W_A = 2+ 0 = 2, W_B = 2 + 0 = 2$$

Step 12: (Final score of Table-B)

For determining overall REBA score we have to refer to Table 10.10. The score of arms and wrist (Table 10.9) of both the workers is used to select required column and score of neck, trunk and legs (Table 10.8) is used to select row in Table 10.10. Thus the final score is obtained for (W_A) and (W_B).

$$W_A = 10 \ W_B = 6$$

Step 13: (Activity score)

Since both workers are engaged in pouring of molten metal and they require to hold one or more body parts in static position for longer 1 minute, therefore we have to add the activity score +1 in basic score.

$$W_A = 10 + 1 = 11 \qquad W_B = 6 + 1 = 7$$

By adding activity score the Final REBA score for both the workers as

$$W_A = 11 \qquad W_B = 7$$

Table 10.10: Overall grand score category-C (Table-C) of REBA.

Score A	Score C											
	Score B											
	1	2	3	4	5	6	7	8	9	10	11	12
1	1	1	1	2	3	3	4	5	6	7	7	7
2	1	2	2	3	4	4	5	6	6	7	7	8
3	2	3	3	3	4	5	6	7	7	9	9	9
4	3	4	4	4	5	6	7	8	8	9	9	9
5	4	4	4	5	6	7	8	8	9	9	9	9
6 (W$_B$)	6	6	6	7	8	8	9	9	10	10	10	10
7	7	7	7	8	9	9	9	10	10	11	11	11
8	8	8	8	9	10	10	10	10	10	11	11	11
9	9	9	9	10	10	10	11	11	11	12	12	12
10 (W$_A$)	10	10	10	11	11	11	11	12	12	12	12	12
11	11	11	11	11	12	12	12	12	12	12	12	12
12	12	12	12	12	12	12	12	12	12	12	12	12

W$_A$ = 10   W$_B$ = 6

Source: McAtamney, L. (2000)

Example-2

A worker in casting industry is engaged in lifting of the molds from the machine and placing them in queue near the induction furnace, he has complain of suffering from lower back and knees pain. Therefore, two postures are captured one W$_D$ for lifting the rammed mold and W$_C$ for placing the mold on the floor, REBA score for both the postures is obtained, let us analyze the same.

Posture W$_C$

Let us first analyze the posture 'W$_C$' of the worker when he is placing the mold on the floor near the induction furnace. The REBA score for W$_C$ was found 11, which consists of score - A (Table 10.8) was as very high at neck 2, trunk 4, legs 3, and load force was also 2 because he was carrying more than 22 kg load. Whereas the score– B (Table 10.9) includes, upper arm 3, lower arm 2, wrist 3 and coupling score 1, which was fair. If the worker tries to keep his trunk straight as much as he can while working, then the trunk score would have reduced significantly and also while placing the box on the floor, if legs were kept relatively straight, the score would have also reduced from 3 to 1. Also the arms could be kept about 20° forward from the centre body line; the reduction in score would have been from 3 to 1. Hence improvement in posture of these body parts would have reduced the overall REBA score. Thus the posture W$_C$ makes the worker to fall under medium risk of MSD.

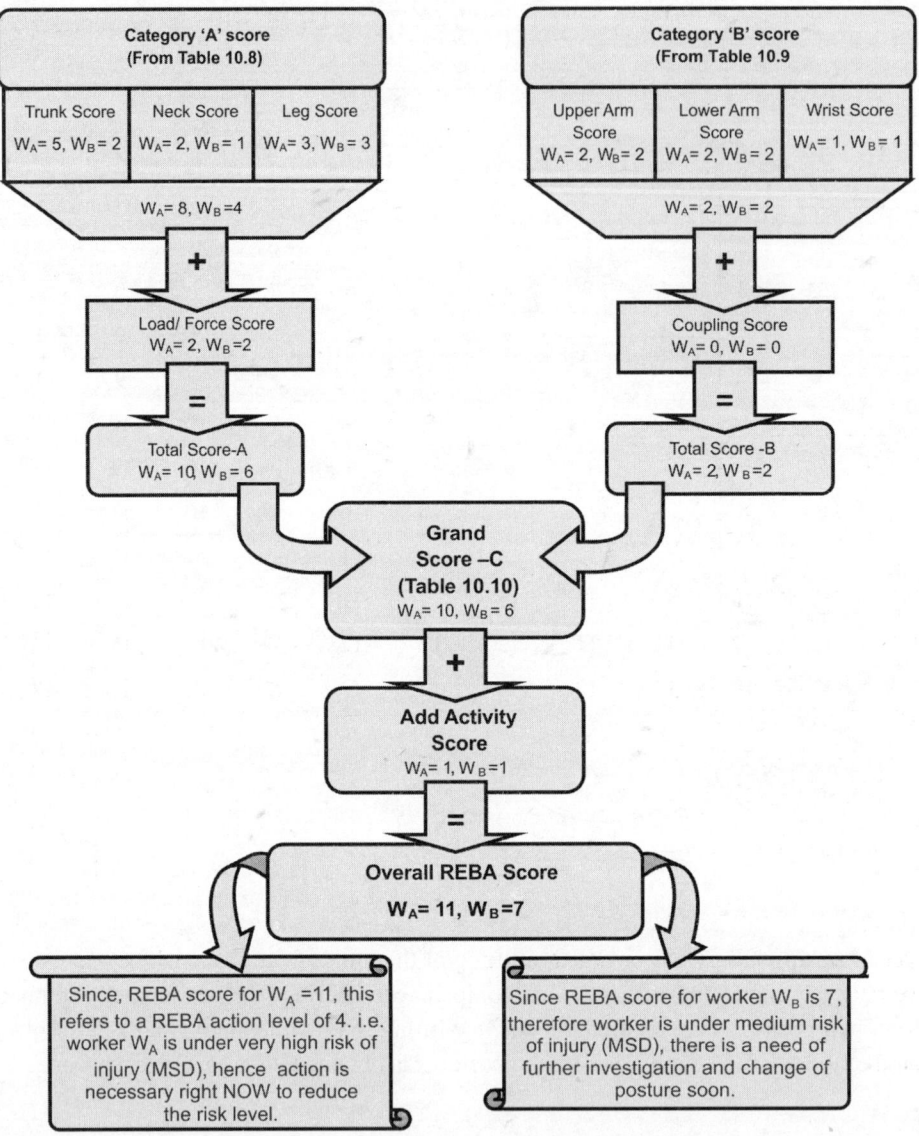

Fig. 10.19: Calculations of REBA score for workers W_A and W_B engaged in pouring operation.

Posture W_D

Overall assessment score for posture 'W_D' was 10; breakup of score reveals that score-A constituted of neck score 2, trunk 3, legs 3, and load force was also 2, and this is because he was carrying load more than 22 kg. On the other side, the scores for upper arm 2, lower arm 2, wrist 2 and coupling score was 1, which was fair. The risk level can be reduced by avoiding the forward bending of trunk and thus trunk score would have reduced significantly. At the same time if legs were kept straight and parallel score would have also reduced from 3 to 1. Also the arms could be kept below 20° forward and backward from the frontal plane line; the score would have been reduced from 2 to 1.

Fig. 10.20: Work postures W_C and W_D of a worker engaged in molding operation.

It is also recommended that the worker should be provided with some platform so that picking of mould becomes comfortable from the machine and worker may need not lift his heals. Improvement in these five factors would reduce the overall REBA score leading the worker to the medium risk range. Workers will be safe from risk of MSDs.

A 29 years old worker is engaged in metal purring in a casting industry. Let us asses his work posture.

Metal pouring is one of the main jobs, which leads to the MSDs among the workers working in the casting industry. In this process of molten metal is poured into the mould cavities with the help of ladle and crucible. This job involves high risk because workers carry molten metal at 1400 °C from the furnace to the moulding boxes. The weight of the ladle and crucible filled with molten metal varies from 60 kg to100 kg. This job requires a fine pouring skill of the worker. This worker has repeatedly complained about lower back and shoulder pains. Let us analyse the posture of this worker.

REBA score was calculated on the posture of this worker while he was pouring the molten metal from the ladle with the help of crucible in the gate of moulding box. The REBA score came out 11, which included score for, neck 2, trunk 4, legs 3 and load force 2, because he was carrying load more than 22 kg. On the other side, the scores of upper arm 3, lower arm 2, wrist 1 and coupling score was 1, which was fair. If the worker tries to keep his trunk straight as much as he can while working, then the trunk score would have reduced significantly and also while pouring legs were kept straight and parallel, then the score would have reduced from 3 to 1. Moreover, the arms could also be kept between 20° forward from the centre body line, and if he stops leaning while pouring then reduction in upper arm score would have been from 3 to 1. By improving these three factors he would have reduced the overall REBA score, leading him to the medium risk range, consequently the worker will be relatively safe from MSDs complaints.

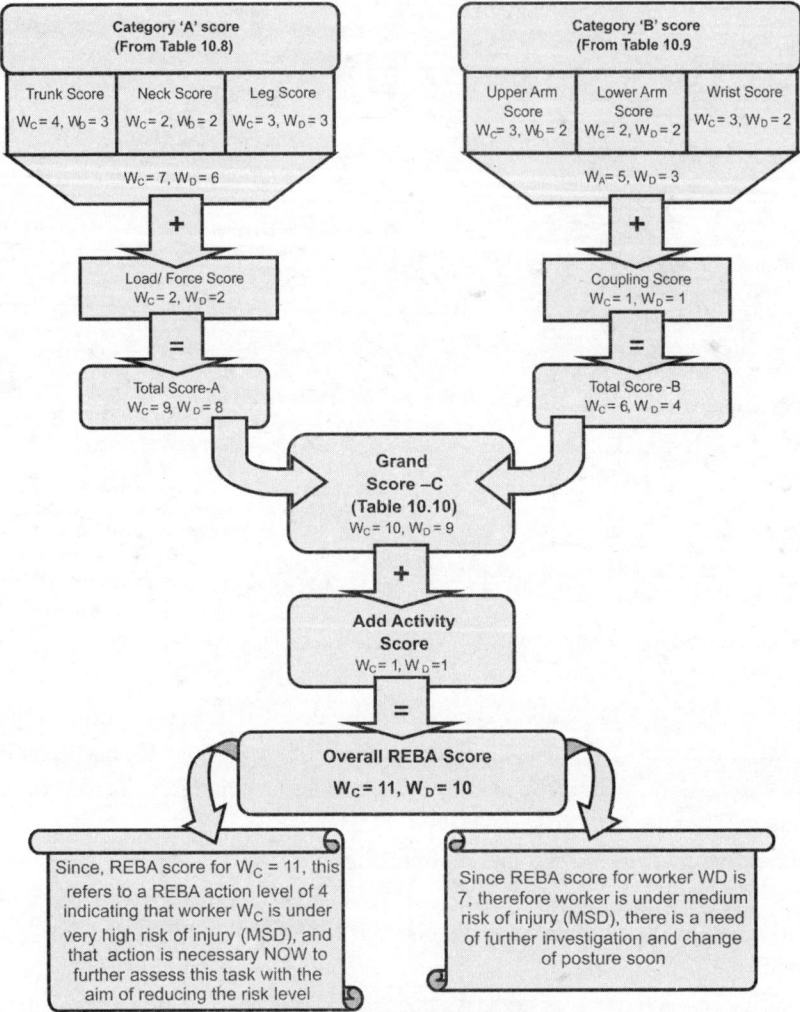

Fig. 10.21: Calculation of REBA scores for W_C and W_D.

Fig. 10.22: A worker engaged in pouring operation with relatively safe posture.

Multiple Choice Questions

1. MSDs stand for
   (a) Musculoskeletal disorders
   (b) Muscular disorders
   (c) Mental stress disorders
   (d) None of these
2. The term VDT termed as
   (a) Visual display technique
   (b) Video display terminals
   (c) Video display technique
   (d) Visual display terminals
3. RULA stands for
   (a) Rapid upper limb assessment
   (b) Repetitive upper limb assessment
   (c) Rapid upper ligament assessment
   (d) None of these
4. REBA stands for
   (a) Rapid ergonomics body assessment
   (b) Rapid entire body assessment
   (c) Repetitive entire body assessment
   (d) None of these
5. Who was the developer of RULA?
   (a) M. J. Griffin
   (b) F. W. Taylor
   (c) McAtamney and Corlett
   (d) G. W. Thomson
6. An ergonomics tool RULA was developed in
   (a) 1998   (b) 1993   (c) 1994   (d) 1989
7. RULA is an ergonomic tool for quick assessment of
   (a) Neck exposure   (b) Trunck   (c) Upper limbs   (d) All of these
8. If load/force description is 5–20 lbs then RULA score is
   (a) 1 or 2   (b) 2   (c) 0   (d) 3
9. If load/force description is more than 10 kg then RULA score is
   (a) 2   (b) 3   (c) 4   (d) 5
10. If load/force description is 2 kg then RULA score is
    (a) 0   (b) 1   (c) 2   (d) 3
11. What would be the risk level if RULA score is 3–4?
    (a) Negligible   (b) Low   (c) Medium   (d) High
12. What would be the action if RULA risk score is 5–6?
    (a) Investigate further
    (b) Investigate further and change soon
    (c) Change immediately
    (d) All above
13. Who was the developer of REBA?
    (a) M. J. Griffin   (b) F. W. Taylor
    (c) Hignett and Corlett   (d) G. W. Thomson
14. REBA was developed in
    (a) 2003   (b) 2004   (c) 2000   (d) 2001
15. If neck is bent in forward position at ≤20°, REBA score is
    (a) 1   (b) 2   (c) 3   (d) 4

Answers

1. (a)   **2.** (b)   **3.** (a)   **4.** (b)   **5.** (c)   **6.** (b)   **7.** (d)   **8.** (a)   **9.** (b)   **10.** (a)
11. (b)   **12.** (b)   **13.** (c)   **14.** (c)   **15.** (a)

Office Ergonomics

11.1. Introduction

In current scenario every one working in the office is greatly dependent upon computers to perform work. Millions of people work with computers every day. Depending upon the type of job, some people perform dedicated computer work, while others do multiple tasks throughout the day. As people are diverse and ergonomics strives to fit the task to the person doing it, therefore, office ergonomics is concerned with the designing of office work place and is deliberated to help an individual for the assessment of the ergonomic design of computer workstation. Therefore office ergonomics help us to setup the job place as per the need of human comfort and provide basic tools to maintain a healthy workspace in the office. Some of the helpful tips show how to identify and correct ergonomic problems to prevent repetitive strain injury, eye strain, fatigue and discomfort.

11.2. Issues in Workstation Design

Whenever a work station or office is designed, most importantly the following issues are considered:

1. Body posture, back, neck, upper arms, lower arms, wrist extension and ulnar deviation
2. Keyboard and mouse
3. Typing speed
4. Hours spent using keyboard and mouse
5. Rest breaks
6. Overtime
7. Unaccustomed work
8. Lighting and glare

Body posture is the most significant aspect when considering the workstation design. Besides that, seat, job facing (working on VDT) aides, observations in maintaining the fine posture are also important. An ergonomically well-designed seat does not assure for accurate position although it provides some help to be in unbiased posture. It is up to the individual to learn and practise appropriate posture. The working posture can be analyzed using various ergonomic tools available, like RULA, which is described in the previous chapter.

11.2.1. Sitting posture

To identify the sitting posture, whether it is poor posture or good, one must be capable to reply a query; 'what do you mean by good posture'? When using a computer, one may start with correct posture, but after some time change to slumping or forward bending posture. At a computer workstation, perfect posture is identified by observing the following points:

1. Head is upright and above the shoulders.
2. Eyes are in downward direction (30° from horizontal line of sight) without bending the neck.
3. Back must be supported by back-rest of the chair which encourages normal curvature of the lower back in lumbar region.
4. Elbows are bent at 90°, forearms horizontal, shoulders are relaxed, but not down.
5. Thighs are horizontal with a 90–110° angle at the hip.
6. Feet are fully supported and flat on the floor, or by a foot-rest.
7. Wrist is straight.

These are the points which represent perfect sitting position; however, no posture is ideal indefinitely. One must change his/her posture and position as often as possible if sitting on a chair working while working. This will ensure proper blood circulation and minimize the occurrence of any hazards, a typical work station is exhibited in Figure 11.1 as given below.

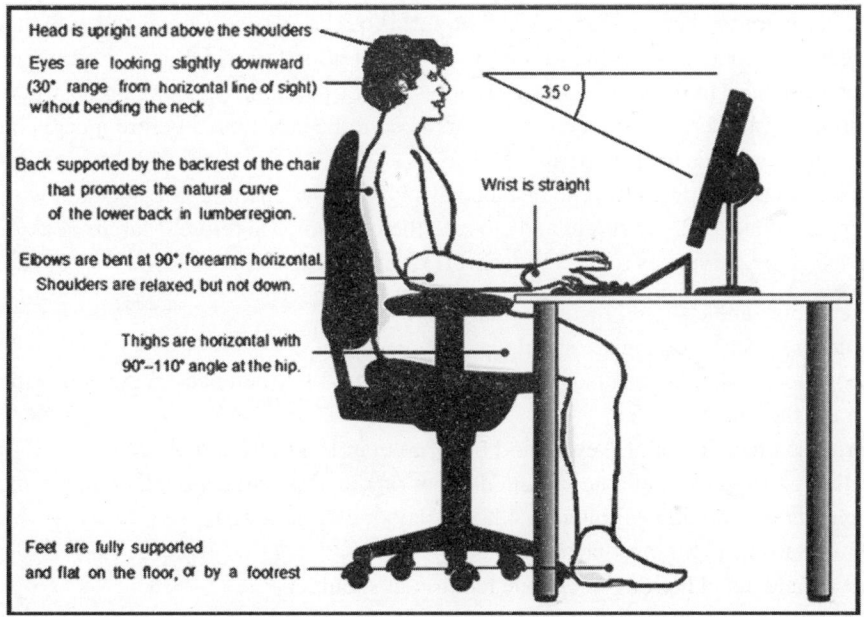

Fig. 11.1: A workstation for computer user.

While considering the dimensions and arrangement of the workplace, a crucial factor to be decided early in the design stage is, whether it is better for the operator to perform his tasks sitting down. Seating has many advantages, indeed Grandjean (1973) illustrated sitting down as being a natural human posture. If an operator is allowed to, he will get rid of the need to maintain an upright posture, which decreases the overall static muscular workload required to lock the joints of the foot, knee, hip, and spine, and consequently reduces his energy consumption. Grandjean had also pointed out that sitting is better than standing for the circulation. In standing position the blood and tissue fluids tend to accumulate in the legs. However, this tendency is reduced in sitting position, because the relaxed musculature and the lowered hydrostatic pressure in the veins of the legs offer less resistance to the return of the blood to the heart. Moreover, seating also helps the operator to adopt a more stable posture, which might help the operator to perform tasks requiring fine or precise movements, and it produces a better posture for the operation of foot controls. But, in spite of these advantages, still, the seated operator remains at inconvenience in some aspects like his mobility is severely restricted. If an operator needs to move up from the seat very repeatedly, fatigue will certainly occur.

A good seat is that, which enables the sitter to stabilize his body joints so that he can maintain a comfortable posture. Although the seat helps the operator to maintain a posture, which is appropriate for fine manipulation, this type of posture is unlikely to be useful if need his hands, arms to operate controls with large forces, or torques. Under such circumstances the operator's normal behavior will be to rise from the seat to enable him to take up the necessary posture, which again aggravates fatigue if repeated too frequently. A third disadvantage of being seated arises in case of the workplace happens to be enough vibrating that may be transmitted through the seat also reduces his accuracy of manipulative performance.

Ultimately, a prolonged sitting may itself cause health problems. For example, Grandjean (1973) pointed out that a sitting posture causes the abdominal muscles to slacken and curves the spine, in addition to impairing the function of some internal organs–particularly those of digestion and respiration. Furthermore, Pottier, Dubreuil and Mond (1969) have demonstrated that prolonged sitting (over 60 minutes) produces swelling in the lower legs of all sitter, which is caused by an increase in hydrostatic pressure in the veins and by compression of the things causing an obstruction in the returned blood flow. Therefore, numerous factors need to be considered before a decision is made whether an operator needs to sit to perform his task.

Having the keyboard low and some distance away from the operator is associated with reduced risk of injury than one at elbow height and close to the operator. Therefore, Marcus et al. (2002) has recommended some guidelines for computer workstation, these are as follows:

1. Inner elbow angle should be more than 121°.
2. Mean shoulder flexion angle should be 38°.
3. Elbow height should be at par with 'J' key height; 'J' key height >3.5 cm above the table surface should be avoided.
4. Horizontal location of 'J' key should be more than 12 cm from table edge.
5. Arm/wrist support should be provided either on the desk surface or chair arm rests.
6. Typing hours should be limited to 4 hours/day or 20 hours/week.
7. Downward head tilt must be avoided.
8. One should avoid holding telephone handset at shoulder.
9. While using the mouse, radial deviation should be limited to 5°.
10. Key activation force should be limited to 48 g.

11.2.2. Computer workstation recommendations

An office can be considered perfect if it provides adjustable chair, foot rest, adjustable work surface, split keyboard, wrist/arm support, document holder, augmented lighting, top of the screen at eye level, frequent, short rest breaks, worker or employee involvement in selecting equipment, setting up workstation and assistive devices. At the same time the office must avoid production incentives, production pressure, unrealistic deadlines, overtimes, supervisory and peer pressure and psychosocial stresses. A computer work station can be improved by providing these recommendations. Figure 11.2 exhibits the workstation with wrist support, a standing workstation, a split keyboard and a traditional keyboard.

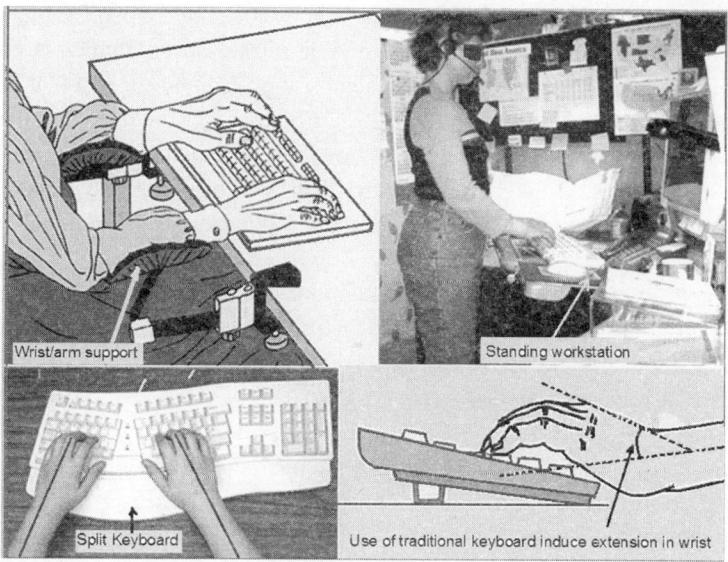

Fig. 11.2: Workstation with arm/wrist support, standing workstation, split and traditional keyboard.

11.2.3. Orthopaedic aspects of sitting and standing workstation

When seated, the primary support structures of the body are the spine, the pelvis, the legs and feet. The spine consists of 33 vertebrae joined together by multiple ligaments and intervening cartilages, the functions of which were discussed in chapter 8. For the convenience of description, the vertebrae are divided into four areas which correspond roughly to the changes in the shape of the spine. These areas are the topmost seven cervical, twelve thoracic and five lumbar vertebrae, followed by five fused sacral and four fused coccygeal vertebrae. From the point of view of seating design, the orientation of the lumbar and sacral vertebrae are important. Although when viewed from the front or back the normal, relaxed spine appears vertical, when viewed from the side its curved nature can be seen as shown in Figure 11.3.

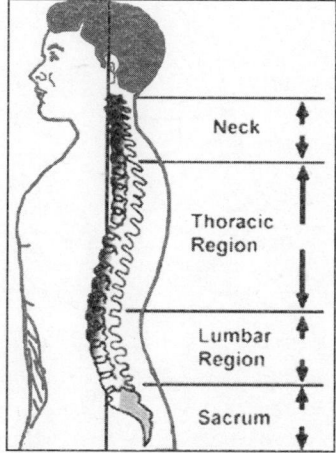

Fig. 11.3: Human spine viewed from the left side.

The top cervical curve bends forwards leading into a convex backward bend throughout the thoracic region. The lumbar region bends forward again, ending in the sacrum, which is positioned on the pelvis. Since the spine has evolved to his shape it seems reasonable to suggest that this natural shape is one which produces both the optimum pressure distribution over the discs and the optimum level of static load on the inter-vertebral muscles. It follows therefore that a seat in which the sitter has to adopt a different spinal posture is likely to cause mal-distributions in the disc pressure and will result in lumbar complains over a period of time. Supine curvature produced in different postures adopted by an individual is shown in Figure 11.4.

Vertebrae are kept in position by muscles and tendons. Therefore, any alteration to the natural shape of spine will produce corresponding stresses on the spinal musculature. This increase in muscle activity can be demonstrated and measured by recording the electrical potentials produced by the different muscles using electromyography (emg).

A study of this nature was carried out by Floyd and Ward (1968), who recorded the activity of four groups of muscles (in the neck, collarbone, in the back just below the shoulder blades and the lumbar region). Therefore again sitting up straight without any backrest produces a fair degree of activity in the lumbar curve (Figure 11.4).

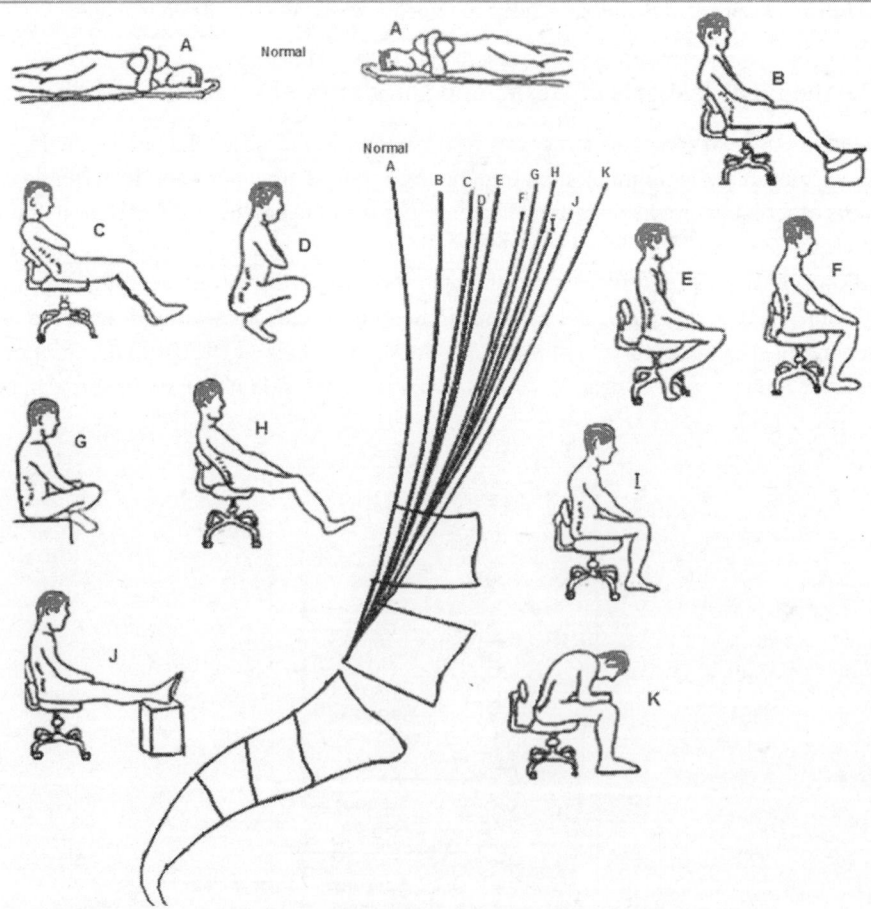

Fig. 11.4: Supine curvature produced in different postures adopted by an individual.

In case of typing or clerical job a forward, hunchback posture causes most activity to occur in the upper back and shoulder regions, again substantiating the orthopaedic observations. Floyed and Ward data indicate that providing a table support on which the author in able to rest his arms, does not reduce muscle activity. Indeed shoulder muscle activity on the side of the writer's preferred hand increased dramatically. Thus both the orthopaedic and muscular evidence suggest that:

1. An upright and forward leaning posture will cause fatigue,
2. The provision of a backrest will reduce some of the lumbar fatigue, and also obtuse angled backrest helps to stabilize the pelvis rotation.

11.2.4. Muscular aspects of sitting

When a workstation is designed wrong, the very first thought that comes in the mind of a person is about the chair. Therefore, a few common guiding principles for an appropriate office chair cover the following parameters:

1. The chair should have provision for lumbar support.
2. Height must be adjustable, so that chair can be comfortably used by persons of varying anthropometric parameters.
3. Chair should have appropriate width, so that large/thick persons can be accommodated comfortably or without any feeling of compactness.
4. Backrest should be adjustable, so that persons of a variety of sitting height can be accommodated in terms of lumbar support.
5. Seat depth should be well fitted or adjustable so that a large percentile of population can be accommodated.
6. Chair should have adjustable or removable armrests, so that can be used for other than computer job.
7. Chair having strong bases with five prongs is much suitable to maintain the sitting balance.
8. Air permeable fabric and cushioning using foam of appropriate density.

11.3. Seat Design

A number of principles for seat design can be extracted from the above discussion. These include:

1. The type and dimensions of the seat are related to the reason for sitting.
2. The dimensions of the seat should fit the appropriate anthropometric dimensions of the sitter.
3. The chair should be designed to provide support and stability for the sitter.
4. The chair should be designed to allow the sitter to vary stability for his posture but the fabric needs to resists slipping when there is fidgeting.
5. Backrests, particularly prominent in the lumbar region, will reduce the stresses on this part of the spinal column.
6. The seat pan needs sufficient padding and firmness to help distribute the body weight pressures from the Ischial tuberosities.

With regard to the motivation for sitting, seats may be divided just into the three groups: Easy or comfortable chairs for relaxation, in these chairs the criterion for an efficient chair ought to be, a loss of awareness of the seat and minimal discomfort of any part of the body's supporting structure. The second type of chairs is used for work, in which stability is an important consideration, requiring adequate support of the lumbar area and distribution of the body weight over the seat pan. The third type of chairs is known as multipurpose chairs. These may be essential for a variety of different function; for example, they could be used at a table, occasionally for working, or as spare chairs, and frequently need to be stacked one over another.

11.3.1. Anthropometric considerations

Apart from the function of the seat, it is necessary that its dimensions must fit the user. This aspect is already clear, and suitable anthropometric data also exist. Normally, these facts all relate to a naked sitter. So the presence of clothing and footwear will increase the dimensions by a proportional amount. Therefore, for designing an office chair, the anthropometric parameters of an individual are required to be considered, out of which some related to the seat design are shown in Figure 11.5 and described as below.

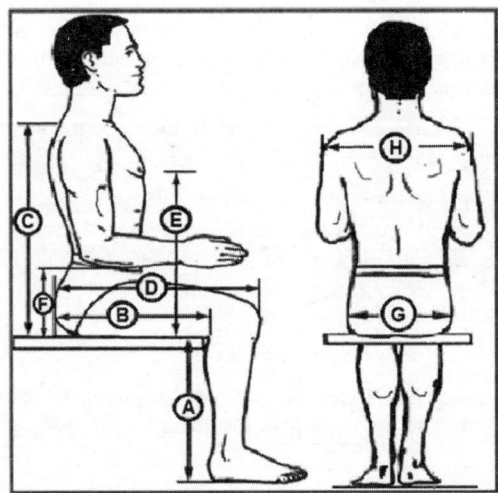

Fig. 11.5: Various anthropometric considerations while designing a seat or chair for workstation.

A. **Popliteal height**: It is the vertical distance between a footrest surface and the lower lateral surface of the thigh, just behind the knee, when the subject is seated with the knee flexed at 90°. This parameter is required to decide the height of a chair. Therefore in order to accommodate a population with larger range of popliteal height, nowadays chairs are provided with height adjustment.

B. **Buttock–knee length**: It is the horizontal distance between the maximum protrusion of a buttock and the anterior point of the knee of a seated subject when the knee is flexed 90°. This parameter is significant for deciding the depth of seat. The depth of a seat should be such that the backrest is properly supporting the low back, especially the lumbar region of the spine. In order to accommodate large number of population having large range of buttock–knee length, chairs are provided with adjustable seat depth.

C. **Cervical height, sitting**: It is the vertical distance between the sitting surface and cervical. This parameter helps to design the backrest height. Depending upon the cervical height, the backrest height is provided in the chair. However, an adjustable backrest is an appropriate solution to accommodate higher range of population.

D. **Buttock–popliteal length**: It is the horizontal distance between the maximum protrusion of a buttock and the posterior surface of the knee of a seated subject when the knee is flexed 90°. This parameter is helpful to decide the distance between the two successive seats in an aircraft, train, bus, car, etc. This parameter is very critical in case of cabin design, since passengers may have to travel long distance for a long duration. If the leg space is less, passengers will have very uncomfortable journey that may affect their performance at the destination.

E. **Chest height, sitting**: It is the vertical distance between the sitting surface and the level of the nipple.

F. **Elbow rest height**: It is the vertical distance between the sitting surface and the bottom of the elbow with the upper arm hanging freely and the forearm flexed at 90°. This parameter is very critical from the arm rest point of view, whether the seat is to be utilized in an office, at computer or in a cabin like aircraft or bus etc.

G. **Hip breadth**: It is the maximum horizontal breadth of the hips. This parameter is required to decide the width of a seat such that a large-sized person can also be accommodated.

H. **Bideltoid breadth**: It is the maximum horizontal distance across the shoulders at the level of the deltoid muscles. It is required to decide the width of the backrest so that majority of population can be accommodated.

After measuring the above parameters, various dimensions of a chair as shown in Figure 11.6 are decided subsequently, which are described as follows:

Seat height: Seat height is normally considered from 38 to 45 cm for easy chair and 43 to 50 cm for work chair. The seat height is adjusted accurately when the thighs of the sitter are horizontal, the lower legs are vertical and the feet are flat on the floor. This is due to reason that the soft undersides of the thighs are not suitable for sustained compression, and pressure from the front edge of the seat pan may become painful. Consequently, the seat height may become a limitation for a short legged person who would not be able to rest his feet on the floor if the seat height is greater than length of his lower legs. The reason for recommending the different seat heights for easy and working or multipurpose chairs is the manner in which they are likely to be used. The height of an easy chair should allow the legs to be stretched well forward since this is one of the preferred relaxing postures for the feet. However, in case of a working chair the sitter is likely to be in a more upright position with his feet flat on floor. If the sitter feels pressure near the back of the seat, one should lift the seat. If you feel pressure near the front of the seat, one should lower down the seat. The objective is to assure the evenly distribute the body weight.

Seat width: Seat width varies from 43 to 45 cm, because, starting from a small to a large-sized person is required to be accommodated. To decide the seat width, hip width is the appropriate dimension and since major gender differences are evident in this dimension, hence restriction may be for the female sitter with upper range of hip width.

Seat depth: Seat depth varies from 40 to 43 cm for easy chair and 35 to 40 cm for work chair. The importance of an appropriate seat depth is to ensure that all possible sitters find support in the lumbar area from the backrest. If the seat is any deeper than the thigh length of an individual sitter, the front edge of the seat will restrict him causing his lumbar area to approach to the backrest, consequently lumbar region remains unsupported. In addition the pressure sensitive areas at the rear side of the knee will compressed against the seat. Therefore, for a work chair, which will be used by a larger proportion of the population, the recommendation is to make the seat depth accommodate even shorter people. However, the only limitation of such a seat is that if a taller person is sitting on it, his knees will overhang slightly at the front. Providing the appropriate seat height and making the feet flat on the floor, avoids the compression fatigue in the thighs.

Seat Angle: This refers to the angle of the seat pan to the horizontal. If a seat pan is tilted backwards, it helps the sitter in two ways. First, by the force of gravity the sitters back moves towards the backrest, thereby so supports the back and reduces statics load on the back muscles. Secondly; the slight backward inclination of the seat pan helps to prevent the gradual slippage out of the seat. The same effect was observed in a study by Branton and Grayson (1967) over a long period sitting. Normally the seat angle varies from 19 to 20 degrees for an easy chair and about 3 degrees for a work chair.

Backrest height and width: Backrest height and width varies from 48 to 63 cm high and 35 to 48 cm wide respectively. The proposed dimensions of the backrest relate, quite simply, to the distance from the shoulder to the underside of the buttock (height) and to the shoulder width as shown in Figure 11.5. The height dimension certainly extends from the compressed seat in case of padding is present. As it has become clear, however, the linear dimension of the backrest is the only component of the query. Since the function of backrest greatly concerned with maintaining a relaxed spinal posture, hence the shape and the angle of the backrest are extremely important. In addition, since the curvature of the spine varies greatly from one person to another, a complex relation between height and shape arise. In another cases the sacrum and flashy parts of the buttocks which overhang or extend behind the sitter can be accommodated, while maintaining the curvature of the lumbar region by fitting with the backrest. Many researchers recommended that the backrest should have an open area or should recede (move away) just above the seat pan. Therefore a space of 15–20 cm is required to accommodate the buttocks in this way. Moreover, a chair with high backrest may not be suitable for some kind of jobs, for example, for typing at computer it may prevent full mobility of the arms. Therefore, in such type of jobs some researchers have suggested a chair with small backrest which supports only at the lumbar region only. On the other hand, the seat should also be sufficiently narrow, so that sitter is able to reach the armrests when they are properly adjusted.

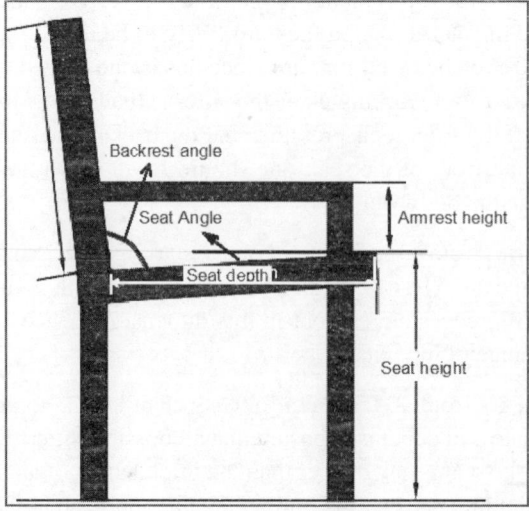

Fig. 11.6: A work chair as per the anthropometric considerations.

Backrest angle: This is the angle of backrest with respect to seat pan. It is extremely important from the posture point of view; it normally varies from 90° to 112°. Just like an angled seat pan, the angle of the backrest to the seat pan serves two purposes. First, it prevents the sitter from slipping forwards, and second, it causes him to lean against the backrest, which enables the lumbar and his sacrum parts of his back being supported properly. However, from an orthopaedics viewpoint, the appropriate angle would be about 115°. For example, Kegan and Radke (1964) established and observed that angle nearest to a natural lumbar shape. However, the sitting comfort responses obtained from seated laboratory subjects suggested that a less obtuse angle has consistently been found to be more comfortable. Overall, the tilt of the back support should allow one to sit with his/her upper body slightly tilted back (110° is usually recommended). Seat tilt can be adjusted to improve one's comfort, which will also affect the distribution of weight. A tilt of five degrees is usually recommended

Armrest height: It ranges from 21 to 22 cm above the compressed seat for easy chair. The primary function of armrests is to rest the arm in order to lock the body in a stable position. In an easy chair this is often accomplished by arm being used to support the head. Armrests are also found to be helpful to change position or as an aid to getting up from the chair. However, it should be considered that armrests may prove to be restrictive to the free movement of the arms and shoulders if they are incorporated in a working chair.

11.3.2. Cushioning and upholstery

The importance of cushioning was demonstrated by Branton and Grayson in the observational study of sitters in two types of train seats. Although the dimensions of the two seats were approximately similar, the type and subjective feel whereas the other appeared subjectively firm. After analyzing the number of jerks observed in the sitters and the length of time for which stable postures were maintained. Cushioning performs two functions, first it helps to distribute the pressure on Ischial Tuberosities and on the buttocks caused by the sitter's weight. Secondly, it allows the body to adopt a stable posture, as the body will be able to sink into the cushioning which then supports it. However the researchers also warn about too soft cushioning.

11.4. Engineering Anthropometry and Workspace Design

In our daily life, we do not like to wear garments or footwear that do not fit to our body size. If we wear an undersized shirt or jeans, it becomes terrible and suffocating, and work performance is affected badly. Similarly, if we use shoes of a wrong size (under- or over-sized), it becomes very uncomfortable for us to walk even for a few minutes. We feel uncomfortable and horrible when we sit on a chair that is too narrow. In another case one cannot reach and grasp an object if it is too high on a wall or too far across a table. In case of classroom benches if the gap between the seat and table is much larger than the buttock–popliteal length, the student will need to bend forward for writing during the exam, which causes low back fatigue and ultimately the pain. It is well known by all that the physical dimensions of a product or workplace should fit the body dimensions of the user. However, most of us may be surprised to know that inadequate dimension (mismatch) is the most common cause of error, fatigue and discomfort, because designers often ignore or forget this requirement or do not know how to put it into design.

Therefore, to match the physical dimensions of workplaces and product with the body dimensions of intended user, basic fundamental and quantitative data is required, which is provided by a scientific discipline called anthropometry. Anthropometry is the study and measurement of human body dimensions. Anthropometric data are used to develop design guidelines for heights, clearance, grips and reaches of workplaces and equipment for the purpose of accommodating the body dimensions of the potential workforce. Anthropometric data are also used in biochemical models in combination with information about external load to assess the stress imposed on workers joint and muscles during the performance of work. Example includes the dimensions of workstations for standing or seated work, production machinery, supermarkets checkout counters, passageway and corridors. Another examples are; the cabin inside dimensions of army tanks, cockpit area in aircrafts. The work force includes men and women who are tall and short, large or small, strong or weak as well as, those who are physically challenged or have health conditions that limit their physical capacity. Anthropometric data are also applied in the design of customer products such as clothes and footwear, automobiles, bicycles, furniture and hand-tools. Because products are designed for a variety of consumers, an important design requirement is to select and use the most appropriate anthropometric database in design. Since anthropometric data is also affected by gender, hence the products designed on the basis of male anthropometric data would not be appropriate for many female customers. When designing for an international market, applying the data collected from one country to other regions with significant size differences is inappropriate.

11.4.1. Factors affecting the variation in human dimension

There are number of factors responsible for the variation in human dimensions, however some of these are described as follows:

Age: It is quite obvious that the stature of a person changes w.r.t. age like; like from childhood to teen age, adolescence and adult. A number of studies have been conducted to compare the stature of people at each year of age. The data shows that stature increase maximum up to age of 25 years and starts decreasing after age about 40 years. Women show more reduction as compared to men. Contrary to stature, some other body dimensions for instance weight and chest circumference may increase through age 60 before reducing.

Gender: On an average, adult men are taller and bigger than adult women. On the other hand, average teen ager (12 year) girls are taller and heavier than their male counterparts. This is because; girls witness their maximum growth rate from ages 10 to 12 years, whereas boys observe their growth between ages 13 and 15 years. Girls continue to show clear growth each year until about age 17, whereas the growth rate for boys sustains gradually until age 20. On an average, adult female dimensions are about 90 per cent of the corresponding adult male values. However, significant differences exist in the magnitude between a male and a female on various dimensions.

Geography, culture and community: Body size and proportions vary greatly between different cultural, regional and national clusters. An anthropometric survey of black and white male in the U.S. Air force showed statistically matched average height, however blacks be apt to have longer arms and legs and shorter torsos than whites. Another comparison of data between the U.S. Air Forces with the Japanese Air Force established that the Japanese were shorter in stature height, but a very less difference was found for their average sitting height as compared to the American data. Similar differences were also found between the American, the French and the Italian anthropometric data. Similarly, a significant difference has been found for the stature height of male soldiers of Indian army. For example, average height of the Sikh regiment jawans is significantly higher than that of the Gorkha regiment jawans.

Occupation: Body size and dimensions also differs for people working in different occupational groups. For example, professional basketball players are much taller than most normal population. Similarly, wrestlers and power lifters are found much heavier than the normal. Data also show that truck drivers tend to be taller and heavier than average, and coalminers appear to have larger torso and arm circumstances. Variation due to occupation can be caused by a numbers of factors, such as nature and quantity of physical activity involved in the job, the special physical requirements of certain occupations and the self-evaluation and self-selection of individuals in making career choices.

Other factors: Other factors include generational or worldly variability. The author Annis (1978) studied the trend of change in stature of the American population since 1840 and noted that there has been a growth in stature of about 1 cm per decade since the early 1920s. Some of the possible reasons for this growth may be improved nutrition and living conditions. Other researcher like Kroemer (1968) noted that a person's body weight varies by up to 1 kg per day because of changes in body water content. Subsequently, the stature of a person may be reduced by up to 5 cm at the end of the day, mostly because of the effects of gravitational force on a person's posture and the thickness of spinal disks. Measuring posture in different positions may also influence results. For example, leaning erect against a wall may increase stature by up to 2 cm as opposed to free standing. Chest circumference changes with the cycle of breathing. Clothes can also change body dimensions of an individual.

11.5. A Case Study: An Investigation on Passenger Seat Design (Lower Side Berth) in Sleeper Class Coaches in Indian Trains

There is huge number of people who travel by trains from one place to another, for short or long distances. A number of activities are done by travellers, for instance consuming meals, liquids, gossips or chatting, resting and other entertainment doings while travelling. Generally, travellers spend the journey in reading and taking nap while resting on their respective berths. The traveller becomes weary and uneasy during journey if occupied berth is not comfortable. The different berth categories are: upper berth, lower berth, middle berth, upper and lower side berth adjacent to the window. The present study specifically focuses on lower side berth which forms by joining two halves that can be used both as a chair and a berth for sleeping by joining the halves. Usually, there occurs some uneven joining of these halves (Figure 11.7) in Indian trains and it puts a significant pressure on the back area of human body which may result into spinal and back ache issues. Hence, lower side berth is optimized at different dimensions of travellers travelling in A/C tiers and sleeper coaches and designed as per the easiness of human body to minimize tenderness of back issues.

Fig. 11.7: Image showing offset along thickness, gap along length and offset in width of two halves of the seat.

11.5.1. Qualitative data

The data pertaining to physical characteristics of a sample of 51 travellers and seat parameters were recorded by using a questionnaire. The most of information has been collected by direct observations and interviews as shown in snaps in which some back ache issues were highlighted by a number of travellers [Figure 11.8 (A) and (B)].

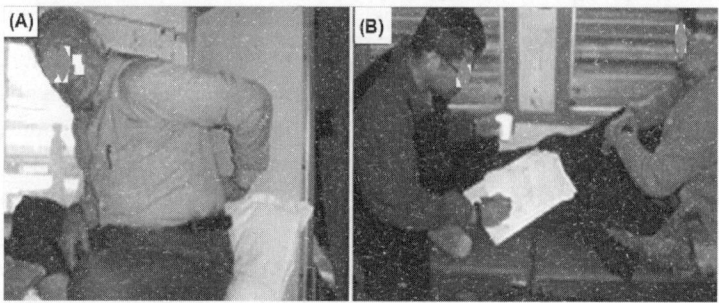

Fig. 11.8: (A) One senior citizen indicating back pain due to design of the berth, and (B) a young passenger filling feedback questionnaire.

All the information gathered from questionnaire revealed that there is a need to redesign the seat as per anthropometry of population. Firstly, all major parameters which influenced while designing a seat was studied thoroughly according to various established standards. Hence, a three dimensional representation of lower side berth was made in both of the forms, i.e., with and without joining the two halves (Figure 11.9 (A) and (B)).

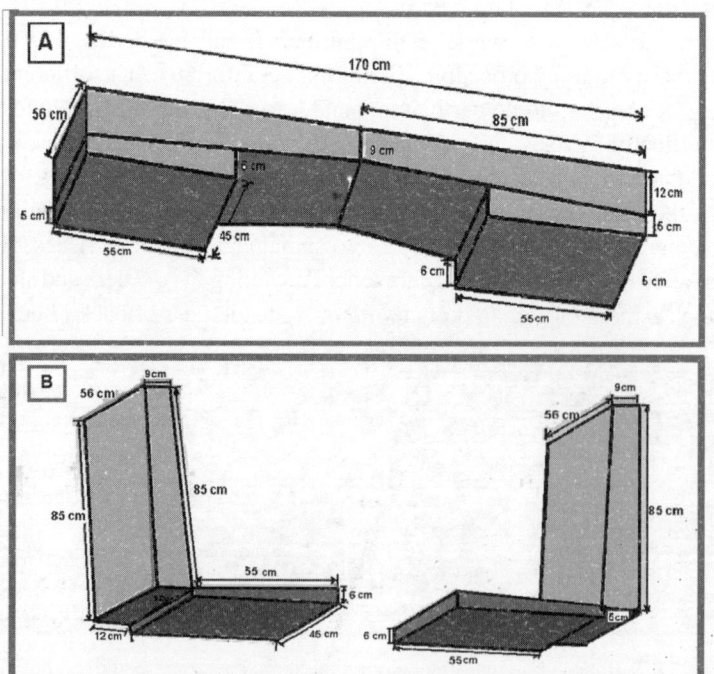

Fig. 11.9: Present design 3-D model of lower side berth (A) while sleeping and (B) while sitting.

11.5.2. Subjective response

The average age (in years), body weight (kg) and stature (cm) of 50 passengers were recorded as (38.61±13.09), (68.1±4.94), (163.4±3.41), respectively. When the train comes in motion and pick up the speed which results into vibrations which enters directly into human body and after continuous exposure it may cause some physical impact on body of travellers such as back ache, fatigue, uneasiness, etc. As per the results obtained for the back ache issues due to lower side berth, 85 per cent of A/C coach travellers and 90 per cent of sleeper coach travellers reported pain in their back during the journey. The reason specified was the non-aligned joint of two halves of lower side berth which makes the laying posture uncomfortable and make excessive pressure on the spine (lumber region) due to vibration while train is in motion. This continuous phenomenon may result into severe pain that can be a source of permanent injury of supine vertebras.

Therefore, about 75 per cent of subjects reported the need of some modification in the seat design as per their expectation of comfort. The percentage of suggestions provided by the passengers of AC and non AC coaches is given in (Table 11.1). About 59 and 50 percent passengers of AC and non-AC coaches respectively wanted increased length of the berth. Similarly 57 and 45 percent asked for more seat breadth, 78 and 75 percent demanded for increased cushioning. Whereas, 40 and 50 percent uttered upon improvement in locking system of the berth, 35 and 39 percent suggested to provide support under the seat pan and 7 and 5 per cent passengers of AC coach and non-AC coach respectively suggested to increase the sitting height of the cabin. These changes are only possible by making some strategic modifications in the internal layout of the coaches. Therefore, subsequently, dimensions and gap along length and breadth of side lower berth were measured on a sample of 50 berths for AC and non-AC coaches and these parameters are exhibited in Tables 11.2, 11.3 and 11.4. Subsequently the anthropometric data of a sample of 51 passengers who were travailing by occupying the side lower berth were taken (Table 11.5), so as to compare the same with the dimensions of the berth and design the berth accordingly.

Table 11.1: Suggestions offered by the passengers for improvements in the seat dimensions and design.

Feedback by Passenger in Coach	Suggestions Offered by Passengers to Improve the Existing Design of Side Lower Berth					
	Increase in Length	*Increase in Breadth*	*Improvement in Cushioning*	*Improvement in Locking System*	*Need Support under the Berth*	*Increase in Height*
AC	59%	57%	78%	40%	35%	7%
Non-AC	50%	45%	75%	50%	39%	5%

Table 11.2: Dimensions of side lower berth while seat is laid for sleeping ($N = 50$).

			Thickness (Tapered Seat)	
Sr. No.	*Length (cm)*	*Breadth (cm)*	*Starting (cm)*	*Ending (cm)*
1	85	54	9	12
2	84.5	53	8.6	12.5
3	85	55	10	13
4	84.4	53.4	8.7	13.1
5	84.8	54.1	8	12.6

Contd.

Contd.

Thickness (Tapered Seat)				
Sr. No.	Length (cm)	Breadth (cm)	Starting (cm)	Ending (cm)
6	84.7	52.4	9.3	12.9
7	85.3	55	9.2	13.7
8	84.5	54.7	9.4	13.1
9	84.6	55.8	9.1	12.9
10	85.1	55.6	8.9	13.7
11	84.9	54.8	8.2	12.8
12	84.6	52.1	8.2	12.8
13	84.1	53.9	8.1	13.6
14	84.3	55.1	9.9	13.8
15	84.1	55.2	9.4	12.7
16	84.6	54.2	9.2	12.8
17	84.9	53.7	8.6	13.6
18	84.7	52.8	8.5	13.9
19	84.5	54.1	8.1	13.7
20	84.2	53.5	9.5	12.9
21	84.5	54.7	9.4	13.1
22	84.6	55.8	9.1	12.9
23	85.1	55.6	8.9	13.7
24	84.5	53	8.6	12.5
25	84.7	52.8	8.5	13.9
26	84.5	54.1	8.1	13.7
27	84.1	53.9	8.1	13.6
28	84.3	55.1	9.9	13.8
29	84.9	54.8	8.2	12.8
30	84.8	54.1	8	12.6
31	84.4	53.4	8.7	13.1
32	84.8	54.1	8	12.6
33	84.7	52.4	9.3	12.9
34	85.3	55	9.2	13.7
35	84.5	54.7	9.4	13.1
36	84.6	55.8	9.1	12.9
37	85.1	55.6	8.9	13.7
38	84.9	54.8	8.2	12.8
39	84.6	52.1	8.2	12.8
40	84.1	53.9	8.1	13.6

Contd.

Contd.

Thickness (Tapered Seat)				
Sr. No.	*Length (cm)*	*Breadth (cm)*	*Starting (cm)*	*Ending (cm)*
41	84.3	55.1	9.9	13.8
42	84.9	53.7	8.6	13.6
43	84.7	52.8	8.5	13.9
44	84.5	54.1	8.1	13.7
45	84.2	53.5	9.5	12.9
46	84.5	54.7	9.4	13.1
47	84.5	54.1	8.1	13.7
48	84.1	53.9	8.1	13.6
49	84.3	55.1	9.9	13.8
50	84.9	54.8	8.2	12.8
Mean	84.61	54.19	8.80	13.22
SD	0.32	1.01	0.62	0.49

Table 11.3: Dimensions while seat is folded for sitting ($N = 50$).

Sr. No.	*Length (cm)*	*Breadth (cm)*	*Thickness (cm)*		*Height from Bottom (cm)*
			Start	*End*	
1	45	54	6	4.7	38
2	45.7	55	6.3	4.5	38.5
3	46.2	54.5	6.2	4.1	38.6
4	47	54.7	7.1	4.6	38.7
5	46.8	54.2	7.3	4.1	38.1
6	46.7	55.7	6.9	4.3	39
7	45.9	56	6.8	4.9	39.1
8	42.6	54.6	6.4	4.8	38.6
9	43.8	55.6	6.5	4.7	39.7
10	44.9	55.1	6.1	4.8	39.9
11	44.9	54.3	7.1	5.2	40
12	45.8	56.5	7.2	5.8	39.8
13	44.7	55.7	6.4	4.7	38.2
14	46.9	53.2	6.2	4.8	38.7
15	46.2	53.8	6.3	4.2	38.2
16	46.5	54.7	6.4	4.3	39.7
17	46.7	53.8	6.2	4.9	38.4
18	47	53.8	6.7	4.7	38.8
19	46.7	53.9	7.1	4.8	39.5

Contd.

Contd.

Sr. No.	Length (cm)	Breadth (cm)	Thickness (cm)		Height from Bottom (cm)
			Start	End	
20	47	55.4	7.2	5.2	33.7
21	46.8	54.2	7.3	4.1	38.1
22	46.7	55.7	6.9	4.3	39
23	45.9	56	6.8	4.9	39.1
24	42.6	54.6	6.4	4.8	38.6
25	47	53.8	6.7	4.7	38.8
26	46.7	53.9	7.1	4.8	39.5
27	45	54	6	4.7	38
28	45.7	55	6.3	4.5	38.5
29	44.9	54.3	7.1	5.2	40
30	45	54	6	4.7	38
31	45.7	55	6.3	4.5	38.5
32	46.2	54.5	6.2	4.1	38.6
33	47	54.7	7.1	4.6	38.7
34	45.8	56.5	7.2	5.8	39.8
35	44.7	55.7	6.4	4.7	38.2
36	46.9	53.2	6.2	4.8	38.7
37	46.2	53.8	6.3	4.2	38.2
38	46.5	54.7	6.4	4.3	39.7
39	46.7	53.8	6.2	4.9	38.4
40	47	53.8	6.7	4.7	38.8
41	46.7	53.9	7.1	4.8	39.5
42	43.8	55.6	6.5	4.7	39.7
43	44.9	55.1	6.1	4.8	39.9
44	44.9	54.3	7.1	5.2	40
45	45.8	56.5	7.2	5.8	39.8
46	47	55.4	7.2	5.2	33.7
47	42.6	54.6	6.4	4.8	38.6
48	47	53.8	6.7	4.7	38.8
49	46.7	53.9	7.1	4.8	39.5
50	45	54	6	4.7	38
Mean	45.81	54.66	6.63	4.74	38.72
SD	1.22	0.87	0.43	0.40	1.22

Therefore, re-designing of side lower berth based upon anthropometric data of passengers was proposed to overcome the problem. The mean ± SD (cm) of sample for anthropometric parameters were; age = 40.61 ± 14.09, stature = 166.4 ± 4.41 cm, sitting height = 82.00 ± 2.78 cm, popliteal height sitting = 43.44 ± 2.23, buttock–popliteal length = 43.074 ± 2.72, hip breadth sitting = 33.83 ± 2.54 cm, interscye breadth =33.27 ± 2.02. The stature height of sample of passengers varied form 159 to 174 cm as recorded in the data is shown in Table 11.5.

Table 11.4: Gap along: length, breadth and thickness.

Sr. No.	Gap Dimension (cm)		
	Towards Length	*Towards Breadth*	*Thickness Offset*
1	2	1	0
2	1	0	0.5
3	2.5	1	0
4	0	1.5	1
5	4	0.5	6
6	0	0.5	0
7	2	1	0
8	2.7	0.8	1
9	0	0	0.4
10	1.5	0.4	0.5
11	1	0	0
12	0.5	0.9	0.8
13	0	0.8	0
14	1.5	1	0
15	0	1	1
16	1.3	1.5	1
17	2.4	1.8	0.8
18	0	1	0.4
19	2.4	1	0
20	0	0.5	0
21	2	1	0
22	2.7	0.2	1.5
23	0	0	0.4

Contd.

Contd.

Sr. No.	Gap Dimension (cm)		
	Towards Length	*Towards Breadth*	*Thickness Offset*
24	1.5	0.4	0.5
25	1	0	0
26	0.3	1.5	1
27	1.4	1.8	0.8
28	0	1	0.4
29	2.4	1	0
30	4	0.5	6
31	0	0.5	0
32	2	1	0
33	2.7	0.8	1
34	0	0	0.4
35	1.5	0.4	0.5
36	1	0	0
37	0.5	0.9	0.8
38	0	0.8	0
39	2.4	1	0
40	0	0.5	0
41	2	1	0
42	2.7	0.2	1.5
43	0	0	0.4
44	1.5	0.4	0.5
45	0	0.8	0
46	1.5	1	0
47	0	1	1
48	1.3	1.5	1
49	2.4	1.8	0.8
50	0	1	0.4
Mean	1.23	0.76	0.65
SD	1.14	0.51	1.19

Table 11.5: Anthropometric dimensions for seat design according to NASA athropometric source book (NASA, 1978).

S. No.	Subject Age (yrs)	Weight (kg)	Stature Height (cm)	Sitting Height (cm)	Popliteal Height Sitting (cm)	Buttock Popliteal Length (cm)	Hip Breadth Sitting (cm)	Interscye Breadth (cm)
1	34	71	167.4	80.1	43.2	35.4	34.3	34.2
2	43	68	159.3	81.3	44.2	43.6	33.2	33.6
3	23	63	161.8	78.5	45.1	39.3	32.6	33.5
4	42	61	165.3	83.3	45.1	42.2	29.1	27.2
5	56	67	166.8	82.9	44.8	42.2	29.8	31.3
6	28	96	175.5	86.4	52.4	53	33.1	34.6
7	34	62	159.2	77.1	42.9	43.5	28.6	31.7
8	21	63	162.1	82.5	42.1	44.1	33.4	31.8
9	38	64	162.8	81.3	41.4	42.7	37.3	33.4
10	58	67	171.1	83.3	44.1	43.5	32.2	31.1
11	64	62	164.3	77.8	42.7	43.1	32.1	31.1
12	68	64	166.1	83.3	44.1	42.9	33.3	31.4
13	24	67	169.3	83.2	41.6	42.7	34.1	33.2
14	46	69	162.1	80.7	42.3	40.4	32.4	34.1
15	42	71	167.3	81.9	43.3	42.4	35.1	33.6
16	37	71	166.1	84.3	41.6	45.6	34.2	33.2
17	52	67	166.4	83.6	41.4	39.1	38.2	35.1
18	34	73	174.1	84.2	41.5	42.2	33.5	35.7
19	22	72	163	83.4	42.6	42.1	37.1	36.4
20	24	69	169	83.1	44.4	42.6	38.9	37.1
21	44	69	169	81.2	41.7	43.8	36.1	36.2
22	33	71	171	71.2	43.7	42.7	35.6	35.1
23	38	67	163	82.2	41.9	42.1	31.2	30.1
24	36	66	161	82.3	44.3	43.2	31.1	32.2
25	35	76	171	84.8	44.3	44.1	34.6	33.1
26	38	73	169	83.6	44.6	45.2	36.1	34.6
27	23	76	171.4	87.1	46.6	45.7	33.8	32.2
28	48	76	161	80.1	43.1	42.3	34.1	33.2
29	44	76	159	78.2	42.7	40.7	33.8	35.1

Contd.

Contd.

S. No.	Subject Age (yrs)	Weight (kg)	Stature Height (cm)	Sitting Height (cm)	Popliteal Height Sitting (cm)	Buttock Popliteal Length (cm)	Hip Breadth Sitting (cm)	Interscye Breadth (cm)
30	22	76	173.5	84.4	44.3	46.1	31.9	33.4
31	56	67	166.8	82.9	44.8	42.2	29.8	31.3
32	28	96	175.5	86.4	52.4	53	33.1	34.6
33	34	62	159.2	77.1	42.9	43.5	28.6	31.7
34	21	63	162.1	82.5	42.1	44.1	33.4	31.8
35	38	64	162.8	81.3	41.4	42.7	37.3	33.4
36	58	67	171.1	83.3	44.1	43.5	32.2	31.1
37	64	62	164.3	77.8	42.7	43.1	32.1	31.1
38	68	64	166.1	83.3	44.1	42.9	33.3	31.4
39	38	64	162.8	81.3	41.4	42.7	37.3	33.4
40	58	67	171.1	83.3	44.1	43.5	32.2	31.1
41	64	62	164.3	77.8	42.7	43.1	32.1	31.1
42	68	64	166.1	83.3	44.1	42.9	33.3	31.4
43	24	67	169.3	83.2	41.6	42.7	34.1	33.2
44	46	69	162.1	80.7	42.3	40.4	32.4	34.1
45	42	71	167.3	81.9	43.3	42.4	35.1	33.6
46	37	71	166.1	84.3	41.6	45.6	34.2	33.2
47	52	67	166.4	83.6	41.4	39.1	38.2	35.1
48	34	73	174.1	84.2	41.5	42.2	33.5	35.7
49	22	72	163	83.4	42.6	42.1	37.1	36.4
50	24	69	169	83.1	44.4	42.6	38.9	37.1
51	44	69	169	81.2	41.7	43.8	36.1	36.2
Mean	40.61	69.1	166.4	82.00	43.44	43.07	33.83	33.27
SD	14.09	6.94	4.41	2.78	2.23	2.72	2.54	2.02

In the laying position of existing design of the berth, the joint was coming exactly under the lumber region of the spine, therefore it was required to be shifted, so as to minimize its influence on human body, hence need some design modifications. Therefore, proposed design of lower side berth was made with a seat lengthwise of 115 cm and other 55 cm so as to move the joint from lower back (lumber region) to some other less sensitive body portion like hamstring muscles (Inner Thais) Figure 11.10(B). Also, the material of seat must be appropriately cushioning so as having better damping and compression qualities to absorb vibrations.

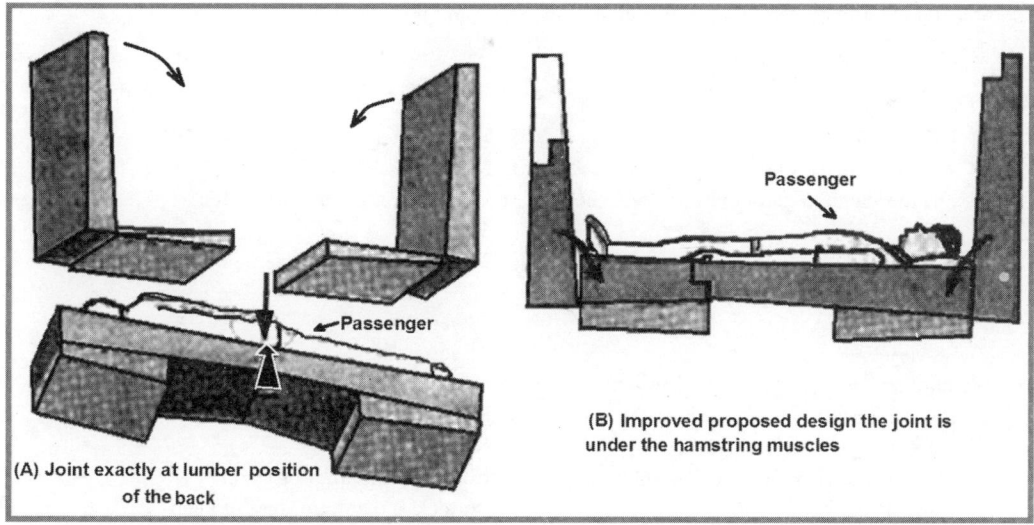

Fig. 11.10: (A) Position of joint exactly at lumber position of the back and (B) improved proposed design, the joint is under the hamstring muscles.

11.5.3. Recommendations

The various recommendations included that at the time of designing and fabrication of seats. It is required that the anthropometric parameters of normal population be considered in the software while reserving the lower side berth. The reservation should only be made as considering age, weight and height of individual so that allocation to adolescents, kids and people with height up to 170 cm. Overweight individuals should not be allowed for reserving the side lower berth to avoid any damage to locking system.

11.5.4. Conclusions

The present case study concluded that the side lower berth occupied travellers suffer from spinal issues generally while resting or sleeping. The joining of two halves made a joint under lumber region of spine of individuals and it is a severe factor of causing low back pain. Hence, there is a need to re-design the seat to change joint location to minimize the back pain issues. The joining region should be covered by a superior quality of cushioning material and provide some locking facility for all lower side berths. Also, a dense cushioning should be made to eliminate or significantly reduce hard feeling of steel reinforcement.

Multiple Choice Questions

1. For an ideal posture the thighs should be horizontal at the hip w.r.t. verticle trunk position.
   - (a) 70°–90°
   - (b) 90°–110°
   - (c) 80°–90°
   - (d) 90°–100°

2. At a computer workstation, ideal posture of elbows should be bent at
   (a) 90°   (b) 80°
   (c) 70°   (d) 95°

3. Prolonged sitting refers to a time
   (a) Up to 40 minutes   (b) >60 minutes
   (c) 40–50 minutes   (d) None of these

4. What is the standard guideline for mean shoulder flexion for computer workstation
   (a) 38°   (b) 48°
   (c) 58°   (d) 28°

5. The human spine is made up of
   (a) 36 vertebrae   (b) 33 vertebrae
   (c) 38 vertebrae   (d) 34 vertebrae

6. The popliteal height is distance between
   (a) Footrest surface and lower lateral surface (b) Knee and footrest
   (c) Knee and lower lateral surface   (d) None of these

7. The buttock knee length is the horizontal distance between maximum protrusion of
   (a) Buttock and knee   (b) Buttock and anterior part of knee
   (c) Buttock and footrest surface   (d) None of these

8. The vertical distance between the sitting surface and the bottom of the elbow is termed as
   (a) Elbow rest height   (b) Cervical height
   (c) Hip breadth   (d) None of these

9. The maximum horizontal distance across the shoulders at the level of the deltoid muscles is
   (a) Cervical height   (b) Hip breadth
   (c) Bideltoid breadth   (d) None of these

10. The seat height for a work chair should be
    (a) 43–45 cm   (b) 43–55 cm
    (c) 41–50 cm   (d) 43–50 cm

11. The armrest height is recommended as
    (a) 23–25 cm   (b) 21–22 cm
    (c) 24–26 cm   (d) None of these

12. Which of the following factors affect s variation in human anthropometry
    (a) Age   (b) Weight
    (c) Gender   (d) Region
    (e) All of these

Answers

1. (b) **2.** (a) **3.** (b) **4.** (a) **5.** (b) **6.** (a) **7.** (b) **8.** (a) **9.** (c) **10.** (d)
11. (b) **12.** (e)

Chapter
12

Physical Stresses

12.1. Introduction

The present chapter considers the aspects of physical stresses due to ambience, the omnipresent sensations to which worker is exposed from diverse sections of machinery or from other systems in his workplace. These ambiance pollutants mainly include vibration, the noise, the temperature, the illumination, etc. But the current chapter is confined to a brief description.

12.2. Vibration

Vibration is the most fundamental environmental factor. Therefore, it is essential to understand the definitions and parameters of structure-borne vibration, and it will help us understand the processes of other environmental parameters – particularly noise and, to a lesser extent illumination. A body is said to vibrate when it describes an oscillating motion about a fixed position. As like sound, the number of times a complete motion cycle takes place during the period of one second is referred to as the frequency, which is measured in hertz (Hz). The motion can consist of a single component occurring at a single frequency, as with a tuning fork, or of several components occurring at different frequencies simultaneously, e.g., with piston motion of an internal combustion engine.

Vibration signals in practice usually consist of many frequencies occurring simultaneously, so that it is not possible to see immediately, just by the examination of; the amplitude–time pattern, how many components there are and at what frequencies they occur. These components can be revealed by plotting vibration amplitude against frequency, the process being known, as in case of sound as frequency analysis. The graph showing the vibration level as a function of frequency is known as frequency spectrogram, and the vibration amplitude is the characteristic which describes the severity of vibration.

12.2.1. Frequency ranges of significance

(a) The principal hazards associated with vibration are cold body vibration and condition known as vibration induced white finger (VWF), the physiological aspects of which are discussed in another subsequent section. The human body is most sensitive to vibration in the frequency range 1–80 Hz and is principally subjected to vibration in 3 supporting surfaces, viz., the feet of a person while standing, the buttocks of a seated person, and the supporting areas of a person lying down. In the longitudinal direction, i.e., feet to head, the human body is most sensitive to the vibration in the frequency range 4–8 Hz. In the transverse direction, however, it is the most sensitive to the frequency range 1–2 Hz. Vibration induced white finger is a condition generally associated with the use of vibratory hand tools, the frequency of the hand tool being the significant factor.

12.2.2. Definition of vibrations

The movements which a body makes about the fixed point are called vibrations. These movements can be regular, like the motion of a weight on the end of spring, or it can be random in nature. The vibrations which are experienced from machines are generally very complex but regular motion. However using appropriate analysing technique, any complex motion can be defined in terms of a number of simple components. The movement of a vibrating body can normally be defined in terms of two parameters: the vibration frequency and intensity.

The frequency is essentially an indication of the speed of movement, and is measured in cycles per second or hertz (1 cps = 1 Hz). Therefore, vibrating body is said to have moved through one cycle when it has moved from its fixed point to its highest deviation to its lowest deviation, and then returned to the position of the original fixed point. The number of times it does this in a specified time (usually 1 sec) is its frequency of motion. The most basic of all motions, known as sinusoidal motion, are shown in Figure 12.1 given below.

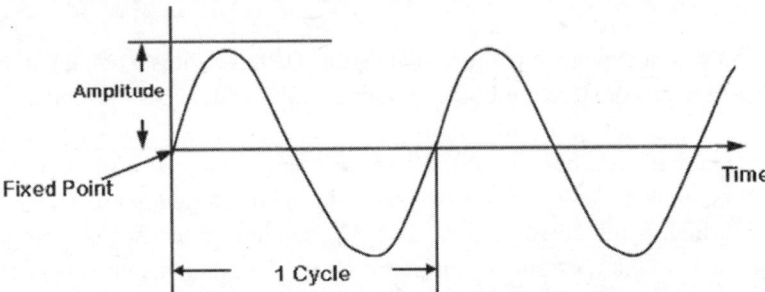

Fig. 12.1: Frequency and amplitude of vibration in sinusoidal form.

There are a number of ways through which the vibration intensity can be measured. However, amplitude and acceleration are normally the used units. The amplitude is the maximum distance that the body moves from its starting point, and it is expressed in the normal units of distance, i.e., inches, feet, cm or mm. But, nowadays it is common practice the intensity is expressed in terms of the body's acceleration, the units being meters per second or 'g' units. Since it is the acceleration which a body needs before it can overcome the force of gravity and lift from the earth's surface, therefore, it is expressed in 'g' units (where, 1 g = 9.81 m/sec^2). The measurement is normally made using a small acceleration sensor, called an accelerometer, placed on the vibrating body. Each of these parameters is related, hence;

$$\text{Acceleration in 'g' units} = [4\pi^2 f^2 / 981]$$

where, 'f' is the frequency of vibration, and 'a' is the vibration amplitude in cm. The direction of vibration can be defined in terms of three coordinates; horizontal frontal or fore-aft (x), sideways or lateral (*y*) and vertical (*z*). For a human being these coordinates are supposed to pass through the body and are related to the back, chest, sides, feet and head. It is important to notice that the physical directions of motion for an operator who is lying down are different from those of a standing man (Figure 12.2).

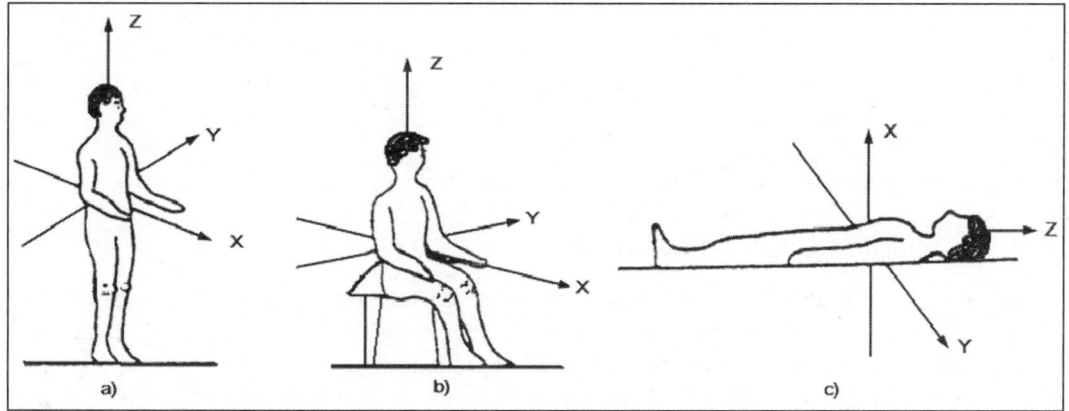

Fig. 12.2: Vibration coordinates in (a) standing, (b) seated and (c) laying position.

12.2.3. The limits of vibration perception

Before considering the different ways in which vibration can affect the operator, it is useful to have some idea of the upper and lower limits (of tolerance and perception) which the human body can accept. In the past investigations of how humans respond to vibration, many researchers attempted to obtain authoritative data concerning the lowest level of vibration needed for perception. But when all of these data were collated, it demonstrated that the threshold of the perception of acceleration varies over different frequencies. These threshold curves were spread over an intensity range of about 10:1. This may be partially influenced due to the use of different apparatus, experimental methods, instructions, subject populations and statistical analysis.

12.2.4. The effect of vibration on health

The human body is an extremely complex structure composed of different organs, bones, joints, muscles, etc. Different parts have different resonant frequencies. Therefore, if the body is vibrated by strong vibration stimuli with frequencies close to the (Frequencies at which the response amplitude is a relative maximum are known as the system's resonant frequencies) structural damage due to vibration amplification may soon occur. A basic indication of the primary resonance characteristics of different parts of the human body is exhibited in Table 12.1. These data, however, should only be taken as a rough guide since body resonances are likely to be affected by many factors such as the muscle flexibility, the dimensions of the bones and the amount of fatty tissue.

Mechanical vibration can cause health injuries which broadly fall into two categories: the first includes changes due to the vibration frequency, which occurs as a result of different body structures being excited at or near their resonant frequencies. The second category is more related to the 'impact' of the stimulus on the body, i.e., the vibration intensity and duration.

Table 12.1: Resonant frequencies for different parts of seated bodies.

Sr. No.	Body Part	Approximate Resonant Frequency (Hz)
1.	Head to Upper Torso	20
2.	Arm Shoulder System	5
3.	Thoracic Abdominal Complex	3
4.	Spinal Column	5
5.	Pelvis (Hips to Knees)	5
6.	Pelvis (Knees to Ankles)	9

12.2.5. Frequency-dependent health effects

Injuries which are caused by vibration frequency usually occur after expanded exposure to the vibrating stimulus, mainly in the higher frequency ranges. The type of appliances which could cause such effects include many of the kinds of powered tools used in industry, for example, chipping hammers, screwdrivers, rivet-drivers, road-drills, stone-breakers and the forester's chain saw. Examples of some of these effects are described in Table 12.2. Most injury reports indicate that the damage occurs to the peripheral blood and nervous stems in the exposed part of the body. Intense vibration from hand tools can be transmitted to the fingers, hands and arms of the operator both from the handles of the machine and from structures which are held or steadied by the hand and vibrated by the appliance. The complaints usually include many of the symptoms like intermittent numbness and clumsiness in the fingers, intermittent blanching of either all or part of the extremities and a temporary loss of muscular control of exposed part of body (Agate, 1949). These symptoms are commonly referred to as white finger disease or Reynaud's disease, and are associated with cumulative exposure to vibration with hand held tools. Relief from all or some of these symptoms can be gained after prolonged abstention from exposure to the vibration stimulus. However, they are likely to re-appear rapidly with restored exposure. Spinal damage and gastro intestinal refers to the chronic disorders associated with repeated exposure to whole body vibrations.

Table 12.2: Frequency-dependent health effects.

Sr. No.	Range of Frequency (Hz)	Health Effect
1	0.1–2.0	Motion sickness
2	4–90	Speech interference
3	3–140	Blurring of vision
4	3–1000	Interference with tasks
5	20–100	Spinal damage *
6	10–140	Gastrointestinal disorders
7	120–12500	Reynaud's disease

   It is difficult to estimate the extent of such damage in industry. However, based upon the assessment studies of large groups of different types of workers using vibrating tools over periods of time, Guillemin and Wechsberg (1953) reported that the incidence of neurovascular damage due to vibration exposure varied between 89.5, 66.2, and 11.6 per cent and as low as 0.2 per cent from different groups. This wide range probably exists due to various factors, such as a) differences between actual daily exposure times, b) large variations in the frequency and intensity of the vibration stimuli, c) the use of various types of materials and d) varying individual susceptibility to the vibration. From their review of the available information Guillemin and Wechsberg (1953) concluded the following: Mechanical vibration endured repeatedly over long periods of time by human subjects produce disabilities that differ in nature and extent in three broad frequency ranges. Those below 1,000 per minute (approximately 16 Hz), occurring with large amplitudes (up to an inch or more), typically produce injuries in the bones, joints and tendons. In the range 2,000–10,000 cycles per minute (33–166 Hz), with amplitudes of tenths or hundredths of an inch, the symptoms are mainly cardiovascular. However above 20,000 cycles per minute (over 300 Hz), where amplitudes are measured in thousandths of an inch, neurovascular disturbances are accompanied by continuous burning pain which may be the predominant symptom. These conclusions have yet not been altered drastically in the light of any new research information.

12.2.6. Physiological effects due to intensity and duration

Intensity dependent effects of vibration occur mainly as a result of body parts moving against one another. In addition, large organ pull on supporting ligaments and it is also possible for soft tissues to be crushed. Different cases of these injuries can be seen in normal working life. For example, Rosegger (1960) carried out a comprehensive survey of the health complaints of 371 tractor drivers and confirmed two forms of damage resulting from long period exposure to vibration as tractor driver; stomach complaints and disorders of the spine (particularly in the lumbar and thoracic regions). Both the number and severity of such cases improved proportionately with the length of service as a tractor driver. Traces of blood in the urine have also been found in the drivers of these vehicles. This is probably due to kidney damage, although this breakdown could be caused as much as by vibrating the kidneys as by damage through high intensity vibration as such.

The effects of these types of stimulation have been investigated at a descriptive level by Magid et al. (1962), who asked ten subjects to describe their feelings about different intense vertical stimuli in the frequency range 1–10 Hz. Many of their subjects reported chest pain between 4 and 8 Hz stimulation (at about 0.5 g) with all subjects reporting this type of symptom pain were produced at lower frequencies (3–6 Hz) but the number of reports was fewer than for chest pain. At 10 Hz (1.2 g) musculoskeletal and abdominal pain was reported.

12.2.7. The effects of vibration on performance

Until or unless there is appropriate damping arrangement for vibration and because of natural tendency of body parts to vibrate in compassion with any vibrating tool or machinery on which they may rest, the effects of vibration on performance lie mainly in the degradation of motor control. This might be the control of a limb like reduced hand control, e.g., bus conductors or may be of eyeballs causing fixation difficulties and blurring. Little evidences exist to suggest that vibration can detrimentally affect central, intellectual processes.

12.2.8. The effects of vibration on vision

It is very obvious that a clearly formed image of an object can only be perceived in the visual cortex of brain if a stable image falls on the retina. If the image is moving it will fall on varying sets of receptors in the retina, thereby producing the signal of overlapping and confused images, which makes it difficult to detect any of the object's detail. This is particularly so if the image falling on the retina oscillates with a relatively large amplitude. Hence there are three situations that can exist to cause a moving image to appear on the retina:

(a) When the object alone is vibrating while the observer stationary.
(b) When the observer is vibrating while the object is stationary.
(c) When both the object and the observer are vibrating.

In this last case the resultant degree of blurring will be determined by the extent to which the man and object are vibrating in phase (in other words whether they go 'up' and 'down' at the same time), and of-course the frequency of the whole body vibration.

12.2.8.1. *Vibrating the object*

The detrimental effects of vibrating the object alone appear to be related to the frequency at which the object is vibrating. At low enough frequencies (<1 Hz) eyes are able to compensate for the movement by tracking the object and are thus able to produce relatively stable images on the retina. However, over time, this is likely to cause fatigue in the muscles which control the eye movements. As the frequency of the vibrating object increases, performance usually measured in terms of reading or of tracking error is likely to deteriorate because, even if the eye muscles attempt to do so, they are unable to make this tracking adequately. This occurs up to some critical frequency, after which performance appears to improve slightly. This noticeable improvement in performance is considered to be due to the eye being unable to track the movement of the object efficiently, so producing a slightly blurred image. Until the object movement is excessively great, however, the blurring is unlikely to cause performance decrements. The precise critical frequency at which this occurs is uncertain. However, various authors have suggested this between 3 and 4 Hz.

At higher frequencies and intensities any performance reduction is due solely to blurred image being produced on the retina. For example, Griffin and Lewis (1978) reported a study which indicated that the number of reading errors, the reading time and the subjective report of reading difficulty increased as the object frequency increased from 5 to 30 Hz, although the details of the vibration intensity were not reported.

Since the degree of blurring is related to the amount of image movement on the retina, it could reasonably be expected that the amplitude of the object vibration will also have a direct effect on performance. Experimental results suggested, however, that this is only the case in particularly unfavourable circumstances. For example Griffin and Lewis (1978) report a series of experiments carried out by Crook et al. (1974) who vibrated their objects at different amplitudes but at a constant frequency. In one experiment, when vibrating an object at 17.5 Hz, with amplitude between 0.15 and 0.75 mm, the author found no significant change in either reading errors or in reading time over the different amplitudes. In later experiments, however, they showed the amplitude of the vibrating object to be an important determinant of performance when either the type or the level of room lighting was poor. They concluded that, providing other environmental conditions are optimal, vibration amplitude of up to 0.5 mm did not weaken performance. The extent to which the amplitude of the object vibration affects performance will be *diminished* somewhat by the distance of the observer from the object. Because the image of an object which is nearer to the observer appears larger than an object more away, if the vibration amplitude remains constant as the object moves away from the eye, the amount of blurring is likely also to be reduced. For this reason the effect of vibration amplitude is often discussed in terms of the angle which the object makes at the eye (which takes account of both the object's apparent size and its distance from the observer). In this respect, Griffin (1976) suggests that a vibration intensity that produces a blur of plus or minus 1 minute of arc at the eyes represents the 'threshold of blur' level. It would probably be over-restrictive in all but the most critical conditions the author suggested that displacements of greater than plus or minus 2 minute of arc may, depending on the difficulty of the task, start to affect reading ability.

12.2.8.2. *Vibrating the observer*

In the case when only the observer is vibrating, the visual effects produced are frequently alike to those resulting from vibrating the object alone. The blurred image in this case, is because of the movement of the eyeball itself relative to the object. At low frequencies (less than about 2–3 Hz) the whole body moves in compassion with the motion, consequently the head and eyes move in a similar fashion. At higher frequencies, however, the head and eyes are likely to vibrate and thus produce blurring effect which is stronger than would have been predicated from the intensity of the vibration input alone. For this reason, therefore, it is important to understand the resonance characteristics of the head and eye relative to the rest of the body.

The vibration effects were briefly described at the beginning of this section. Depending on the frequency, any part of the body may treat the incoming vibration in one of three ways: (a) the internal structure can be such that it reduces the level of vibration as it is transmitted through it (as, for example, would a soft cushion), (b) it can transmit the vibration in a 1:1 ratio (in other words, the input level to the body is not affected) and (c) it can highlight the vibration (this occurs at the resonant frequencies of the structure). The resonance characteristics of the body can be measured, therefore, by placing two vibration sensing devices at each end of the object, one measuring the level of vibration input (I) and the other the level of vibration which has been transmitted through the body (T). *The ratio, T/I, provides an index of the object's transmissibility. Thus a ratio of greater than 1 indicates resonance, while a ratio of less than 1 indicates attenuation or reduction.*

Using this procedure on a human body, Lee and King 1971 placed a vibration sensor at the feet (input) and on the head (output) of a standing man, and plotted a transmissibility curve over different frequencies which indicate peak head resonance at about 8–10 Hz as shown in Figure 12.3 (A) whereas Figure 12.3 (B) demonstrates eye-to-head transmissibility. From these two sets of results, a clear eyeball resonance can be seen which starts at about 20 Hz, while Thomas (1965) has demonstrated a slightly higher eyeball resonance at about 30 Hz. Taken together, these data suggest that the eyeball will begin to accentuate the input vibration if the vibration frequency is above about 20–25 Hz. At lower frequencies, between about 8 and 10 Hz, the head and neck resonance is also likely to cause the eyes to move at higher intensities than those which enter the body, as long as the movement is not too great. However, this may still be compensated by the eye muscles to allow a relatively stable image to fall on the retina. In order to ascertain which is better or worse to vibrate the object or the observer, Dennis (1965) investigated the effects of two vibration conditions (frequency range 5–37 Hz) on a task of reading different numbers on a display. In the first condition the numbers were vibrating and the observer or subject remained stationary; in the second the subject was vibrating with the display stationary. Dennis's results signified that below 6 Hz vibrating the display produced higher visual impairment than when the subject was vibrated. However above 14 Hz, the outcome was just reverse. This implies, therefore, that, above approximately 14 Hz the decrement in performance is primarily subject-based (probably resulting from eyeball resonance), whereas below this level any decrement is likely to be display-based.

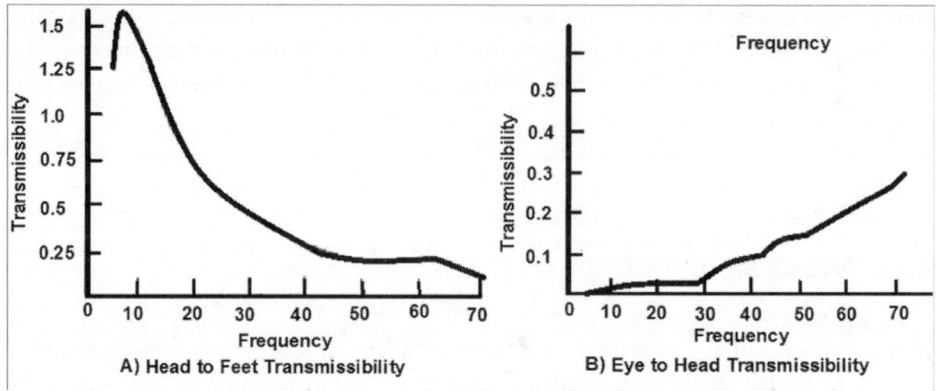

Fig. 12.3: (A) Head to feet transmissibility, (B) eye to head transmissibility.

12.2.8.3. *Combined effect when both object and observer are vibrating*

The case of combined object and observer vibration exist in many practical situations. A very ample example is transport means, the vibration are being applied to both the observer and to the object which he is required to view. In such cases if the eyes remain in a fixed position in the head and the vibration frequency is low enough, the motions of the object would be accompanied by similar motions of the man. In this case the image of the vehicle interior, for example, would maintain a fixed position on the retina. Unluckily, though, vibrations are not always of low frequency. Because of different resonance characteristics of the man and parts of his machine or vehicle there are differences in the amplitude, phase and perhaps direction of the observer and object motions which cause complications. For example, even if the observer and object are vibrating with the same frequency and amplitude, if they are vibrating out of phase (that is, if one 'goes up' while the other 'goes down') their relative amplitude will be twice that of the individual amplitude. Because of these problems, very few well-controlled studies have been carried out to investigate combined object and subject vibration effects, but a few studies which were accessible have been reviewed by Griffin and Lewis (1978).

12.2.9. The effects of vibration on motor performance

Where the operator can normally balance for small amounts of vibration when carrying out a visual task, decrements in motor performance are less easily conquered. Thus is due to the reason that the vibration tends physically to move the operator's limbs, possibly out of phase with the rest of his body. Any balancing has to be done by tensing different muscles to steady the limb just like the seated body seeks stability. Even if the operator is able to balance for the motion, consequently, it is likely to cause fatigue. The task most commonly used to investigate the effects of vibration on motor performance has been some type of tracking task. The subject is given a control which operates part of a display such as a spot on an oscilloscope, and he is required either to keep the otherwise moving spot stationary or to make it follow some predetermined pattern. Clearly, this is a type of activity which is used constantly in different working situations, particularly by such operators as drivers, pilots and astronauts. It is also a task which has been consistently shown to be detrimentally affected by vibration. When the vibration stimuli is below 20 Hz, some evidences revealed that any decrease in performance is related to the amount of vibration which is transmitted through the body, e.g., Buckhout (1964) has shown that for vertical vibration at 5, 7 and 11 Hz, tracking error was highly and positively correlated with the amount of vibration transmitted to the sternum. Vibrations transmitted through the seat back to the shoulder and the chest areas are likely to play a part in reducing an operator's performance on a motor tracking task. This effect will be *emphasized* if the operator is using a seat belt, which tends to force the body back into the seat and into the backrest (Lovesey, 1975). With regard to specific intensities there is good agreement amongst investigators that, for a given vibration spectrum performance is progressively degraded as the level of vibration is increased above a threshold. The threshold level, however, is extremely variable and depends, amongst other factors, on the type of task (primarily its difficulty) and on the operator's motivation and workload. These factors affect his ability to compensate for the effects of the vibration on his limb control. The frequencies which effect motor performance lie between 4 and 5 Hz with decrement in performance becoming progressively smaller as the frequency becomes higher or lower (Huddleston, 1970). In addition to the frequency, the direction of the vibration stimulus will also determine the direction of the control task which is affected. For example, although vertical vibration has clear effects on a vertical tracking task, it has little effect if the operator is required to track an object in the horizontal axis. This has obvious implications for the design of a control system in a vibrating environment.

12.2.10. Permissible exposure limits.

The daily exposure limit is has two aspects; a) Daily Exposure limit value (ELV), and b) Daily Exposure action value (EAV), these are briefly described as follows:

(a) Daily Exposure limit value (ELV) – It is the maximum amount of vibration to which an employee may be exposed in any single day. the daily exposure limit value for hand-arm vibration is 5 m/s2 A(8); whereas for whole body vibration the daily exposure limit value is 1.15 m/s2 A(8);

(b) Daily Exposure action value (EAV) – It is that level of daily exposure to vibration above which the employer is required to take actions to reduce exposure. Therefore, for hand-arm vibration the daily exposure action value is 2.5 m/s2 A(8), whereas for whole body vibration - the daily exposure action value is 0.5 m/s2 A(8).

There are mainly two types of vibration exposures. One is whole body vibration; vibration transmitted through a supporting area to whole body like sitting on a chair, standing on vibrating ground 0.5 - 80 Hz; ISO 2631-2, sea sickness 0.1- 0.5 Hz, ISO 2631-1. The second is hand arm vibration; vibration transmitted through hand just in case of holding a vibrating tool like hand drill, 5 - 1500 Hz, ISO 5349-1. However a number of attempts have been made since 1964 to combine all the preceding data on the effects of vibration on health, performance and comfort, to compose an accepted standard for human exposure to whole body vibration. Such a standard was finally produced in 1974-ISO 2631: Guide for the Evaluation of Human Exposure to Whole Body Vibration. This guide laid down limits of exposure to both vertical and lateral vibration under the three principle criteria: a) the preservation of health (exposure limit) working efficiency (fatigue-decreased proficiency, FDP) boundary and comfort (reduced comfort, RCB) boundary. As originality the levels for each criterion, within the frequency range 1–80 Hz, are defined in terms of the maximum time for which an operator should be exposed to the vibration from 1 minute to 8 hours. For any exposure time, exposure limit = 2 × FDP, and RCB = FDP/3.15. Some of the levels produced for vertical vibration are illustrated in Figure 12.4.

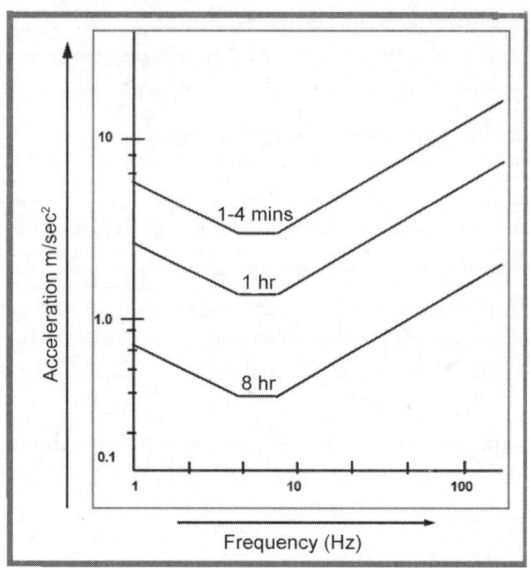

Fig. 12.4: Vertical vibration exposure limits for fatigue decreased proficiency (FDP).

12.3. Occupational Noise Exposure

Occupational noise exposure refers to the exposure to noise when a person is occupied in some activities that are performed mainly at work palace, at home or at some public places. Since one third time of the day is spent by an individual at work place, occupational exposure to noise is an important issue. If we consider a person employed at an industry which manufactures some auto-parts and hand tools. The manufacturing processes in such an industry generally include moulding machines, forging presses, punching/blanking presses, drop hammers and grinders, etc. Since noise is an integral part of machinery in operation, therefore it is being accepted as a part of job. Hence noise is perceived as a necessary evil in such small scale industrial units. This attitude persists in the management as well as the employees and workers of majority of firms where the traditional methods and machinery are still being utilized. Moreover, the small and medium enterprises (SMEs) especially casting and forging units are lagging far behind in noise control efforts. Since the workplaces in most of the SMEs are very congested due to lack of availability of space. Hence simultaneous running of moulding machines, blanking and punching presses, drop hammers, grinders and barrels, etc., make the work places very noisy. All continuous, intermittent, and impulsive noise between the levels of 80 and 130 dB (A) must be included in the exposure assessment (OSHA's Noise Standard, 2000). Continuous noise is a noise whose maxima occur more often than once per second. Thus impulsive noise is assumed to have peaks occurring less often than once a second, and is limited to peak sound pressure levels of 140 dB (OSHA's Noise Standard, 2000). The impact produced by a hammer falling on a plate can generate an impulse with peak level of 137 dB and duration range of 140–470 ms (Tremolieres and Hetu, 1980). In developing countries like India besides the economic growth and technological changes, the companies are striving for increasing the sales turnover; thus the workers are more stressed for the productivity along with the additional burden of noise exposure. The small scale casting and forging units of developing countries are lagging far behind in implementing hearing conservation, occupational noise control and occupational health and safety programmes. Therefore consequently the workers of these units are more exposed to the noise levels beyond the permissible limits. High noise exposure not only affects the communication among the workers and Noise Induced Hearing Loss (NIHL) but also lead to, physiological and psychological effects.

12.3.1. Hazardous effects of occupational noise exposure

A number of studies have been carried out in last few decades regarding the assessment of occupational noise exposure and its auditory and non-auditory health effects. The auditory effects include mainly the hearing disability whereas the non-auditory effects include; cardiovascular symptoms, BP, noise annoyance and other psychosocial effects. In a study it was found that chronic exposure to noise levels typical of many workplaces is associated with excess risk for acute myocardial infarction death (Davies et al., 2005). Another study by Lercher et al. (1993) investigated the effect of occupational noise annoyance on blood pressure and also its combined effect with social support at work, nightshift work, and work satisfaction. A significant effect of noise annoyance on rising diastolic blood pressure (DBP) was observed; whereas Lang et al. (1992) reported that a long exposure to noise over 85 dB (A) could be a risk factor for high BP, and possibly inducing major increases of BP among sensitive individuals. Noise may not risk employee's life immediately but might be the cause of neurobehavioral change, psychological stress and unhappiness in daily life without showing the symptoms of chronic /acute diseases. Thus quality of life of industrial worker is one of the prime factors which cannot be ignored (Kisku et al., 2006). The same study recommended the revision of occupational Indian Noise standard at par with International/European standards [Kisku et al., 2006]. Rai et al. (1981) reported that exposure to high intensity noise of 88–107 dB (A) 6–8 hour/day for long duration (10–15 years) brings about biochemical changes which make anyone prone to cardiovascular pathology. Levels of free cholesterol and cortisol were significantly high in exposed group. Rai et al. (1981) have also reported a dose–response relationship between noise intensity and hypertension incidence in the occupational population. Workers, who perceived noise exposure at workplace, were found with a higher prevalence of myocardial infarction. Perceived noise exposure at work potentially increases the risk of death due to intracerebral haemorrhage among subjects with hypertension, but not among the subjects without hypertension, which may draw attention for occupational health policy (Fujino et al., 2007).

The other researchers have investigated the auditory effects of noise, as its exposure has been known to cause hearing loss for over a century. However the exposure limits have been legally adopted recently, but strict enforcement of current permissible exposure levels would not eliminate occupational hearing loss from industry (Raymond, 1994). Chen et al. (2007) assessed the temporary threshold shift (TTS) recovery time using an audiometric test at post exposure times of 2, 20, 40, 60, 80 and 120 minutes. Also TTS driven by noise exposure is enhanced by heat and workload. On the other side, Smeatham and Wheeler (1998) compared industrial and military impulsive noise to categorize the different types of impulsive noise. The categorization enabled to define laboratory impulsive noise sources which could be used in a laboratory environment. David et al. (1996) reported that, noise exposure was potentially hazardous to hearing and proposed to give serious consideration to work environment of truck drivers and programme for the prevention of permanent hearing damage. Before that, Mantysalo and Vouri (1984) had reported that exposure to high levels of impulse noise despite the use of ear protectors could be more detrimental to hearing than are high levels of continuous noise (event continuous with slightly impulsive features). Group with shortest duration of exposure and with intermediate duration of exposure to impulse noise had the highest threshold at 6.0 KHz in both ears while the group with longest exposure to impulse noise had the highest threshold asymmetrically at 4.0 KHz in left ear and at 6.0 KHz in the right ear (Mantysalo and Vouri, 1984). Another study

(Khuranet and Attar, 2000) reported that, the workers operating machinery of two food manufacturing companies, a cement–brick making plant, a furniture factory and a car washing facility in Kuwait were exposed to high noise levels without any kind of ear protection. The same study recommended that, the concerned environmental and health organizations in Kuwait should be compelled to establish noise level standards and controls to safeguard the health of workers and public at large. Workforce of two textile plants was found at high risk of developing NIHL and other associated ailments. Author recommended that exposure limit stipulated by OSHA of 90 dB (A) for 8 hours/day shall be followed with caution. It was also recommended to establish a hearing conservation programmes including the noise assessment, awareness among workers and audiometry (Raman Bedi, 2006). Indian factory act - 1948 lay down a limit of 90 dB (A) for 8 hour/ day but the Indian working hours are 48 hour/ week which leads to high noise exposure. Most of the workers were illiterate and semiliterate, they were not aware of the exposure norms and health hazardous effects of noise (Bedi et al., 2004, IS: 7194 - (1994), IS: 9989 - (1981) Indian Standard). OSHA's Noise Standard (2000), recommends that workers must be allowed to observe the monitoring procedures and must be told about their exposures. Hearing protection must be provided at no cost to employees and must be worn by all workers exposed to a TWA of 90 dB (A) and above. Also hearing protection is mandatory for those exposed to 85 dB (A) and above if they have not yet had a baseline audiogram.

Some of the researchers have investigated the chronic effects of occupational noise on brain stem auditory evoked potential response, i.e., latency time. A study by Murata et al. (1990) reported that, the combined stressors of local vibration, noise, climate and heavy work, affected not only the peripheral nervous system but also the brainstem portion of the auditory pathway, the brainstem effect. Murata et al. (1990) reported V^{th} peak latency of BERA was significantly prolonged in chain saw operators. Another study by Thakur et al. (2004) reported that, high noise exposed subjects working at Mumbai airport showed alteration in BERA indicating altered auditory conduction up to the level of the brainstem.

Legislative Criteria for Occupational Exposure with respect to Noise as per Indian Factories Act, 1948

The permissible limits for occupational exposure with respect to noise, heat stress and dust exposure is described as follows:

Permissible Limits for Exposure with Respect to Occupational Noise

Permissible limits of exposure have been prescribed in the Model Rules framed by the Directorate General Factory Advice Service and Labour Institutes (DGFASLI), Ministry of Labour and Government of India under the Factories Act 1948 for continuous noise. These limits are laid down on the basis of the duration of noise exposure at various sound levels in a day. The permissible limits of exposure emphasize to reduce employees' exposure to noise by adopting suitable engineering or administrative controls, otherwise. PPEs must be provided to the workers along with effective Hearing Conservation Programmers in the factory. The permissible limits of exposure is 90 dB (A) measured with a frequency weighting for 8 hours exposure and the higher level permitted is 115 dB (A) for 15 minutes (Ref. Chapter 13). As per the permissible limits of noise exposure, no exposure in excess of 115 dB (A) is to be permitted. However the noise dose (D) criteria recommend only 100 per cent dose in 8 hours of noise exposure noise dose D, in per cent, is given by:

$$D = 100 \ C/T;$$

Where, C is the total length of the work period in hours, and T is the reference duration corresponding to the measured sound level, L, can be calculated by using the following formula:

$$T = 8/2^{\frac{(L-90)}{5}}$$

Any unwanted sound that creates annoyance to the human operator, is generally treated as noise, *thus any unnecessary sound is known as noise* which results in trouble and lowers quality of life consequently may leads to sleep and mental/auditory fatigue. Assessment of occupational noise exposure and noise induced hearing loss is described in the subsequent chapters 13 and 14.

Multiple Choice Questions

1. A body is said to vibrate when it describes an oscillating motion about
   - (a) a fixed position
   - (b) a free position
   - (c) Partially fixed position
   - (d) Both fixed and free position
2. The term VWF referred as
   - (a) vibration induced white finger
   - (b) vibration influenced white finger
   - (c) vibration impact white finger
   - (d) None of these
3. The human body is most sensitive to vibration in the frequency range
   - (a) 1–80 Hz
   - (b) 81–90 Hz
   - (c) >90 Hz
   - (d) None of these
4. Vibration induced white finger is a condition generally associated with the use of
   - (a) vibratory hand tools
   - (b) frequency of the hand tool
   - (c) Type of hand tool
   - (d) Both vibratory hand tools and frequency of the hand tool
5. The vibrations which are experienced from machinery are generally very complex but
   - (a) Regular motion
   - (b) Irregular motion
   - (c) Both regular and irregular motion
   - (d) None of these
6. Acceleration in 'g' units is
   - (a) $4\pi^2 f^2/981$
   - (b) $2\pi^2 f^2/981$
   - (c) $4\pi f^2/981$
   - (d) $5\pi^2 f^2/981$
7. If the frequency of vibration is in the range of 3–140 Hz then there will be
   - (a) Motion sickness
   - (b) Blurring of vision
   - (c) Spinal damage
   - (d) Speech interference
8. If the frequency of vibration is in the range of 20-100 Hz then there will be
   - (a) Motion sickness
   - (b) Blurring of vision
   - (c) Spinal damage
   - (d) Speech interference
9. The frequency of vibration of an object at which human eyes can make more stable image on retina is.
   - (a) <4 Hz
   - (b) <8 Hz
   - (c) <2 Hz
   - (d) <1 Hz
10. The eyeball will begin to accentuate the input vibration if the vibration frequency is above about
    - (a) 20–25 Hz
    - (b) 10–15 Hz
    - (c) 20–35 Hz
    - (d) 20–45 Hz

11. The term FDP is termed as
    (a) Fatigue decreased proficiency
    (b) Fatigue disorder proficiency
    (c) Fatigue decease proficiency
    (d) None of these
12. For any exposure time, exposure limit
    (a) $3 \times$ FDP
    (b) $2 \times$ FDP
    (c) $4 \times$ FDP
    (d) $5 \times$ FDP
13. The term DBP is termed as
    (a) Diastolic blood pressure
    (b) Diabetic blood pressure
    (c) Decreased blood pressure
    (d) None of these
14. The permissible limits of exposure is
    (a) 90 dB (A)
    (b) 90 dB (B)
    (c) 90 dB (C)
    (d) None of these
15. The formula of noise dose (D) criteria is
    (a) $D = 100 \, C/T$
    (b) $D = 90 \, C/T$
    (c) $D = 800 \, C/T$
    (d) $D = 100 \, T/C$

Answers

1. (a) **2.** (a) **3.** (a) **4.** (d) **5.** (a) **6.** (a) **7.** (b) **8.** (c) **9.** (a) **10.** (a)
11. (a) **12.** (b) **13.** (a) **14.** (a) **15.** (a)

Chapter

13

Occupational Noise Exposure

13.1. Introduction to Noise

Any unnecessary sound which creates disturbance to human is generally treated as noise. Thus any unnecessary sound is known as noise that results in disturbance and lowers quality of life which leads to sleep and mental/auditory fatigue. The annoying effect of noise, whether from a factory, motorway, marriage palace or nightclub, varies from person to person. Secondly, it can divert concentration and audible caution signals or interfere with occupation that may result into accidents. Finally exposure to excessive noise can result in hearing loss, provided the exposure period is of sufficient duration; for example, loud music may lead to hearing disorders.

13.2. Sound

The pressure difference in air, water or any other medium that is detected by human ear is known as sound. Physically, sound is a vibration of gas, liquid or solid medium particles.

13.2.1. The nature of sound

Sound is a mixture of many waves or fluctuations (sound waves) impinging on the eardrum. It is generated from any energy source setting up hasty and varying pressure in the nearby air. The rate at which variations occur, is called frequency or pitch, and is expressed in hertz (Hz), i.e., cycles per sec. One hertz means the number of complete air waves passing a fixed point per sec. The normal human ear is sensitive to frequencies between about 20 and 20,000 Hz being particularly sensitive in the range 2,000–6,000 Hz. (with maximum sensitivity at 4000 Hz.) that is less perceptive at elevated and lower frequencies.

13.2.2. Characteristics of sound waves

Sound having one frequency only is called as pure tone, such as sound produced by the tuning fork (see Figure 13.1 A). As components distributed over a wide range of frequencies, most of industrial noise are very complex, called broad band (see Figure 13.1 B). Noise produced by looms, an air jet industrial machinery or printing presses, are few examples.

There is a significant noise due to impact between metal parts in steel industry setup. Due to a wide range of impacts per sec, noise is treated as broad band noise, as in a riveting machine. On the other hand noise having widely spaced impacts termed impulse noise, as drop hammer-operated hand tool (see Figure 13.1 C).

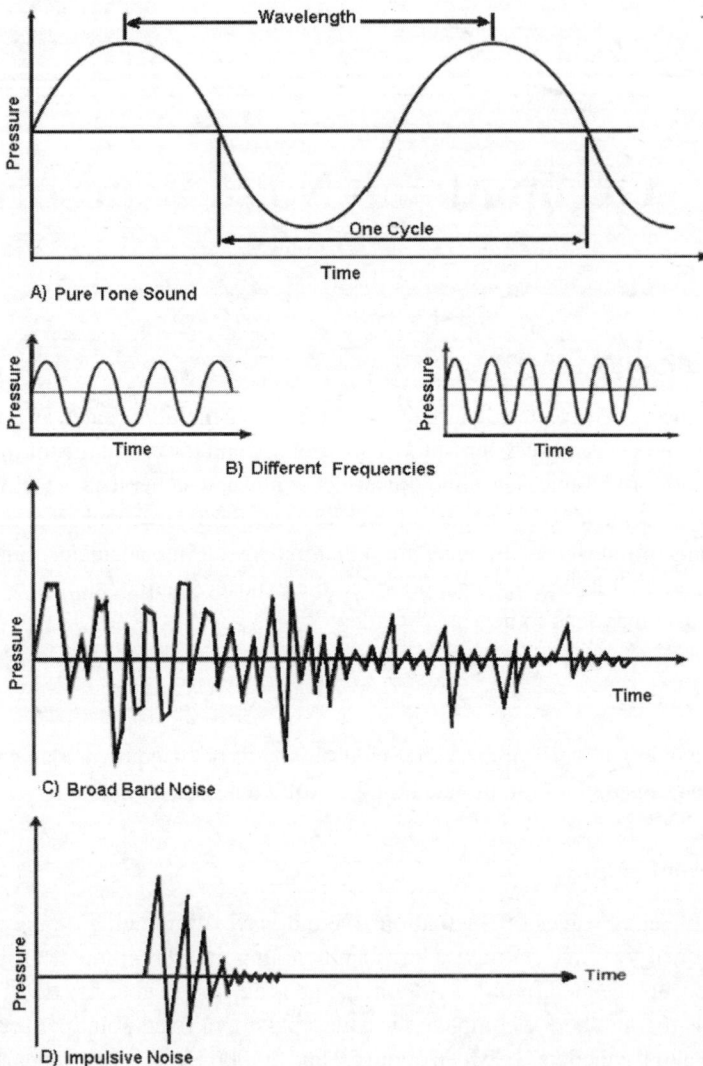

Fig. 13.1: (A) Pure tone sound, (B) different frequencies, (C) broad band noise and (D) impulsive noise.

13.2.2.1. *Wavelength*

Wavelength is the physical distance from one peak of a sound wave to the next in the air. It is ratio of the speed of sound in the medium to fundamental frequency.

13.2.2.2. *Amplitude*

The maximum displacement of a particle from its rest position is known as amplitude of a sound wave. The amplitude of a sound wave determines intensity, although the two are not directly correlated.

The different parameters of sound are as follows:

- Energy in the form of pressure waves
- Waves can be described by frequency (f), speed (c), and wavelength (λ)
    - $c = f\lambda$
- Sound moves at 344 m/sec in air, 6100 m/sec in steel
- Some materials will amplify or reflect sound
- Frequency (f) is related to pitch
    - A healthy, young person can detect 20–20,000 Hz (cycles/sec)
    - This declines with age and exposure history

13.3. Sources of Noise and Vibration

Noise and vibration in working environment can include the following:

(a) Noise produced as a result of vibration in machines;
(b) Noise taking a structure borne pathway;
(c) Radiation of structural vibration into the air;
(d) Instability formed by air or gas flow;
(e) Noise taking an air borne pathway;
(f) Noise produced from vibratory hand tools, e.g., chain-saws.

13.3.1. Important aspects of sound and noise

All the sources of noise mentioned in the previous section, result in the vibration of the ear drum. This vibration is amplified in the middle ear by the ossicles (malleus, incus and stapes) and vibrations are transmitted, further from nerve impulses to brain. It is in this way that a sound is heard.

Sound intensity

It is the particular power/level of the sound energy with which it confronts the ear, and describes rate of flow of sound energy. High-intensity sound has more energy than low-intensity sound.

Sound pressure level

The intensity of sound is difficult to measure directly, but the passage of sound energy through air is accompanied by fluctuations in atmospheric pressure. These fluctuations can be measured and related to the amount of sound energy that is flowing. Therefore, it is usual to measure sound pressure level, which is a measurement of the magnitude of air pressure variations or fluctuations which make up sound. So, for the pressure variations root mean square value is used and generally expressed in terms of decibels (dB).

Frequency

Frequency is the number of complete pressure variations passing a fixed point per sec. It is measured in hertz (Hz), i.e., 1 Hz = 1 cycle per sec; 1 kHz = 1,000 cycles per sec. The frequency of a sound gives it its distinguishing character. For instance, high-frequency sound, such as a train whistle, will sound high pitched, whereas a low frequency sound, such as that from a double bass, will sound low pitched. The more rapidly the vibrations occur, the higher is the frequency and vice versa.

Pitch

This is the subjective quantity of a sound which determines its position in the musical scale. It is determined by frequency.

Loudness

It generally depends upon the intensity and amplitude of the sound waves involved. However, high and low frequencies are less sensitive to human ear and loudness also depends on frequency and the subjective perception of sound by human beings.

13.4. Basic Theory of Noise Measurement

It is the measurement of intensity of sound on a comparative basis. The range of intensities to which ear responds, however, is huge from the threshold of hearing to the threshold of pain. For example, at 1,000 Hz the threshold of pain is 100,000,000,000,000 (10^{14}) times more intense than the threshold of hearing, where sound is just discernible. It is clearly difficult to express such ratios on a simple arithmetic scale so a logarithmic scale is used. The ratio would therefore be expressed as under:

$$\text{Log}_{10}(10^{14}/1) \text{ or } 14, \text{ rather than}$$

$$10^{14}/1$$

The unit used is the bel. Thus 1 bel is $\log_{10}10^1$ (a tenfold change in intensity), 2 bel is $\log_{10}10^2$ (a hundred fold change in intensity) and so on. The bel, however, is a very large unit, so it is further split into tenths, called decibels (dB); 1 bel equals 10 decibels. For example, $10\log_{10}10^{14}$ equals 140 dB. Thus 1 dB equals a change in intensity of 1.26 times, since $10^{1/10}$ is 1.26 (or 1.26^{10} is 10). Also, a change of intensity of 3 dB = $1.26^3 = 2$, so that doubling the intensity of a sound gives an increase of 3 dB.

If there are two sounds of intensities I_1 and I_2 and they differ by N dB, then

$$N = 10 \log_{10} (I_1/I_2)$$

It is normal practice to relate intensity to a standard reference level, so that

$$N = 10 \log_{10} (I_1/I_0)$$

and I_0 is taken as 10^{-12} watts per square meter.

However, as intensity is proportional to pressure squared,

$$N = 10 \log_{10}(P^2/P_0^2)$$

$$= 10 \log_{10}(P/P_0)^2$$

or

$$N = 20 \log (P/P_0) \text{ dB}$$

where, P is the standard reference level of 2×10^{-5} Newton/square meter (Pascals) and N is sound pressure level in dB. Pressure is the easiest quantity to measure, hence the use of dB sound pressure level. The standard reference level of 2×10^{-5} Newton/square meter is chosen since it is the average threshold of audibility at 1,000 Hz (i.e., It is 0 dB).

Note: Under the SI system, sound pressure is expressed in pascals. A pascal is a unit of pressure corresponding to a force of one Newton acting uniformly upon an area of 1 square meter. Hence 1 Pa = 1 N/m².

The use of a logarithmic scale in sound measurement has a further advantage, as the evaluation of intensities is simplified by the replacement of multiplication with addition and of division with subtraction. Furthermore, the response of the ear tends to follow a logarithmic scale.

The addition of decibels is carried out on a ratio basis, rather than an arithmetic one and Table 13.1 may be used to simplify the procedure. To add two sound pressure levels, take the difference between the two levels and add the corresponding value in the right-hand column to the higher sound pressure level.

Table 13.1: Addition of decibels.

Difference (dB)	Add to Higher (dB)
0.0–0.5	3.0
1.0–1.5	2.5
2.0–3.0	2.0
3.5–4.5	1.5
5.0–7.0	1.0
7.5–12.0	0.5
Over 12.0	0.0

13.4.1 Other units used in noise measurement

Phons

All frequencies are not equally responded by the human ear. Sounds of different frequency at a constant sound pressure level do not evoke equal loudness sensations. This phenomenon is linear with neither amplitude nor frequency and loudness level is measured in phons, the sound being compared again to a standard reference signal of 1,000 Hz. The loudness level in phons of any sound is taken as that which is subjectively as loud as a 1,000 Hz tone of known level. 0 phon is 0 dB at 1,000 Hz. 50 phon is the loudness of any tone which is as loud as a 1,000 Hz tone of 50 dB. This can be demonstrated by equal loudness curves for pure tones shown in Figure 13.2. Maximum sensitivity occurs between 1 and 5 kHz. The curves are obtained by finding the sound levels at different frequencies which seem equally loud to the listener in comparison with a reference sound at 1KHz – a scale designed to provide scale numbers having linear unit of loudness. The scale is precisely defined by its relation to the phon scale.

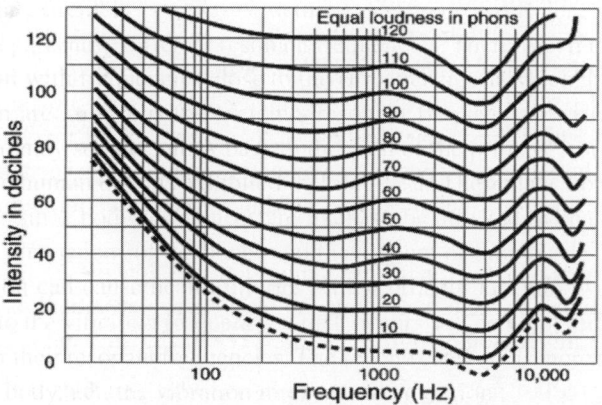

Fig. 13.2: Equal loudness curves for pure tones.

13.4.2. Octave bands and octave band analysis

It is possible to make a single measurement of the overall sound pressure of the entire range of audible frequencies. However, this measurement, if taken in linear decibels, is of limited use since the ear is more sensitive to some frequencies than others. Use of a weighted dB scale provides a reasonable means of assessing likely risk to hearing, but knowledge of the way in which the sound is distributed throughout the frequency spectrum provides a much more accurate picture. This can be obtained by dividing the noise into octave bands and measured by the sound pressure level at the central frequency of each band. (An octave represents a doubling of frequency, so that the range 90–180 Hz is one octave, as in the range 1,400–2,800 Hz.) The octave bands are usually identified by their geometric central frequencies, e.g., geometric central frequency of octave band 90–180 Hz is approx. 125 Hz. The octave bands with standard ranges having geometric central frequencies are shown in Table 13.2.

Table 13.2: Standard range of octave bands.

Limits of Band (Hz)	Geometric Centre Frequency (Hz)
45–90	63
90–180	125
180–355	250
355–710	500
710–1400	1000
1400–2800	2000
2800–5600	4000
5600–11200	8000

13.4.3. The sound pressure level meter

A sound pressure level meter is an instrument which measures linear sound pressure level in the human audio frequency range unless provided with and set to various weighting networks. The objective measurement of sound pressure level in accordance with the manner of response of human ear is given by weighted network.

The microphone senses the air pressure fluctuations and converts mechanical vibration to an electrical signal containing frequency and amplitude signals. The amplitude increases the weak signal from the microphone and incorporates gain adjustment, which enables the instrument to cope with the wide range of pressure amplitudes which the ear can sense. The output socket as a sound signal may be fed to external instruments such as recorders or noise dose meters.

Since an accurate response from the sound level meter is necessary, provision is made to calibrate it for accurate reasons. This is best done by these of the portable acoustic calibrator placed directly over the microphone. The calibrator is basically a miniature audible signal generator giving a precisely defined sound pressure level to which the sound level meter can be calibrated. Electronic oscillators are most commonly used.

When the sound level fluctuates, the meter reading should follow these variations. However, if the level fluctuates too rapidly, the meter needle may move so irregularly that it is not possible to obtain a significant value. For this reason, two meter response characteristics are used:

(A) Fast: This gives a fast reacting indicator response which enables the user to follow and measure noise levels which are not fluctuating to rapidly.

(B) Slow: This gives a damp response and helps average out meter fluctuations which would be otherwise impossible to read.

The following instruments are used for occupational noise exposure assessment.

(a) Sound Level Meters (SLM)
Used for area surveys
Settings for average, peak, impulse, ABC scales

(b) Noise Dosimeters
Used for individual monitoring
Clip microphone near the ear
Wear all day

13.5. Noise-measuring Meters

These devices are generally called the Sound Level Meters (SLM), particularly based on the safety regulations provided by occupational safety and health administrator (OSHA). The specific category and manufacturing detail must be mentioned of any noise measuring devices so as to be used in any noise detecting surveys.

13.6. Basic Sound Level Meters

The time constant remains standard as taken as 1/8 and 1 sec, i.e., fast and slow, respectively, that provide limited averaging just for steady signals of noise. On the other hand, these time constants become very small than the required time to calculate the RMS level of various types of firms. The error increases with the amount of variation of sound level more than 6 dB around the mean.

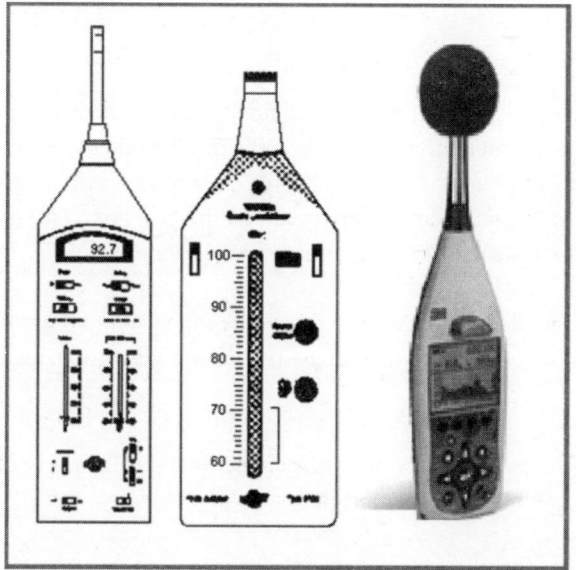

Fig. 13.3: Type 1 Integrating Sound Level Meter (right) and
Type 2 Integrating Sound Level Meter (left).

13.7. Integrating Meters

These are also referred as *integrating–averaging sound level meters* and *noise dosimeters which* are mostly used for noise surveys with Type 2 Classification or better.

13.7.1. Noise dosimeters

These are known as noise integrating devices used for individual sample of noise for a long span of time which are small enough to be worn by workers. It indicates the noise exposure dose acquired during that time at the end of the sampling time. Noise dosimeters are normally set as follows:

Criterion level: Lc = 85 dBA
Threshold level: Lt = 80 dBA or Off
Exchange rate: q = 3 dB
Time constant = 'Slow'
Dosimeters must have the following minimum specifications:
Classification: Type 2
Weighting: A-weighting
Dynamic range: 50 dB
Crest factor: 30 dB

13.7.2. Weighting networks

This sound level meter incorporates electrical circuits known as waiting networks. These provide for various sensitivities to sounds of different frequencies, the original objective being to simulate the response characteristics of the human ear at different frequencies. These weighting networks are known as A, B, C, D weighted decibel scale operating condition and can be selected on the sound level meter.

(A) **scale**

The 'A' scale is normally used for industrial noise measurement designated as dB(A). This scale makes the instrument more sensitive to the middle range of frequencies as compared to lower and higher range. It is one which most closely approximate to the ear response. In this text mainly different examples are given on the basis of 'A' type sound, if the noise level of a particular place is 95 dB then the same is denoted as 95 dB(A).

(B) **scale**

This scale was intended for measuring of middle range sound pressure levels between 55 and 85 dB. It is not commonly known, as it does not give good correlation to subjective test of hearing perception.

(C) **scale**

This scale gives more sensitivity in low frequencies and is therefore of limited use as with the B scale it does not give good correlations with subjective test.

(D) **scale**

This scale is generally limited to the measurement of air craft noise and has little or no application in the measurement of industrial noise.

13.8. Noise Control

As per the legislative requirements, work place noise exposure must be restricted permissible limits, therefore mainly two factors must be considered to reduce work place noise:

First, the source of the noise and secondly, the actual adopted noise to the receiver. Personal protective equipment e.g., for preventing the individuals form hearing disorders like ear plugs, ear defenders, or acoustic wool can be used. The better way of preventing risk of persons sustaining noise induced hearing loss is, if practicable, to tackle the potential problem at the decision stage, to a certain extent. There are a number of suitable ways to control the level of noise with various sources and possible levels in the path to the receiver. These are mentioned in the following table. The noise pathway for vibration-induced noise has three distinct stages:

(A) Structure-borne noise emission
(B) Radiation of the noise from the structure into the air
(C) The actual air borne noise pathway

In this case, the way to reduce structure borne noise is to isolate the noise-generating parts of the machine from the radiating surface with the vibration isolators, consisting of steel springs. The amount will be increased if the surfaces resonate or oscillate. In this case it would be appropriate to apply a damping treatment using a spray on, stick on or magnetic damping compound. Panel damping materials can be extremely effective in a wide range of applications, some of the noise control methods are given in Table 13.3.

Table 13.3: Methods of noise control.

Sr. No.	Sources and Pathways	Control Measures
1.	Vibration by a machines	Reduction at origin e.g., replacement of nylon particles for metals as tapering equipment
2.	Vibration due to structure	Vibration isolation, by using damping devices
3.	Emission of structural vibration	Reduction of vibration at the origin by implementing silencers
4.	Instability due to flow of air or gas	Noise insulation–reflection; heavy barrier
5.	Noise path as carried by the air	Noise absorption without any reflection; porous lightweight barriers

Once the noise has been radiated into the air, two completely different methods of control are possible. The recipient can be insulated from the noise with the heavy limp airtight screen (barrier), which simply acts as an impervious barrier to prevent noise transmission. Alternatively, noise can be absorbed in porous materials mounted on walls or suspended from a ceiling. Absorption of noise only will be helpful if most of the noise which reaches the subject has been reflected off walls and ceilings.

13.9. Permissible Limits for Exposure with respect to Occupational Noise

Occupational noise exposure standards for 8 hours/day are shown in Table 13.4. Permissible limits of exposure have been prescribed in the model rules framed by the Directorate General Factory

Advice Service and Labour Institutes (DGFASLI), Ministry of Labour, Government of India under the Factories Act 1948 for continuous noise. These limits are laid down on the basis of the duration of noise exposure at various sound levels in a day. The permissible limits of exposure emphasize to reduce employees' exposure to noise by adopting suitable engineering or administrative controls, otherwise. PPEs must be provided to the workers along with effective Hearing Conservation Programmes in the factory. The permissible limits of exposure is 90 dB (A) measured with a frequency weighting for 8 hours exposure and the higher level permitted is 115 dB (A) for 15 minutes. The permissible limits of exposure for noise are presented in Table 13.4 (DGFASLI, 1987). No exposure in excess of 115 dB (A) is permitted. The noise dose (D) criteria are shown in given in Table 13.5, and noise dose D, in percent, is given by:

$$D = 100 \ C/T;$$

Table 13.4: Various occupational noise exposure standards.

Noise Exposure Standard	Maximum Permissible Limit of Noise Level
ACGIH	85 dB(A)
NIOSH	85 dB(A)
OSHA	85 dB(A)
PEL	90 dB(A)

Table 13.5: Permissible limits for occupational noise exposure as per Indian Factory Act 1947.

Sound Pressure Level in dB (A)	Duration of Exposure per Day (hours)
90	8
92	6
95	4
97	3
100	2
102	1
110	½
115	1/4

(Schedule – xxiv, Indian Factory Act 1947)

where, C is the total length of the work period (noise exposure) in hours, and T is the reference duration of exposure corresponding to the measured sound level, L, as given in Table 13.6, or by the following formula:

$$T = \frac{8}{2^{\frac{(L-90)}{5}}}$$

Table 13.6: Occupational noise dose criteria.

A-weighted Sound Level, L (dB)	Reference Duration, T (hour)	A-weighted Sound Level, L (dB)	Reference Duration, T (hour)	A-weighted Sound Level, L (dB)	Reference Duration, T (hour)	A-weighted Sound Level, L (dB)	Reference Duration, T (hour)
80	**32.0**	93	5.3	118	0.16	117	0.19
81	27.9	94	4.6	119	0.14	**120**	**0.125**
82	24.3	**95**	**4.0**	106	0.87	121	0.11
83	21.1	96	3.5	107	0.76	122	0.095
84	18.4	97	3.0	108	0.66	123	0.082
85	16.0	98	2.6	109	0.57	124	0.072
86	13.9	99	2.3	**110**	**0.5**	**125**	**0.063**
87	12.1	**100**	**2.0**	111	0.44	126	0.054
88	10.6	101	1.7	112	0.38	127	0.047
89	9.2	102	1.5	113	0.33	128	0.041
90	**8.0**	103	1.3	114	0.29	129	0.036
91	7.0	104	1.1	**115**	**0.25**	**130**	**0.031**
92	6.1	**105**	**1.0**	116	0.22		

OSHA limits workers to 100 per cent of the daily dose or 90 dB(A) for 8-hour TWA

$$D = 100 \times \left[\frac{C_1}{T_1} + \frac{C_2}{T_2} + \dots + \frac{C_n}{T_n} \right]$$

D = daily nose dose, in per cent

C = total time of exposure at the measured noise level

T = reference allowed duration for that noise level from Table 13.6.

Example:

An occupational hygienist visited casting unit and his exposure was monitored for 2 hours at 80 dB(A), 2 hours at 95 dB(A), and 4 hours at 103 dB(A). Let us estimate the DND (Daily Noise Dose) received by the hygienist.

The reference time can be calculated as

$$T = 8/2^{\frac{(L-90)}{5}}$$

Reference time for exposure at 80 dB:

$$T_{80} = 8/2^{\frac{(80-90)}{5}}$$

$$T_{80} = 8/2^{-2} = 8 \times 4 = 32 \text{ hours or from Table 11.3}$$

$$T_{95} = 8/2^{\frac{(95-90)}{5}} = 8/2^1 = 4 \text{ hours}$$

Similarly

$$T_{103} = 8/2^{\frac{(103-90)}{5}} = 8/2^{\frac{13}{5}} = 8/2^{2.6} = 1.3195 = 1.3 \text{ hours (approx.)}$$

Therefore: $DND = 100 \times \left[\dfrac{C_1}{T_1} + \dfrac{C_2}{T_2} + \ldots + \dfrac{C_n}{T_n} \right]$

$$D = 100 \times \left[\frac{2}{32} + \frac{2}{4} + \frac{4}{1.3} \right] = 363.94\%$$

13.10. Assessment of Exposure

The regulations bring in the concepts of daily personal noise exposure and action levels when undertaking assessment of exposure. It is the individual noise exposure with part 1 of the plan to policies without considering the use of ear preventive measures. There must be more than one noise exposure levels to consider the combined effect at various stages.

The mixed exposures are taken as more than limit value when $C_1/T_1 + C_2/T_2 + \ldots + C_n/T_n$ exceeds unity. *Here, C_1, C_2,* , , up-to *Cn* are the actual durations of noise exposure at various noise, and T_1, T_2, , , *up-to* T_n are permissible duration of exposure at respective levels. The peak sound pressure level of impulsive or impact noise level exposure should not be more than 140 dB. The permissible exposure limits for impulsive or impact noise (Schedule – xxiv, Indian Factory Act 1947) is exhibited in Table 13.7.

Table 13.7: Permissible exposure levels of impulsive or impact noise
(Schedule – xxiv, Indian Factory Act 1947).

Peak Sound Pressure Level in dB	*Permitted Number of Impulses or Impact per Day*
140	100
135	315
130	1000
125	3160
120	10000

13.11. Adding Decibels

Often there is a need to combine two or more noise sources. Since decibels are logarithms, they cannot be added directly.

$$80 \text{ dB} + 85 \text{ dB} \neq 165 \text{ dB}$$
$$80 \text{ dB} + 85 \text{ db} = 86 \text{ dB}$$

$$SPL_{total} = 10 \log \left(\sum_{i=1,2,\ldots}^{N} 10^{L_i/10} \right)$$

Where, L_i is the sound pressure level of ith location

Example 1

The sound pressure level of three machines installed in a room M1, M2 and M3 was measured at 80, 85, and 87 dB, respectively, find out the total sound pressure level in dB (A) of the room.

Solution:

$$SPL_{total} = 10 \log (10^{80/10} + 10^{85/10} + 10^{87/10})$$
$$SPL_{total} = 89.6 \text{ dB (A)}$$

Example 2

In a hand tool manufacturing company, 4 hammers are running simultaneously in the forging section. According to the sound level meter time weighted averages of sound level for these hammers are: H1-103, H2-101, H3-102, H4- 103 dB (A). Calculate the net noise level in the section.

Solution:

$$SPL_{total} = 10 \log (10^{103/10} + 10^{101/10} + 10^{102/10} + 10^{103/10})$$
$$SPL_{total} = 108.35 \text{ dB (A)}$$

Otherwise the method of estimation can be used to calculate the total noise level for two or more sources of noise. The criterion for estimation is given in Table 13.8.

Table 13.8: Estimation method for calculation of noise level for two or more sources of noise.

Excess of Stronger	Add to the Larger Value to Get Combined Level
0	3.0
1	2.5
2	2.1
3	1.8
4	1.5
5	1.2
6	1.0
7	0.8
8	0.6
9	0.5
10	0.4

For example; if for two sources of noise 62 dB + 62 dB difference is zero. Hence add 3 to the larger vale i.e., 62 dB = 62 dB + 3 dB = 65 dB(A)

Let us try to find out the net noise level in a grinding section with three grinding G1, G2, and G3 machines running simultaneously at 80, 85, and 87 dB respectively. Using the estimation method;

Start with 85 and 87; difference is 2 dB, so select 2 dB from the table 13.8.

87 + 2 = 89 dB(A)

Add 89 and 80, difference is 9 dB, so select 0.5 dB

89 + 0.5 = 89.5 dB(A) comparable to 89.6 dB(A)

13.12. Hearing Conservation Program (HCP)

After measuring the hearing threshold using formal audiometry, if standard threshold shift (STS) is identified, then it is the duty of the employer to notify the employee in writing. In addition, to it provide additional training to the employee and the adequate hearing protection must be provided. Workers' Compensation (WC) laws for identifying and compensating STS may vary across the states.

Hearing Conservation Program (HCP) is required whenever employee exposures exceed 85 dBA 8-hour TWA. This is half the allowable noise exposure for an 8 hour day or 50 per cent Daily Noise Dose (DND). An exposure at 90 dB (A) for an 8-hour TWA is 100 per cent DND. The major components of hearing conservation programmes are:

- Exposure monitoring
- Audiometric testing
- Hearing protective devices
- Training program
- Access to the written standard
- Record keeping

13.13. Noise Control Programmes

In considering how an organization should approach compliance with the noise regulations at work place it is appropriate to consider the elements of occupational hygiene practice, i.e., identification/recognition of the health risk, measurement, evaluation against current standards (legal and otherwise) and control.

Identification/recognition

This may be achieved through routine observation in a working area. Generally, if it is necessary to shout to make one understand, it is fairly certain that sound pressure levels are around or above 90 dB(A). Alternatively, there may have been complaints from safety representatives, shop stewards or employees, claims against the company for occupational deafness, or action by the local environmental health officer in the event of noise nuisance to local residents. A preliminary survey, using a sound pressure level meter, will indicate variations in sound pressure level from one part of the premises to another.

Measurement

The sec stage of the operation is the carrying out of a full-scale noise survey of the premises using a precision grade sound level meter with the facilities for octave band analysis. Such a survey will produce an indication of those items of plant and machinery producing unacceptable noise emissions, the frequency ranges involved and risk of occupational deafness to operators.

Evaluation

Reference to action levels specified in the regulations, the actual results of the noise survey, the number of people exposed generally and in specific locations, and to the transmission pathways of the noise, will make it possible to decide on the relative urgency of action necessary. Therefore, to eliminate or reduce such exposures there must be some short-term, medium-term and long-term measures.

Control strategies

Many options are available based on the degree of risk, cost and sheer practicability of implementation. The ultimate objective must be controlled through recognized methods as opposed to the provision of ear protection. The options are:

(a) Control

1. Reduction at source through proper preventative maintenance (PM) on machines parts such as motors, bearings, drive belts and pumps.
2. Adjustments, lubrication, replacement, vibration isolators, think outside the box for new designs and work with suppliers.
3. One of the best engineering controls is to provide distance between source of noise and the receiver (person), because the relationship between noise and distance follows the inverse square law. Doubling the distance reduces the noise by $1/4$ and tripling the distance reduces the noise by $1/9$.
4. Installation of sound proof enclosures and closed shields; or devices, insulating curtains; coverings for noise-reflective floors, ceilings, walls; vibration isolation devices.
5. Some surfaces absorb the sound energy, or do not allow it to reflect effectively.
6. Noise control curtains; fiber-filled cloth, office partitions.
7. Installation of noise refuges/protection.

(b) Prevention

1. New plant specification, including liaison with designer, manufacturers and suppliers of plant and machinery;
2. Preventive maintenance schedules which incorporate attention to existing and future potential noise emissions;
3. Designation of demarcated ear protection zones;
4. Provision and use of ear protection by all persons who may be exposed;
5. Staff training to recognize the hazards from noise exposure;
6. Propaganda aimed at informing operators as to the causes and effects of exposure to noise; and
7. Health surveillance through audiometric testing of new employees and of existing employees on an annual basis as part of an annual health examination.

13.14. Administrative Controls

It is used when engineering controls are exhausted or infeasible. It includes limiting the time in exposed areas; worker rotation; limiting the number of workers in exposed areas (limited access).

It is further recommended that an organization should practice and publish a statement of policy on noise which states the intention of the company to take the above measures with a view to preventing employees going deaf at work.

13.15. Hearing Protective Devices (HPD)

After engineering and administrative controls are exhausted and infeasible, the OSHA standard requires that the employer must make a variety of hearing protection devises available for all workers who are exposed at ≥85 dB(A) for 8-hour TWA i.e., the workers must be provided HPD at no cost. Employers must ensure workers actually wear the HPD. The ability to select the most comfortable device that provides adequate protection will hopefully help and motivate the workers to use hearing protection. The different types of HPD are shown in Figure 13.4.

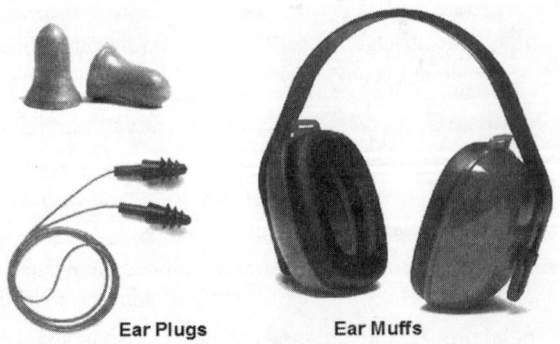

Ear Plugs   Ear Muffs

Fig. 13.4: Different types of hearing protective devices (HPD).

Employers must train workers to use HPD and how to care for them. HPD attenuation must effectively reduce noise exposure to below the OSHA action level (85 dBA). The HPDs are specified with noise reduction rating (NRR) typically ranges from 22 to 30 dB NRR. The numerical attenuation value is determined in a laboratory, when using the A-scale noise, one must deduct 7 dB from the NRR, however no deduction is required for the C-scale.

Example:
An individual is working in a noise level of 98 decibels for an eight hour TWA. Earplugs are available with a 29 NRR and earmuffs are available with a 25 NRR. Let us calculate the NRR of both HPDs.

Since for A-Scale, Earplugs (NRR = 29 – 7 = 22) and Earmuffs (NRR = 25 – 7 = 18)

Earplugs: 98 – 22 = 76 dBA
Earmuffs: 98 – 18 = 80 dBA

Both are below the 85 dBA 8-hour TWA Action Limit.

13.16. Occupational Noise Exposure Assessment: A Case Study

Specially, in iron and steel industries occupational noise is an essential part which may cause severe hearing disorders. The major avoidable cause of hearing disorder is exposure to excessive noise. As reported by Nelson et al. (2005) that 16 per cent of the disorders are due to noise exposure while doing any activity, employees in industry like mining, construction, printing, crushers, drop forging, iron and steel companies, are at high risk of NIHL (Nadi et al., 2008). The present case study revealed that, the workers are exposed to high noise throughout their life time of work. Singh et al. (2009), reported that majority of workers in small scale hand tool forging units are highly exposed to high noise levels as given by OSHA norms [>90 dB (A)], 60–72 hours/week without appropriate personal protective equipment. The employees engaged in various jobs of casting and forging industry are exposed to continuous, intermittent and impulse/impact noise. The noise having maxima more than once per sec is referred as continuous noise and impulsive noise is limited to peak sound pressure levels of 140 dB. The noise in between 80 and 130 dB (A) must be considered as the exposure assessment. Employees exposed to a time weighted average (TWA) of 90 dB (A) and above must be provided some hearing protection (OSHA's Noise Standard Defines Hazard, Protection, 2000 Resource Guide).

In this case Quest Sound Level Meter Model SOUNDPRO SP-DL-1-1/3 was used to assess 'A' weighted (Leq) ambient noise level. In addition, the noise exposure was measured using Noise Dosimeter Model Noise Pro DLX-1 ANSI SI.25-1991 (Figure 13.5). The workers working in various sections were approached and requested to participate in collecting data. The workers who volunteered were instructed about the use of Noise Dosimeter. Dosimeter was attached to the waist and microphone with the collar of the worker. OSHA norms for hearing conservation were incorporated, which include; an exchange rate of 5 dB(A), criterion level at 90 dB(A), criterion time of eight hours, threshold level equal to 80 dB(A), upper limit equal to 140 dB(A) and with F/S response rate. The sound level meter was positioned horizontally on a tripod stand at 1.5 meter height from the ground level. There must be a distance of 1–2 feet radius from the ear of the workers for noise measurement. Equivalent sound pressure level has been measured L_{eq} in various sections of these plants. The sound pressure was recorded for 15 minutes each time on each work station and one long term recording for eight hours was done. Time weighted average dose was calculated under a long term recording of 8 hours. Five observations at different locations in each type of operation were taken.

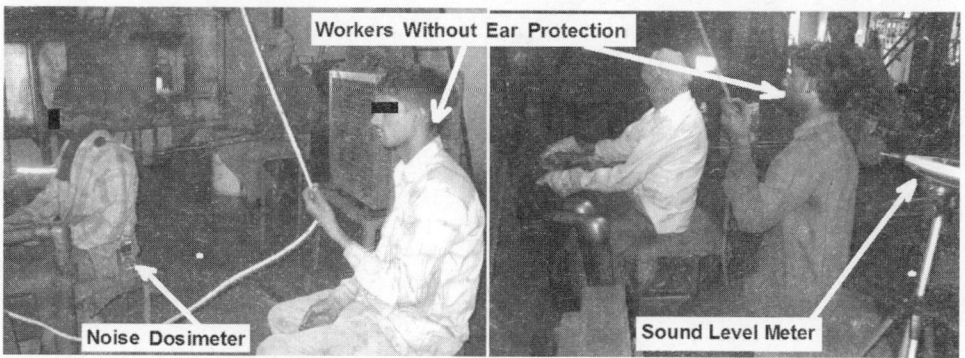

Fig. 13.5: Noise level and noise dose assessment at onsite location.

13.16.1. Occupational exposure with respect to noise levels

Moulding, sand mulling/mixing, grinding, over-head cranes, fettling/shot blasting, cupola–iron feeder and its motor drive, etc., were the noise generating processes in casting units. The sound pressure was recorded for short as well as for long duration. There was hardly a difference of 0.5–1.0 dB (A) between long-term recoding and short-term recording. In the moulding section at different locations, noise was intermittent/impulsive, whereas in grinding and melting sections noise was steady (continuous). In forging industry most of the work locations like; drop forge hammers, punching, blank cutting, and trimming have impulsive noise, whereas the sections like grinding, broaching/machining, tool room, nickel/chrome plating have steady (continuous) noise. Therefore, the workers engaged in different processes were exposed to continuous, intermittent/impulsive/impact noise. The prevailing noise levels (L_{eq}) in various processes of casting and forging SMEs are shown in Tables 13.9 and 13.10. In casting units, the noise level varies from 88.9 dB (A) to 103.5 dB (A) and in forging units ranged from 86.5 to 110 dB (A) L_{eq}. In forging units, noise level at various sections like drop hammer, blanking, punching press, grinding, barrelling, etc., were >90 dB (A), i.e., the prescribed limits of OSHA and the Indian Factory Act 1948. However, the noise levels in other work areas like nickel plating, tool room section, die section recorded less than 90 dB (A), but due to the long working like overtime, the noise exposure in these sections also exceeded the permissible limits. Similarly in casting units the noise levels in all the sections were more than 90 dB (A).

Table 13.9: Prevailing noise levels at various sections of casting units.

M/C Moulding Section	L_{eq} dB (A)	L^{min} dB(A)	L_{max} dB (A)	L_{pk} dB (A)	Wheel Grinding	L_{eq} dB (A)	L^{min} dB(A)	L_{max} dB (A)	L_{pk} dB (A)
Location 1	93.5	68.3	111.4	128.8	Location 1	92.3	81.8	102.1	114.7
Location 2	91.2	67.8	110.2	126.8	Location 2	93.2	80.9	102.4	114.5
Location 3	89.3	67.2	111.0	124.7	Location 3	91.5	81.7	102.7	114.2
Location 4	90.6	70.2	112.0	125.8	Location 4	92.9	81.5	103.8	113.9
Location 5	92.3	72.5	111.7	126.9	Location 5	93.8	82.5	101.9	115.2
Location 6	91.4	71.2	112.4	128.6	Location 6	91.8	82.1	102.7	114.2
Location 7	90.3	73.4	113.5	127.4	Location 7	93.6	81.6	101.8	114.8
Disk Grind Cutting Section	L_{eq} dB (A)	L^{min} dB (A)	L_{max} dB (A)	L_{pk} dB (A)	**Shot Blasting Section**	L_{eq} dB (A)	L^{min} dB (A)	L_{max} dB (A)	L_{pk} dB (A)
Location 1	103.5	81.1	114.1	127.0	Location 1	89.5	75.5	113.0	130.8
Location 2	102.6	80.7	114.0	126.2	Location 2	89.8	81.3	109.9	124.9
Location 3	103.1	81.4	116.7	127.4	Location 3	90.4	78.4	108.7	125.4
Location 4	103.8	80.9	115.3	126.6	Location 4	91.2	80.5	110.9	126.3
Location 5	102.2	81.5	114.7	126.9	Location 5	88.9	78.2	111.4	131.5
Cupola Furnace	L_{eq} dB(A)	L^{min} dB (A)	L_{max} dB (A)	L_{pk} dB (A)	**Induction Furnace**	L_{eq} dB(A)	L^{min} dB (A)	L_{max} dB (A)	L_{pk} dB (A)
Location 1	90.9	88.5	95.2	123.4	Location 1	90.9	88.5	95.2	122.7
Location 2	91.3	89.2	96.9	124.7	Location 2	89.7	89.7	94.9	123.1
Location 3	90.4	89.9	96.3	122.9	Location 3	90.8	87.9	95.1	122.9
Location 4	90.6	88.3	95.8	123.8	Location 4	91.5	88.2	94.8	123.2

Table 13.10: Prevailing noise levels at various sections of forging units.

Blank Cutting	L_{eq} dB (A)	L^{min} dB (A)	L_{max} dB (A)	L_{pk} dB (A)	Belt Grinding	L_{eq} dB (A)	L^{min} dB (A)	L_{max} dB (A)	L_{pk} dB (A)
Location 1	101.9	92.5	115.0	135.2	Location 1	94.9	90.4	98.9	112.9
Location 2	101.2	92.9	116.1	135.4	Location 2	94.5	90.2	98.4	112.5
Location 3	101.6	92.3	115.6	134.9	Location 3	95.5	90.8	99.2	112.9
Location 4	102.4	92.8	115.9	135.1	Location 4	94.9	90.6	99.2	112.6
Drop Forge	L_{eq} dB (A)	L^{min} dB (A)	L_{max} dB (A)	L_{pk} dB (A)	**Barrelling**	L_{eq} dB (A)	L^{min} dB (A)	L_{max} dB (A)	L_{pk} dB (A)
Location 1	102.1	94.8	115.1	135.0	Location 1	99.2	92.1	103.8	118.4
Location 2	101.7	95.5	110.9	129.8	Location 2	101.9	94.7	113	134.6
Location 3	100.6	90.8	113.3	134.8	Location 3	105.3	98.8	111	124.2
Location 4	101.2	94.0	116.0	125.2	Location 4	110.2	98.0	115.0	129.2
Location 5	101.7	92.9	118.6	134.6	Location 5	107.0	93.5	113.5	126.5
Location 6	101.2	82.4	115.7	134.8					
Location-7	103.8	92.2	118.6	140.9					

Contd.

Contd.

Punch Press	L_{eq} dB (A)	L^{min} dB (A)	L_{max} dB (A)	L_{pk} dB (A)	Straightening	L_{eq} dB (A)	L^{min} dB (A)	L_{max} dB (A)	L_{pk} dB (A)
Location 1	101.9	94.7	113.6	134.6	Location 1	93.3	88.1	104.0	120.1
Location 2	99.1	92.1	103.6	118.4	Location 2	92.6	87.1	103.4	119.5
Location 3	96.1	92.4	98.2	112.9	Location 3	93.1	87.9	103.8	121.0
Location 4	96.5	91.1	98.9	118.2	Location 4	92.9	87.9	103.9	119.8
Broaching/ Machining	L_{eq} dB (A)	L^{min} dB (A)	L_{max} dB (A)	L_{pk} dB (A)	**Gauging- Sizing**	L_{eq} dB (A)	L^{min} dB (A)	L_{max} dB (A)	L_{pk} dB (A)
Location 1	101.9	94.7	113.6	134.6	Location 1	93.3	88.1	104.0	120.1
Location 2	99.1	92.1	103.6	118.4	Location 2	92.6	87.1	103.4	119.5
Location 3	96.1	92.4	98.2	112.9	Location 3	93.1	87.9	103.8	121.0
Location 4	96.5	91.1	98.9	118.2	Location 4	92.9	87.9	103.9	119.8
Wheel Grinding	L_{eq} dB (A)	L^{min} dB (A)	L_{max} dB (A)	L_{pk} dB (A)	**Nickel Plating**	L_{eq} dB (A)	L^{min} dB (A)	L_{max} dB (A)	L_{pk} dB (A)
Location 1	96.1	92.4	98.2	112.9	Location 1	88.0	82.8	106.2	121.3
Location 2	96.5	91.1	98.9	118.2	Location 2	87.9	84.7	105.9	119.9
Location 3	94.6	90.2	98.9	113.0	Location 3	88.1	83.8	106.1	120.7
Location 4	95.6	91.2	98.4	114.0	Location 4	87.9	84.5	105.7	119.6
Location 5	96.5	91.2	98.6	116.4	Location 5	86.5	82.4	103.2	118.8
Duplex Grinding	L_{eq} dB (A)	L^{min} dB (A)	L_{max} dB (A)	L_{pk} dB (A)	**Gen Set**	L_{eq} dB (A)	L^{min} dB (A)	L_{max} dB (A)	L_{pk} dB (A)
Location 1	94.7	90.5	104.1	120.0	Location 1	96.1	91.1	102.1	125.6
Location 2	93.8	89.8	99.4	115.3	Location 2	96.5	90.8	102.0	124.5
Location 3	94.7	90.5	104.1	120.0	Location 3	95.9	91.2	100.6	123.6
Location 4	93.8	89.8	99.4	115.3	Location 4	96.8	91.4	102.3	125.5

L_{eq} (dB): The true equivalent sound level measured over the run time, L_{min} (dB): Minimum sound level sampled sound level during the instrument's run time allowing the response that the unit is set for (fast or slow). L_{max} (dB): The highest level sampled sound level during the instrument's run time allowing the response that the unit is set for (fast or slow). L_{pk} (dB): The highest instantaneous sound that microphone detects.

It has been ascertained in chapter17, that majority of the workers were working for overtime (2–4 hours/day) i.e., 12–24 hours/week even without proper ear protection. Therefore, daily noise exposure of the workers in sections like blank cutting, forging, punching, machine moulding, grinding, barrelling, machining/broaching, gauging/sizing exceeds maximum exposure limit of 90 dB(A) specified by OSHA. Moreover, OSHA norms are not valid in Indian SMEs because the Indian Factory Act has prescribed 46 hours/week since the workers work for 10–12 hour/day and six days/week, i.e., exposure time is 60–72 hours per week. Therefore, the workers were under total exposure of 20–32 hour per week (i.e., 50–80 per cent) higher than work exposure norms of 90 dB (a) for 40 hours per week in the USA or the European countries. Thus the workers of casting and forging SMEs were highly exposed to occupational noise.

The prevailing noise levels L_{eq} (A) dB and 8 hours percentage dose is shown in Table 13.11. The noise levels in punching/blanking, forging, moulding, grinding, except gas cutting/welding and tool room sections were more than the prescribed limits, i.e., 90 dB (A). Therefore the percentage dose was also more than 100 per cent. The workers engaged in different activities in forging section were found with high noise dose per cent (like Punching blanking or trimming 367 per cent, forger 811 per cent, furnace job taker 635 per cent and rope puller (hammer operator) 749 per cent). These values are based upon 8 hours of time–weighted average. However, in fact the workers work 10–12 hours per day. Therefore, the actual dose could be rather higher than level shown in Table 13.11.

Table 13.11: Prevailing noise levels L_{eq} (A) dB and mean noise dose (% age).

Occupation	Range of Noise Level $L_{eq}(A)$ dB	Mean Dose %
Punching/blanking/trimming	98.9–102.4	367.5
Forging	101.2–105.1	811
Hot job transferring from furnace to forger	101.2 –103.3	635.3
Rope pulling (Hammer Operating)	101.2–104.5	749.2
Moulding/casting	89.3–93.5	110.4
Grinding	93.8–97.9	302.8
Tool and die making	89.6–92.6	95.1
Gas Cutting/Welding	89.4–60.3	92.0

The noise dose, D, in percent, is given by: $D = 100$ C/T; where C is the total length of the work period in hours, and T is the reference duration corresponding to the measured sound level, L, as given in Table 2, or by the following formula

$$T = 8/2^{\frac{(L-90)}{5}}$$

In most of the casting and forging units, where the main assets are heavy moulding machines, overhead cranes, forging presses, drop hammers, punching/blanking presses, grinding, barrelling, etc., noise is an integral part of machinery in operation. Therefore it is being accepted as a necessary evil. The work place in casting and forging industry of the region is very congested due to lack of availability of space. The simultaneous running of blanking/punching presses, drop hammers, grinders and barrels, machine moulding, fettling and grinding operations, etc., make the working environment very noisy. Hence, the noise exposure is inevitable, there is need to control the noise levels.

13.16.2. Conclusions

On the basis of occupational exposure assessment with respect to onsite hazardous working conditions, the study concluded that majority of the workers were highly exposed to high noise levels >90 dB (A), 60–72 hours/week i.e., the prescribed limits of OSHA as well as the Indian Factory Act 1948. Thus, there is a strong need to strictly enforce the noise exposure norms. The factors like lack of education and awareness about hazardous effects of working conditions, general backwardness in sanitation and lack of control on hazardous working conditions in the industry of this geographic region seem to aggravate the health hazards of workers. Hence, some of the health parameters have further been investigated in the subsequent chapters.

Unsolved Problems

1. In a hand tool manufacturing company 4 hammers are installed in a forging section. According to the sound level meter time weighted averages of sound level for these hammers are: H1 -103, H2-101, H3-102, H4- 103 dB (A). Calculate the net noise level in this section.
2. The time weighted averages of sound level for grinding machines G1,G2,G3,G4,G5,G6,G7 and G8 are 98,96,100,94,96,93,98 and 96 dB (A) respectively. Calculate the net noise level in section.
3. An occupational hygienist visited casting unit and his exposure was monitored for 2 hours at 80 dB(A), 2 hours at 95 dB(A) and 4 hours at 103 dB(A), calculate his the DND (Daily Noise Dose)?
4. A worker engaged in forging job work for 2 hours at ambience noise level of 103 dB without the ear plugs, then he take a tea break of 15 minutes at ambience noise level of 70 dB, after the break he again work for 2 hours in the same forging section at 103 dB (A) but with ear plugs with NNR 22, subsequently he break for lunch for 30 minutes at 70 dB(A). After lunch the worker is shifted to grinding section, there he works for 2 hours without any ear protection, after which he again take a tea break of 15 minutes at 70 dB (A). After the break the worker works for 2 hours at blank cutting machine at 99 dB (A) without ear protection. Calculate the net noise dose received by the worker.

Multiple Choice Questions

1. The pressure difference in air, water or any other medium that is detected by human ear is known as
   (a) Sound
   (b) Vibration
   (c) Air pressure
   (d) All of these
2. The normal human ear is sensitive to frequencies between
   (a) 20 and 20,000 Hz
   (b) 30 and 30,000 Hz
   (c) 10 and 10,000 Hz
   (d) 25 and 25,000 Hz
3. What is the unit of pitch?
   (a) mm
   (b) mm^2
   (c) Hz
   (d) None of these
4. The ratio of the speed of sound in the medium to fundamental frequency is termed as
   (a) Noise
   (b) Wavelength
   (c) Impulsive noise
   (d) Pressure
5. The maximum displacement of a particle from its rest position which determines intensity is
   (a) Amplitude
   (b) Wavelength
   (c) Noise
   (d) None of these
6. What is the speed of sound in air?
   (a) 400 m/sec
   (b) 300 m/sec
   (c) 344 m/sec
   (d) 500 m/sec
7. What is the speed of sound in steel?
   (a) 3000 m/sec
   (b) 4500 m/sec
   (c) 6100 m/sec
   (d) 7000 m/sec
8. Which of the following could be a source of noise in any working environment?
   (a) Noise produced as a result of vibration in machines
   (b) Noise taking a structure borne pathway
   (c) Radiation of structural vibration into the air
   (d) Noise taking an air borne pathway
   (e) All of the above
9. The vibration is amplified in the middle ear by the ossicles and vibrations are transmitted from
   (a) Malleus to incus
   (b) Incus to stapes
   (c) Nerves to brain
   (d) None of these

10. The particular power/level of the sound energy with which it confronts the ear and describes rate of flow of sound energy is known as
    - (a) Sound intensity
    - (b) Noise
    - (c) Vibration
    - (d) Impulsive noise

11. Loudness of the sound wave depends upon
    - (a) Intensity and amplitude
    - (b) Intensity only
    - (c) Amplitude only
    - (d) None above

12. 1 bel is expressed as
    - (a) $\log_{10}10^1$
    - (b) $\log_1 10^1$
    - (c) $\log_{20}10^1$
    - (d) $\log_{10}10^{1.5}$

13. 1 dB equals a change in intensity of
    - (a) 1.36 times
    - (b) 1.46 times
    - (c) 1.26 times
    - (d) 1.16 times

14. The unit of pressure is Pascal and 1 Pa is equal to
    - (a) 1 N/m^2
    - (b) 10 N/m^2
    - (c) 1 N/m
    - (d) 1 N/mm^2

15. The instrument which measures linear sound pressure level in the human audio frequency range is named as
    - (a) Sound pressure level meter
    - (b) Sound intensity level meter
    - (c) Vibration level meter
    - (d) None of these

16. The terms SLM stands for
    - (a) Sound loudness meter
    - (b) Supply loudness meter
    - (c) Sound level meter
    - (d) Sound level measurement

17. The use of Noise Dosimeters include
    - (a) Clip microphone near the ear only
    - (b) Only individual monitoring
    - (c) Clip microphone near the ear and individual monitoring
    - (d) Settings for average, peak, impulse

18. DGFASLI stands for
    - (a) Directorate General Factory Advice Service and Labour Institutes
    - (b) Director General Factory Advice Service and Law Institutes
    - (c) Directorate General Factory Advice Service and Labour Inspection
    - (d) Directorate General Firm Advice Service and Labour Institutes

19. The permissible limits of occupational noise exposure is
    - (a) 70 dB
    - (b) 80 dB
    - (c) 90 dB
    - (d) 95 dB

20. The term noise dose is
    - (a) $D = 100 \times C/T$
    - (b) $D = 90 \times C/T$
    - (c) $D = 80 \times C/T$
    - (d) $D = 100 \times T/C$

Answers

1. (a)	**2.** (a)	**3.** (c)	**4.** (b)	**5.** (a)	**6.** (c)	**7.** (c)	**8.** (e)	**9.** (c)	**10.** (a)
11. (a)	**12.** (a)	**13.** (c)	**14.** (a)	**15.** (a)	**16.** (c)	**17.** (c)	**18.** (a)	**19.** (c)	**20.** (a)

Occupational Noise Induced Hearing Loss (NIHL)

14.1. Introduction

The most common condition associated with exposure to noise is occupational hearing loss. Occupational noise is an integral part of the job especially in iron and steel industries, noise may cause severe hearing loss. The hearing loss occurred due to occupational noise exposure is called as occupational noise induced hearing loss or occupational deafness. Worldwide, about 16 per cent of the hearing disability is caused by occupational noise exposure, according to Nelson et al. (2005). The workers in industries such as mining, construction, printing, crushers, drop forging, iron and steel companies are at high risk of NIHL (Nadi et al., 2008). When an individual is over-exposed to excessive sound levels, sensitive structures of the inner ear can be damaged. This can result in permanent noise-induced hearing loss (NIHL). These structures can be injured by exposure to a brief but intense sound, such as an explosion, or from regular exposure to excessive sound levels over time. Noise may cause hearing loss in three ways.

(a) Temporary threshold shift (TTS) refers to the short-term effect, i.e., a temporary reduction in the hearing ability of an individual that may follow to noise exposure. For example, if a person attend a concert or party function for 1–2 hours, in which DJ system was run at high decibel level, or an individual visit a factory in which various machines such as forge hammer and mechanical presses are running simultaneously and noise level is very high. This noise exposure will consequent in short-term hearing disability which is called temporary hearing threshold shift. The state of temporary threshold shift is usually recoverable, a person has hearing disability for some time, and then recovers the original hearing threshold level, and this is also known as auditory fatigue.

(b) Permanent threshold shift happens where the permissible limit of tolerance is exceeds in terms of; exposure time, noise level and individual vulnerability or susceptibility. The permanent hearing threshold shift (PTS) is result of chronic auditory fatigue, i.e., repeated temporary threshold shifts (TTS). It has been established by Chiu et al. (2007) that TTS driven by noise exposure is further enhanced by heat and work load. In this case recovery from threshold shift does not take place completely, rather will reduce and ultimately stop at some point of time, even after the end of the exposure. The term 'permanent' refers to a condition of no possibility of further recovery from haring loss. The term permanent threshold shift (PTS) denotes the degree of hearing loss (impairment) that remains persist even after 40 hours. However some (partial) recovery of hearing may be found after 2 days away from noise, and even longer.

(c) Acoustic trauma is quite a different condition from occupational noise induced hearing loss (NIHL). It involves abrupt aural damage resulting from short term intense exposure or even from a single exposure. Explosives pressure raises are often responsible, such as exposure to gunfire, major explosions or even fire-crackers.

For most steady type of industrial noise, intensity level and duration of exposure are the main factors in the degree of noise induced hearing loss. Hearing impairment also occurs with age (presbyacusis), therefore it may become very difficult to differentiate between the hearing loss due to noise exposure and normal age induced hearing loss. Research by UK Medical Research Council and the National Physical Laboratory has shown that the risk of noise induced hearing loss can be related to the total amount of energy that is taken in by the ears a working lifetime.

Mild form of noise-induced hearing loss

This is a condition, when a person experiences some difficulty in conversation or discussion with people, the wrong reply may be given occasionally, and difficulty to listen domestic sounds like clock ticking.

Severe form of hearing loss

This refers to a condition of severe degree of deafness; when a person faces difficulty even in face to face conversation with people, they generally complaints that dialogue in the TV and radio are unclear. There is inability to listen the domestic sounds at home and street. It is often impossible to tell the direction from which sound is coming so as to access the distance from the sound. This condition can give rise to inability at common law in an action against an employer, even though the resultant deafness is quite insignificant. Moreover, it is also justifiable to assign an employer's liability for an employee's occupational deafness according to length of time; the employer can be answerable shown to have been in breach of his duty of care.

14.2. Hearing Threshold Shift Due to Occupational Noise Exposure

Occupational noise is an integral part of the job especially in iron and steel industries, and may cause severe hearing loss. As mentioned earlier that 16 per cent of the hearing disability is caused by occupational noise exposure according to a study by Nelson et al. (2005). The workers in industry such as mining, construction, printing, crushers, drop forging, iron and steel companies are at high risk of NIHL (Subroto et al., 2008). Numerous researches reported by various researchers on auditory effects of noise. Taylor et al. (1984) indicated that for long-term exposures of 10 years or more, hearing losses resulting from impact noise in the drop-forging industry are greater than the resulting from equivalent continuous noise. Toppila et al. (2000) also reported that impulsive noise appears to be more harmful for hearing at high noise exposure levels. Krishnamurti (2009) revealed considerable influence of age on noise-induced permanent threshold shifts. On the other hand, gender or ear differences were considerably non-significant. The NIPTS at 2 kHz had been more as compared to NIPTS at frequencies 0.5 kHz, 1.0 kHz, 4.0kHz and 8.0 kHz. Karlsmose et al. (2000) and Helzner et al. (2005) determine the prevalence of and risk factors for hearing loss in a sample of

2,052 older adults (aged 73–84; 46.9 per cent male, 37.3 per cent black) The prevalence of hearing loss was 59.9 per cent; the prevalence of high-frequency hearing loss was 76.9 per cent. Hearing loss was most common in white men followed by white women, black men, and black women. National Academy on Aging Society Washington (1999), reported that, hearing loss was highly associated with aging and According to Kyoko et al. (2005), occupational noise exposure and age both were found to be associated with hearing loss. Ferrite and Santana (2005) also explored that age and occupational noise exposures were, separately, positively associated with hearing loss. Tabuchi et al. (2005) conducted a research work on hearing reduction in 36 small-scale factories where press machinery was actively used. Noise levels in working environments were measured in 34 factories. It was found that as workers' age increased, the percentage of workers having some degree of hearing loss also increased. Ighoroje et al. (2004) also reported that hearing loss was significantly associated with working experience of more than 10 years and overtime. Karlsmose et al. (2000) also reported that, the deterioration increased with age and was higher in males than in females. Other authors like Fransen et al. (2008) had also established that, BMI was associated with hearing loss.

Some studies have shown an association between hearing impairment and chronic alcohol abuse like (Rosenhall et al., 1993). On the other hand, a protective effect of moderate alcohol consumption on hearing had been noted before in some, but not all, studies like; Brant et al. (1996), Popelka et al. (2000), Gates et al. (1993) and Itoh et al. (2001). However in contrast, Starck et al. (1999) could not establish any association between smoking and NIHL.

14.3. Hygiene Standards for Noise-induced Hearing Loss

Standards are based on noise-induced deafness, age-based deafness, and speech range at frequencies of interest, i.e., 500 Hz, 1 kHz, 2 kHz and 3 kHz. The various standards are as follows:

(a) National Insurance Commission: 50 dB hearing loss averaged through 1, 2 and 3 kHz in the ear which hears best.

(b) American Academy of Ophthalmologists and Otolaryngologists: 25 dB hearing loss averaged through 0.5, 1 and 2 kHz in both ears.

(c) British Association of Ophthalmologists: 40 dB hearing loss averaged through 1, 2 and 3 kHz in both ears.

14.4. Cause of Noise-induced Hearing Loss

In order to ascertain how noise induced hearing loss takes place, it is necessary to understand the physiology of the human hearing system. The ear is composed of three specific parts: the outer, middle and inner ears. The anatomy of ear is exhibited in Figure 14.1.

The inner ear contains the vital organ of audibility, called cochlea that comprises a curled fluid-filled pipe, similar in appearance to snail's shell. The cochlea has a basilar membrane, an auditory nerve and frequency responsive hair cells incorporated along its coil. The hair cells run along the length of basilar membrane. A sound entering the cochlea causes vibration of fluid resulting in the hair cells also being vibrated. At the beginning of the cochlean coil the hair cell detects the lower frequencies whilst progressively higher frequencies are detected by hair cells located towards the centre of the coil. The sectioned view of cochlea is shown in Figure 14.2 (A and B). The hair cells effectively turn mechanical energy into electrical impulses which they send to the brain by means of auditory nerve.

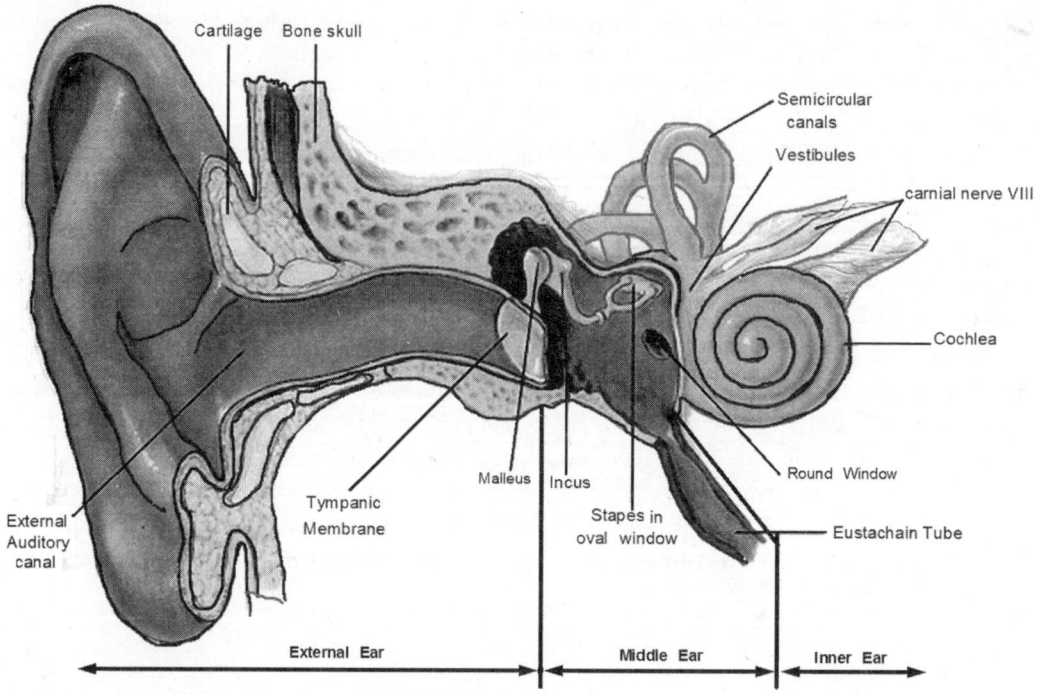

Fig. 14.1: Anatomy of human ear.

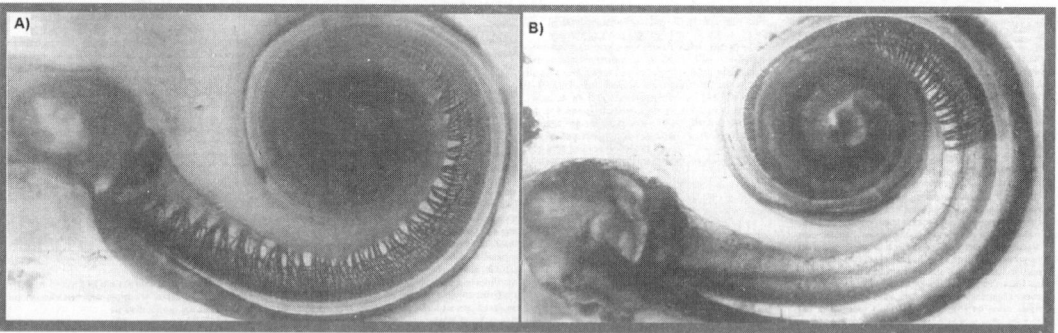

Fig. 14.2: Sectioned view of cochlea with (A) normal hair cells and (B) damaged hair cells.

When the ear-drum is vibrated, the vibration is amplified by the ossicles and transmitted to the cochlea. The cochlea transforms the vibrations into nerve impulses which are sent to the brain via the auditory nerve. If the ear-drum is vibrated between 20 and 20,000 times per sec a sound will be heard. Normally a person can recovers after periods of excessive vibration from noise. However, noise can cause so fatigue to some of these cells that no longer revives and actually dies; thus the row of cells gradually depletes.

In case of noise-induced hearing loss, the projecting hair cells can be compared with a cornfield that has been subjected to storm or people walking through it. Some of the hair cells are flattened and many will be broken off at their base. However in other words it can be explained like; wheat or corn crops survive in a wind up to certain velocity, but when a heavy storm occurs the crops get lay down permanently, and can not revive their position.

14.5. Measuring Hearing Loss-Audiometry

Hearing loss is measured by audiometry, the most widely used technique being pure tone audiometry. This involves the subject sitting in a soundproof booth and listening through earphones to a series of pure tone sounds. Each sound is gradually increased in the intensity until the subject can hear it, whereupon the subject presses a button or raise the index finger of respective side, to indicate that he has perceived that sound. In this way, the hearing threshold, or lowest level at which sound can be heard, is established over range of frequencies.

At the end of the audiometric test, an audiogram is produced which records the hearing of the subject over these frequencies. The audiogram is then compared to an audiogram which, notionally would be produced by perfect hearing, however no one has perfect hearing. The audiogram is used to assess the degree of hearing loss across the frequencies of interest, which are particularly those at which normal speech takes place i.e., 0.5 kHz, 1 KHZ and 2 kHz.

14.5.1. Assessment of hearing threshold using pure tone audiometry

Pure tone audiometry allow us to determine the hearing threshold i.e. the lowest level of sound that a person can hear at given test frequencies. In this method the entire auditory system, from ear canal to auditory cortex, is tested. There are different methods and equipments of testing the threshold levels. Each method involves presenting of sounds to the listener from just audible to just inaudible levels. The listener's threshold is somewhere between these two levels. Whenever a testing program is established, a method is selected and accordingly an audiometer is chosen.

14.5.2. Method for pure tone audiometry

In pure tone audiometry, there are three basic methods that are generally used for testing the hearing threshold; these can be described as follows:

(a) *The method of constant stimuli:* In this method, the listener is given with a series of tones at each intensity and the number of responses is recorded. The intensity at which the number of responses equals half the number of presentations is defined as threshold i.e. the listener should respond at least 50 per cent time of the number of presentations given.

(b) *The method of limits:* In this method, various intensities are presented and how the listener responds at each intensity is recorded. The lowest intensity to which the listener responds at least 50 per cent of the time is recorded as threshold.

(c) *The method of adjustment:* In this method, the listener has control of the signal intensity and sets it to a level so that the signal is just-barely heard, such that if it were less intense it could not be heard at all. This intensity setting is recorded as threshold.

There is a trade-off between time to administer these three methods and the accuracy of the threshold determination. While the most accurate, the method of constant stimuli takes the longest amount of time. The method of adjustment takes the least amount of time, but is the most inaccurate. That leaves the method of limits as the method upon which manual audiometry is based.

The version of the *method of limits* used in manual audiometry is known as the modified Hughson–Westlake procedure (Carhart and Jerger, 1959). This procedure is detailed in ANSI S3.21-1978 (R-1992). In this procedure the signal intensity is first presented at a level the listener can hear clearly. Then the intensity is reduced in fixed-size decrements until the listener no longer responds. The intensity is then increased in smaller fixed-size increments until the listener responds again. From this point on, whenever the listener responds, the signal is decreased and whenever the listener fails to respond the signal is increased. The intensity, when the signal is being increased, to which the listener responds two out of three times is recorded as threshold. A flow diagram for managing the customized Hughson–Westlake method (after Martin, 1986) is shown in Figure 14.3. The tabular audiogram used for recording the hearing threshold is shown in Table 14.1.

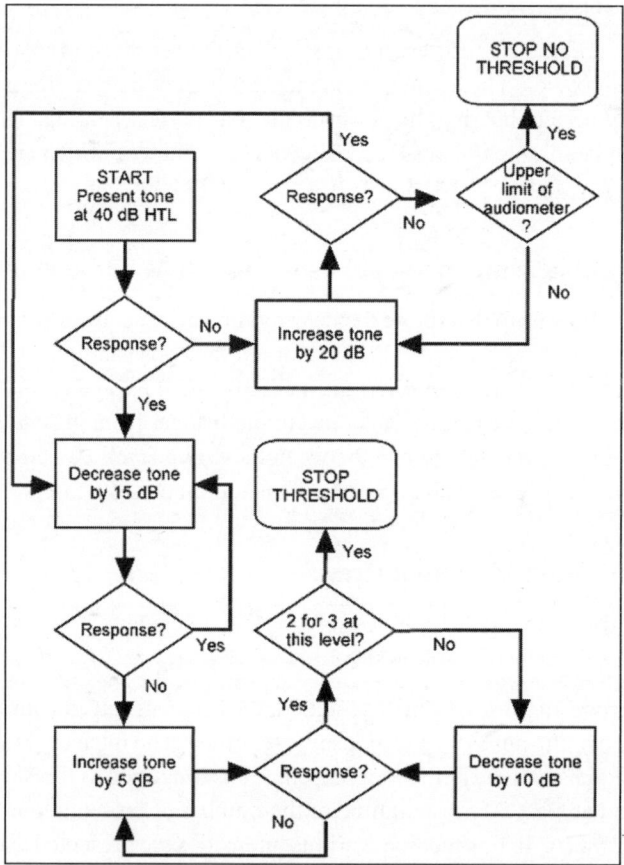

Fig. 14.3: Flowchart of occupational audiometry.

Table 14.1: Tabular audiogram in which hearing thresholds were recorded.

Sr. No.	Frequency (kHz)	0.25	0.50	1.00	1.50	2.00	3.00	4.00	6.00	8.00
1.	Right Ear (dB)	5	10	15	15	20	20	30	20	5
	Left Ear (dB)	10	10	15	15	25	25	35	25	5

14.5.3. Audiometer

A type audiometer has been defined by ANSI S3.6-1996 which have an intensity range from 0 to 90 dB HL for all test frequencies. The Arphi 500 km III portable audiometer works on 220/230 volts AC mains supply as well as on 15 volts DC supply as shown in Figure 14.4. On the right hand side of the audiometer is a battery compartment, which can be accessed by unscrewing the thumbscrews. By a special internal arrangements, the batteries are automatically put off, when the main chord is plugged in. The audiometer is very easy to operate. The earphones have to insert in the jack socket on the side panel. The audiometer can be started by operating switch to 'ON'. The mode can be turned to either 'Right' or 'Left' depending upon the ear to be tested. The frequency dial is adjusted to the test frequency. The examination should be initiated at a frequency of 1000 Hz and increased to 8000 Hz and reset to 1000 Hz down to authenticate feedback from individual's side. The intensity of the tone can be controlled by rotating the hearing loss attenuator (intensity dial). A subject being tested for hearing threshold is shown in Figure 14.5.

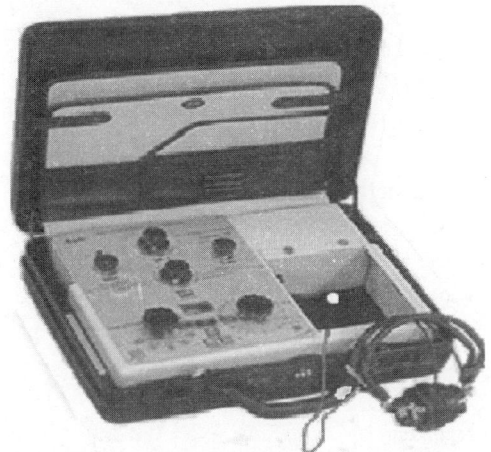

Fig. 14.4: Arphi model 500 MK-III portable audiometer.

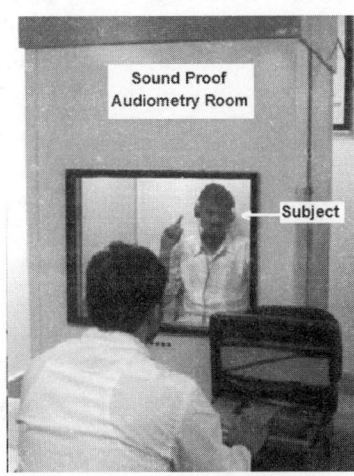

Fig. 14.5: Audiometry test conducted in a sound proof audiometry room.

14.6. Hearing Impairment Criteria

The prevalence of hearing loss is usually determined based on hearing threshold levels (HTL) with a low fence of 25 dB (NIOSH, 1998; Sriwattanatamma and Breysse, 2000, Table 14.2).

Table 14.2: Hearing impairment criteria.

Category	Description of Hearing Impairment	Audiometric Value (Better Ear)	Performance
Zero	No harm	Less than 25 dB	No, or very less hearing problems, able to hear even whispers.
One	Slight impairment	Between 26 and 40 dB	capable for listening
Two	Reasonable impairment	Between 41 and 60 dB	capable for listening words using raised voice at 1 m.
Three	Rigorous impairment	Between 61 and 80 dB	Able to hear some words when shouted into better ear.
Four	Profound harm including deafness	More than 81 dB	not capable for listening

14.7. Protocol and Procedure for Pure Tone Audiometry

The starting of audiometry is the most significant part, which includes positioning of the headphones on the subject. The accurate position is assured by keeping opening of the ear canal to be lined centre to centre with the cut-outs in the headphone. It is very important to assure that left and right side of head phone is positioned on the left and right ear respectively. Among various work-related situations, the headset covering has been highlighted as blue and red on left and right side of ears respectively. Therefore, the person to be tested for audiometry must be given some instructions about the test, these are enumerated as follows:

1. The subject is made to sit in an audiometric room. The subject is asked about the ear in which he hears better, otherwise the test can be started from right ear.
2. Subject is instructed about listening of a few dim voices like beeps.
3. Subject is instructed that, there will be a pulsation of voices for listening to a number of beep which become dim and dim after passage of time and after that quietness.
4. The intensity of sound is fluctuated as in upward range to examine hearing capacity.
5. Whenever the subject listens to some voices of some definite range then he/she must have to halt the process by making any gesture.
6. As and when the subject clearly understands all the instructions, the earphones are put on and the test is conducted. Subject is asked to wait for removal of the earphones after the test was over. An audiometry test being conducted is shown in Figure 14.5. A typical type of audiogram of forge workers versus control group is shown in Figure 14.6.

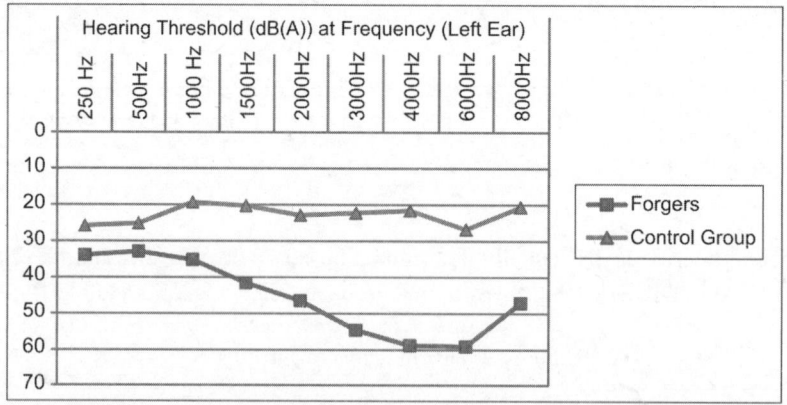

Fig. 14.6: A typical audiogram of a group of workers exposed to high noise in forging section versus control group.

14.8. Effect of Occupational Noise Exposure on Latency Time (Brainstem Auditory Evoked Potentials Response, BERA)

Auditory evoked potentials are scalp recorded electrical responses of the brain generated by acoustical stimuli (Tafti and Ravindran, 1996). Various researchers have investigated the effect of occupational environmental conditions on brainstem auditory evoked potential response. The study of Discalzi et al. (1993) of both mercury and lead exposure to the subjects demonstrated a considerable persistence of wave I–V time. Murata et al. (1990) reported that the combined stress of local vibration, noise, climate and heavy work affected not only the peripheral nervous system but also the brainstem portion of the auditory pathway, the brainstem effect. Murata et al. (1990) reported V[th] peak latency of BERA was significantly prolonged in chain saw operators. Seidel et al. (1992) reported that noise caused an attenuation of the N1 and P2 amplitudes and prolongation of P3 latencies.

Different researchers have investigated the effect of alcohol on latency response. Nicholas et al. (2002) hypothesized that alcoholics would show similar evoked potential changes to those seen in aging. In order to test this hypothesis, Nicholas et al. (2002) studied seven middle-aged abstinent long term alcoholics and eight age-matched normal controls. Differences were noted in either amplitude or latency of the P2 or N350 components, and both groups displayed a prominent LLP potential. Floyd et al. (1997) hypothesized that chronic alcohol ingestion is associated with modifications in components of mid-latency auditory evoked potentials (MAEPs). These findings suggest that chronic alcohol consumption may produce time-dependent structural and/or neurochemical alterations in substrates for cortical information processing, which may be irreversible. In the present paradigm, this irreversibility may occur after 6 or more months of ethanol intake, and may be detected with the use of MAEPs. Keenan et al. (1997) have shown event-related potentials (ERPs) to be different between alcoholics and non-alcoholics. The alcoholic group had significantly longer latencies in P300 measures in both the family history positive and negative groups. Cadaveira and Grau (1990) studied short- and middle-latency auditory evoked potentials (BAEPs and MAEPs) in 15 chronic alcoholic patients after 1 month's abstinence and compared with those of 15 healthy controls, matching the patients pair wise by sex and age. In the alcoholic group, BAEP peak V was significantly delayed and the inter-peak intervals, III–V and I–V, were lengthened. Floyd et al. (1995) also reported ethanol intake associated with increased latencies. Díaz et al. (1990) BAEP peak V was significantly delayed and the inter-peak intervals, III–V and I–V, were lengthened.

Some of the researchers have exclusively investigated the effect of smoking on latency response. The authors like Domino (2003) studied short, middle (EP) and long latency ERP in non-smokers and smokers just after smoking, and after overnight abstinence from tobacco. Short latency potentials were unaffected by tobacco smoking or abstinence. Middle- and long-latency potentials were reduced during abstinence and enhanced immediately after tobacco smoking.

14.9. Assessment of Latency Time using Brainstem Auditory Evoked Potentials Response (BERA)

Brainstem auditory Evoked Potential Response (BERA) are the potentials recorded from the ear and vertex in response to a brief auditory stimulation to assess the conduction through the auditory pathway up to mid-brain. Brainstem auditory evoked potentials comprising of 5 or more peaks within 10 ms of the stimulus. The clinical utility of BAEPs has been established for evaluation of hearing in uncooperative patients and very adolescent, severity of hearing deficits in infants and the middle portion of brainstem function.

14.10. Method for Latency Time (Brainstem Auditory Evoked Potentials)

A Brainstem Auditory Evoked Potential Response Test (BERA) is a recording of the electrical activity coming from the brainstem. The auditory signal when applied to the ear produces action potential. The action potential travels along the auditory pathway. These are described as brainstem auditory evoked potentials and are named from I to VII wave forms. Electrodes are attached to the head of the subjects so that the electrical activity can be measured and recorded. The test is used to evaluate the health of certain brain pathways that cannot be readily accessed by an EEG test (Tafti and Ravindran, 1996). For the measurement of latency time of subjects RMS BERA Mark-II was used, as shown in (Figure 14.7)

Fig. 14.7: Latency time assessed by brainstem auditory evoked potential test using RMS BERA Mark-II.

14.11. Protocol and Procedure for the BAEP

1. The subject is to be instructed not to use hair oil or hair gel before the test if so the skin should be properly cleaned with acetone.
2. The subject is requested to sit on a comfortable chair and four electrodes are glued onto the head. The impedance level of ≤5 kΩ (kilo ohm) must be assured.
3. The earphones are placed over the ears of the subject.
4. The hearing threshold is required to be checked for each ear; e.g., if threshold of either ear is 25 dB, the stimulation is given at (25 dB + 60 dB = 85 dB) for 1500–2000 clicks to each ear. The waves from I to VIIth is plotted as shown in Figure 14.8 (Mishra and Kalita, 2005).

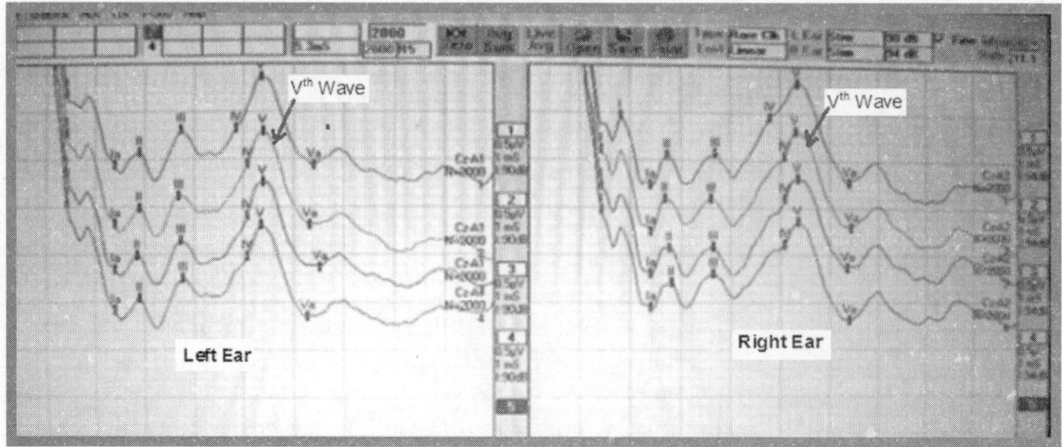

Fig. 14.8: Wave forms produced after recording.

14.12. Case Study: Investigation of Auditory Functions of Steel Industry Workers

14.12.1. Introduction

The steel industry is providing employment to the significant population of the country. The present study is focused to explore the issue of hearing safety, and afterwards determine occurrence of noise at workplace, persuade reduction in hearing capacity along with casting forging firm subjects. Assessment of hearing capacity was performed on a sample of 165 industry workers and compared with sample of 57 controls. Audiometric tests were examined at 1.0–8.0 kHz frequencies. The occurrence of hearing loss was determined based on hearing threshold level (HTL) with a low fence of 25 dB. Student's- test and ANOVA was used for statistical comparison among a variety of groups, a 'p' value of <0.05 was considered as significant. The subjects of steel firms in developing country like India are maximally suffering from noise at work place. The majority of workers are not being protected from occupational noise induces hearing loss (NIHL) as they work without ear protection. Hence, it is an essential duty of every industry to make their employees hazard-free, by offering some preventive measures and safety equipment strictly so as to reduce the health related negative issues due to excessive suffering of noise at workplace.

Alternative hypothesis (H_1): Auditory functions (Hearing Threshold and Latency Time) of exposed group deteriorate significantly.

14.12.2. Results of the case study

The 't' test results of age, BMI and hearing threshold at all frequencies recorded using pure tone audiometric examination of both the groups are shown in Table 14.3. Age of both the groups was statistically matched at 'p' $\leq$ 0.05, whereas the BMI of exposed group was significantly lower than the control group. The industry subjects showed significantly higher hearing threshold level at 1 kHz, 2 kHz to 8 kHz for both ears (at $p < 0.01$). The overall audiograms of both ears; exposed group v/s control group are shown in Figure 14.9.

Table 14.3: The 't'-test results of exposed v/s control group for hearing threshold.

Parameters	Exposed Group (N = 165)	Control Group (N = 57)	Difference of Means	'P' value
Age	30.06 (7.78)	31.98 (8.89)	-1.92	0.123
Frequency	Left Ear Hearing Threshold, Mean (SD)			
1.00 kHz	26.30 (13.59)	19.56 (7.75)	6.74	0.000
1.50 kHz	29.73 (16.15)	20.44 (7.45)	9.28	0.000
2.00 kHz	32.73 (16.36)	22.81 (6.94)	9.92	0.000
3.00 kHz	37.82 (19.04)	21.67 (7.40)	16.15	0.000
4.00 kHz	42.76 (18.05)	21.05 (7.83)	21.70	0.000
6.00 kHz	45.24 (17.45)	25.79 (9.72)	19.45	0.000
8.00 kHz	34.09 (18.46)	19.65 (9.35)	14.44	0.000

Contd.

Contd.

Parameters	Exposed Group (N = 165)	Control Group (N = 57)	Difference of Means	'P' value
Frequency	Right Ear Hearing Threshold, Mean (SD)			
1.00 kHz	27.85 (12.34)	19.82 (7.90)	8.02	0.000
1.50 kHz	31.52 (14.35)	26.49 (31.07)	5.02	0.103
2.00 kHz	34.52 (15.48)	21.93 (7.12)	12.58	0.000
3.00 kHz	39.76 (17.61)	22.54 (7.02)	17.21	0.000
4.00 kHz	43.91 (16.09)	21.67 (7.75)	22.24	0.000
6.00 kHz	48.03 (16.53)	25.79 (9.81)	22.24	0.000
8.00 kHz	34.15 (18.98)	18.07 (10.12)	16.08	0.000

$0.01 > p$ value < 0.05; Significant at 95 per cent confidence interval and p value ≤ 0.01; Significant at 99 per cent confidence interval.

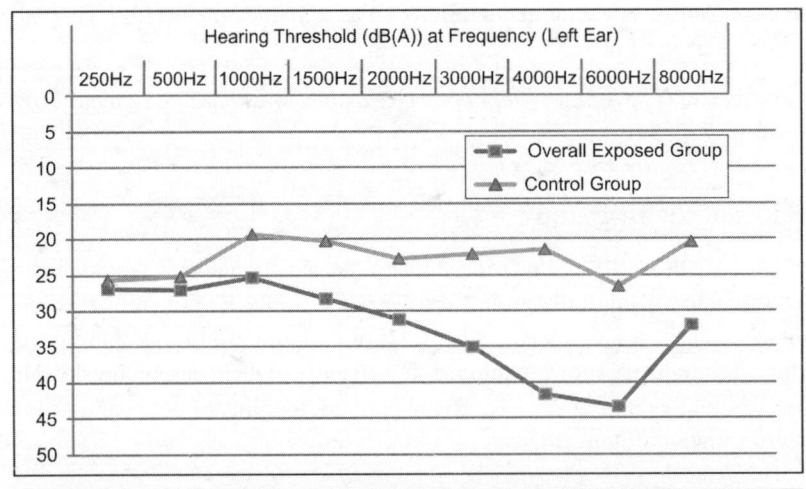

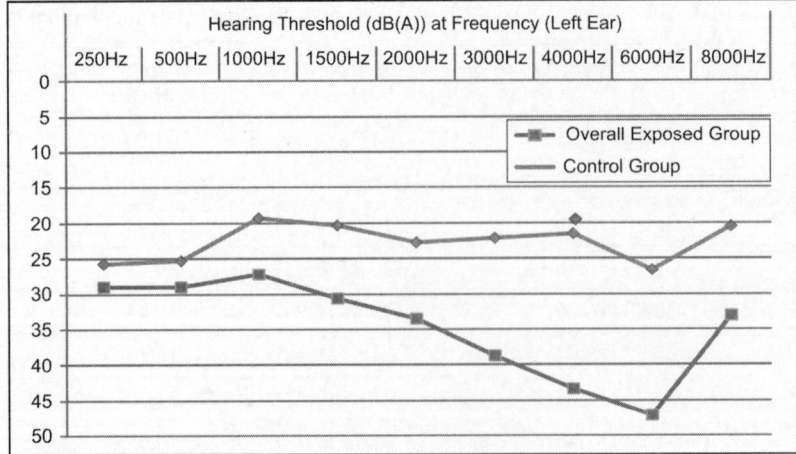

Fig. 14.9: Audiogram of left and right ear of exposed group v/s control group.

14.12.3. Comparison of exposed group v/s control group for latency time

The '*t*' test results of age, BMI and middle ear latency time recorded using brainstem auditory evoked potential response (BERA) of both the groups are shown in Table 14.4. Latency time for both ears of exposed group subjects was significantly higher than that of the control group subjects (at '*p*' ≤ 0.01). The age of both the groups was statistically matched, but BMI of exposed group was significantly lower than the control group at *p* less than 0.05.

Table 14.4: The '*t*'-test results of exposed v/s control group for latency time (LT).

Group Statistics 2	Exposed Group (n= 142)	Control Group (n= 57)	Difference of Means 3	'P' Value
	Latency Time, Mean (SD)			
Age (years)	30.08 (8.51)	29.12 (6.48)	0.95466	0.447
BMI	21.68 (3.460	23.44 (3.56)	-1.75324	0.002
Left Ear LT (ms)	5.68 (0.16)	5.57 (0.10)	0.11958	0.000
Right Ear LT (ms)	5.69 (0.15)	5.57 (0.10)	0.11884	0.000

0.01> *p* value < 0.05; Significant at 95 per cent confidence interval and *p* value ≤ 0.01; Significant at 99 per cent confidence interval.

Since the values of hearing threshold and latency time of exposed group were significantly higher than the control group, hence the hearing capacity and latency response of exposed group were significantly lower.

Alternative hypothesis (H₁): *'Auditory functions (Hearing Threshold and Latency Time) of exposed group deteriorate significantly' is accepted.*

H_1: Auditory functions $_{Exposed\ Group}$< Auditory functions $_{Control\ Group}$ is true.

14.12.4. Effect of workplace conditions on auditory functions (hearing threshold and latency time)

The results of the effect of work place conditions on auditory functions are described as follows.

14.12.4.1. *Effect of workplace conditions on hearing threshold*

Within the exposed group subjects, the work place conditions significantly influenced hearing threshold of the left ear at all frequencies. The right ear hearing threshold was significantly influenced all frequencies, except 1.5 kHz. The workers exposed to impulsive noise, i.e., engaged in forging process were found with significantly higher hearing threshold as compared to the workers engaged in other sections. Process wise, audiograms of left and right ear threshold are shown in Figure 14.10. Workers engaged in forging sections were found with higher loss of hearing as compared to workers engaged in other sections.

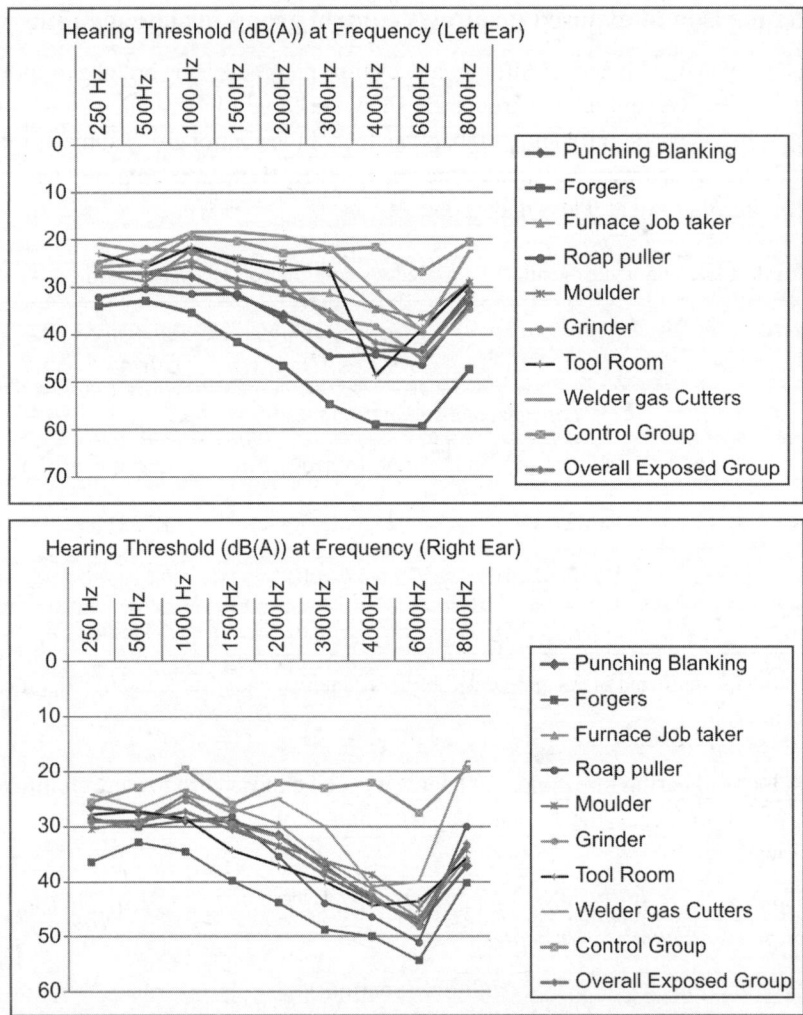

Fig. 14.10: Audiogram of left and right ear of workers engaged in various sections and control group.

14.12.5. Discussion

The results of the present study showed a significant hearing loss, i.e., permanent hearing threshold shift (PTS) in the left ear among the industry subjects at frequencies (1.0 kHz–8.0 kHz) and in the right ear at 1.0 kHz & 2–8 kHz. The industry workers showed a significant hearing loss within less than 9 years of work exposure at the medium (1.5–3.0 kHz) as well as higher frequencies (4.0–8.0 kHz), as compared to the control group. Whereas the previous studies; Celik et al. (1998) reported a significant difference only for the thresholds obtained at 4 kHz ($p < 0.0005$) in a group of noise-exposed industrial workers and the control group with normal hearing. The hearing loss was developed within the first 10 years of noise exposure and associated with slight progress in the following years. The other researchers; Taylor et al. (1984) reported that, for long-term exposures of 10 years or more in the drop-forging industry resulted in hearing losses, as great as or greater than those resulting from equivalent continuous noise. Toppila et al. (2000) also reported that impulsive noise appears to be more harmful for hearing at high noise exposure levels. Since, in the present study, workers were not using proper ear protection and consequently they were highly exposed to continuous as well as impulsive noise levels >90 dB (A), therefore the prevalence of hearing loss was higher. Moreover they were working 60–72 hours/week which was significantly higher than the prescribed working hours of 46 hours/week as per the Indian Factory Act and 40 hours/ week in the USA or European countries. Thus, the hearing loss was associated with occupational noise exposure and other risk factors, i.e., gross occupational exposure to noise demonstrated as a cause of hearing loss in similar to Ahmad et al. (2001).

14.13. Comparison of Exposed Group v/s Control Group for Latency Time

As far as the latency response was concerned, in the present study latency time (for both ears) of industry workers was significantly higher than that of the control group subjects at $p < 0.05$. The past researchers; Murata et al. (1990) reported that, the combined stressors of local vibration, noise, climate and heavy work, affected not only the peripheral nervous system but also the brainstem portion of the auditory pathway, known as the brainstem effect. Murata et al. (1990) reported V^{th} peak latency of BAEP was significantly prolonged in chain saw operators. Similarly, in the present study the industry workers were highly exposed to hostile conditions like high noise level even more than 90 dB (A) as prescribed by Indian standards. Therefore the exposed group showed a significantly prolonged latency response time as compared to the control subjects.

14.14. Effect of Workplace Conditions on Hearing Threshold

Within the exposed group, data analysis demonstrates a prevalence of significant hearing loss amongst the workers exposed to impulsive noise (forging tasks) as compared to the workers associated with the other activities. Therefore it revealed, that the prolonged exposure to impulsive/impact noise at 101.2–105.3 dB (A) for 10 years or more proved to be more hazardous to hearing as compared to continuous or intermittent noise. There was significant difference between hearing threshold level at 1.0 kHz, 1.5 kHz, 2.0 kHz, 3.0 kHz, 4.0 kHz, 6.0 kHz and 8.0 kHz for left ear of forgers and subjects engaged in other job. At the same time, hearing threshold of right ear of forgers was significantly higher as compared to control group at all frequencies and workers of tool room at 2.0 kHz, 3.0 kHz, 4.0 kHz, 6.0 kHz and 8.0 kHz. Hence, PTS was significantly influenced by the type of occupational noise exposure. The permanent hearing threshold shift (PTS) is result of chronic auditory fatigue, i.e., repeated temporary threshold shifts (TTS). It has been established by Chiu et al. (2007) that TTS driven by noise exposure is enhanced by heat and work load. In forging section the workers were exposed to impulsive noise, heat stress and physical work load, thus they were more prone to the hearing loss. The workers of moulding section were under significant hearing loss of both ears at 3–8 kHz, these workers were also under exposure of intermittent/impulsive noise, heat stress and physical work load. Tambs et al. (2006), also compare the frequency-specific effects of noise on hearing acuity across the range 0.25–8.0 kHz. The effect of impulse noise was strongest at 3–8 kHz and varied little within this frequency range. Whereas, Mantysalo and Vuori (1984) had reported, that impulse noise seemed to produce permanent threshold shifts at 4 and 6 kHz after a shorter duration of exposure than continuous steady state noise since, the frequencies most sensitive to impulse noise are 4.0 and 6.0 kHz. The exposure to high levels of impulse noise (despite the use of ear protectors) is more detrimental to hearing than the high levels of continuous noise (Mantysalo and Vuori, 1984). Thus impulsive noise produces permanent threshold shifts at certain frequencies after an exposure of within a decade as compared to continuous noise. Therefore, it is reasonable to consider that, relatively better and superior quality of ear protectors must be provided to protect the hearing of workers exposed to impulsive noise in forging and machine moulding sections as compared to the workers exposed to continuous noise in other sections like grinding, etc. The workers did not use PPEs like ear protectors as they feel uncomfortable with the existing PPEs being provided by the managements of different organizations. Hence, there is a need to provide ergonomically designed ear protectors which can be worn comfortably by workers under extreme weather conditions especially in months of June–August when WBGT and Humidex levels are beyond the comfortable working conditions.

14.15. Effect of Workplace Conditions on Latency Time

Within the exposed group, there was insignificant difference in latency response of workers, engaged in various processes. However, there was significant difference between the latency time of control subjects and the industry subjects engaged in punching, forging, grinding moulding and tool room sections. Post hoc Tukey's test for multiple comparisons among various industry subgroups and control group revealed that latency time of both ears of workers engaged in punching/blank cutting section, forger, grinders and moulder were significantly prolonged than that of the control subjects. Hence, our results are in accordance with Thakur et al. (2004) who reported a considerable decline in the auditory conduction up to the level of the brainstem (BAER).

14.16. Effect of Work Exposure on Hearing Threshold

In a previous study Hong (2005) reported, that hearing loss was particularly higher among workers who reported longer years of working in the construction industry. In the present study within the exposed group, the workers with work exposure 10–15 years and more than 15 years were at significant hearing loss of left ear at all frequencies as compared to the workers with work experience up to 5 years. Hence, hearing loss increased with work exposure. This is in accordance with the previous studies; Kyoko et al. (2005) and Ighoroje et al. (2004), who reported that hearing loss was significantly associated with working experience of more than 10 years and overtime.

14.17. Conclusions

1. The exposed group as compared to the control group, showed significantly lower auditory functions as reflected in higher hearing threshold and latency time. The type of work conditions of specific tasks like forging explained a significant influence on the hearing impairment. Therefore, the PPEs should be provided as per the type of work conditions like noise to which the workers are being exposed.
2. There is a need to consider that, relatively better and superior quality of ear protectors must be provided to protect the hearing of workers exposed to impulsive noise in forging section as compared to the workers exposed to continuous noise in other sections like grinding etc.
3. The workers did not use PPEs like ear protectors as they feel uncomfortable with the existing PPEs being provided by the managements of different organizations. Hence, there is a need to provide ergonomically designed ear protectors which can be worn comfortably by workers under extreme weather conditions especially in months of June–August when WBGT and Humidex levels are beyond the comfortable working conditions.

Multiple Choice Questions

1. The expanded form of NIHL is
   (a) Noise intensity and hearing loss
   (b) Noise intensity and hearing level
   (c) Noise induced hearing loss
   (d) None of the above
2. The term TTS stands for
   (a) Temporary threshold shift
   (b) Tuned threshold shift
   (c) Total threshold shift
   (d) None of the above
3. The chronic auditory fatigue i.e. repeated temporary threshold shifts (TTS) results into
   (a) Predetermined threshold shift
   (b) Chronic threshold shift
   (c) Permanent threshold shift (PTS)
   (d) None of these
4. Human ear can listen to the sound at frequency of
   (a) ≤20 kHz only
   (b) ≥20 Hz only
   (c) Between 20 Hz and 20 kHz
   (d) Any of the above
5. The normal range of hearing is considered when a subject is able to listen at
   (a) ≤15 dB(A)
   (b) ≤35 dB(A)
   (c) ≤25 dB(A)
   (d) None of the above
6. Occupational hearing loss is also affected by
   (a) Occupational noise exposure only
   (b) Duration of noise exposure only
   (c) Occupational heat stress exposure only
   (d) All of the above
7. The spread form of BERA is
   (a) Brainstem Evoked Response of Audiometry
   (b) Brainstem Auditory Evoked Potentials Response
   (c) Brainstem Evoked Potentials Response
   (d) None of these
8. Before starting the BERA recording what level of impedance is required to be maintained
   (a) >5 kΩ
   (b) <3 kΩ
   (c) =5 kΩ
   (d) None of the above
9. If threshold of either ear of a subject is 25 dB, then in BERA testing at what intensity the stimulation should be given?
   (a) 65 dB
   (b) 85 dB
   (c) 95 dB
   (d) None of these
10. In order to investigate the latency time, which wave should be selected
    (a) VIInd wave
    (b) Vth Wave
    (c) IIrd Wave
    (d) All of these
11. Brainstem auditory evoked potentials comprising of 5 or more peaks within
    (a) <10 ms of the stimulus
    (b) >10 ms of the stimulus
    (c) <5 ms of the stimulus
    (d) None of the above
12. The range of hearing frequencies for normal communication is at
    (a) 3–6 kHz
    (b) <4 kHz
    (c) 6 kHz
    (d) None of the above

Answers

1. (c) **2.** (a) **3.** (c) **4.** (c) **5.** (c) **6.** (d) **7.** (b) **8.** (c) **9.** (b) **10.** (b)
11. (a) **12.** (a)

Occupational Heat Stress Exposure

15.1. Introduction

The implementation of occupational hygiene and pollution control measures is still ignored in various developing countries like India. The agenda of occupational safety is not being taken as seriously in small scale industries, this may be due to illiteracy and cheap availability of workers from a large population. The preventive measures and equipment are taken as unwanted and sometimes not provided by the organization. Employees of small scale industries (SMEs) are mostly migrants and they agree with working environment and never consult about preventive measures. Hence, workers engaged in these units suffer from various occupational health hazards and risks. This chapter mainly focuses upon the occupational heat stress assessment and its hazardous effects.

15.2. Heat Stress and Human Performance

All human beings are largely affected by environmental temperature. The physical and mental performance deteriorates because of complex association of physiological and pathophysiological processes. The lack of fluid in human body occurs due to a long exposure to heat stress that lowers patience, possibly affecting mental and psychomotor functions. The assessment of heat stress and physiological response of worker is very vital to maintain optimal working environment for human health and productivity in developing country like India.

Gomes et al. (2002) compared the occupational exposure to heat and protective measures used by the workers of foundry unit and a bottling plant. The foundry workers were found at higher risk of exposure to heat, noise, IR and UV radiations as compared to the workers engaged in the bottle filling unit.

The level of heat stress and noise exceeds the permissible limit. Simultaneously there were observations of high rate of muscle spasm, hearing loss and visual disability (Gomes et al., 2002). The skin temperature is significantly affected by body activities and clothes; thus an index that recognizes work activity and clothing is more appropriate. There must be provision of sufficient rest breaks after a continuous work time so as to minimize the effect of metabolic heat (Onder et al., 2005). The standards of care and awareness will refuse due to any physical discomfort in any working environment which may increase the risk of health and safety in mining (Onder et al., 2005). The systolic blood pressure (SBP) changes with the induced heat stress (Cui et al., 2004). Moreover if a worker doing an activity at a rate of 49 and 98 W/min in front of a usual Soderberg pot, the heat stress individually can put load on the cardiovascular system in a range of 20–25 beats/minute which corresponds to a twenty per cent enhancement in the work load. Therefore, it should be considered to prevent undue fatigue among workers engaged in activities under excessive heat exposure (Rodahl K., 2003). ISO 7243 is a heat stress standard based on the wet bulb globe temperature (WBGT). ISO 7933 gives the applicability of heat stress index in an environment where; mean radiant temperature (t_r) exceeds dry air temperature (t_a). In the warm humid environment, the ISO 7243 standard slightly underestimated and the ISO/DIS 7933 model somewhat overestimated the observed physiological strain. In the hot dry environment, neither of the standards took into account the higher sweating capacities of the physically trained men (Rodahl K. 2003). Webber et al. (2003) stressed the importance of optimizing the quantity of air supplied and the amount of cooling needed and conductive mining methods to establish optimum thermal environments in ultra-deep mining. The effects of ambient temperature on cardiovascular responses in men and women resulted in higher heart rates in the hot room compared to baseline. Systolic blood pressure dropped in the hot condition relative to the baseline. Diastolic blood pressure increased in the cold room relative to the baseline whereas the baseline and the hot room did not differ (Sollers et al., 2002). Forstho et al. (2001) compared sweat rates with recommended rates throughout a re-analysis of 556 climatic chamber experiments on 16 individuals as predicted by ISO standard. As per the obtained results it has been seen non-applicability of ISO 7933 (1989) to climatic conditions with high radiant temperature.

There is a significant effect of heat on the workers' performance that may lead to prospective health hazard and risks. Education of workforce and implementing environmental monitoring with simple interventions could significantly reduce lost time injuries and improve production (Bates and Miller, 2005). Thatcher et al. (2005) reported poor performance correlation between lower productivity and thermal comfort and proposed a total redesign of the thermal environment. There is a series of standards particularly focused upon the estimation of risk level under heat stress environments (Parson, 1999). These include a thermal index to monitor and control hot environments (ISO 7243), an investigation w.r.t. the heat exchange amid a worker and environment (ISO 7933), and a standard describing the basic principles of physiological measurement to establish personal monitoring systems of workers under heat stress (ISO 9886) (Olesen and Parson, 2002). The maximum value of WBGT in heat stress areas for continuous work and interrupted work are identified by international standard. These values are given for metabolic rates ranging from 120 to 340 Wm^{-2} (ISO 7243, 1989). WBGT lowers as 22.5 °C may lead to restrictions of the 'time allowed for work' or the 'work ability' for un-acclimatized people. The restrictions on worktime of individuals acclimatized to hot environments begin at about 26 °C (WBGT) (ACGIH, 2001). The assessment of hot environments, physiological measurements, estimates of metabolic heat production and contact with solid surfaces should be considered (Parsons, 2008).

15.3. Assessment of Heat Stress at Workplace

There are number of equipments available for area heat stress monitoring. However, the current text is confined to the one equipment as has been used by the author. The prevailing onsite heat stress level can be assessed using Area Heat Stress Monitor 'Model Quest Temp 36/6°'. The ambient temperature at each and every work unit should be measured for about 15 minutes at least in different locations. Observations should be recorded at actual workplaces of workers, concurrently on sites of their regular engagements (Figures 15.1 and 15.2). The various categories of job and permissible heat stress criteria are shown in Tables 15.1 and 15.2, respectively (ACGIH).

Fig. 15.1: Heat stress monitoring at onsite locations in forging industry.

Fig. 15.2: Heat stress monitoring at onsite locations in casting industry.

Table 15.1: Categories of job activities on the basis of workload*.

Categories	Example Activities
Light	Light activity with arms while sitting or standing at machine, very light assembly operations, inspecting or monitoring hot processes, using a table saw, standing with light or moderate work at machine or bench and some walking about.
Moderate	Scrubbing in a standing position, carrying or stacking light items, walking about with moderate lifting or pushing, opening valves, walking on level at 6 km/hour while carrying 3 kg weight load.
Heavy	Carpenter sawing by hand shovelling dry sand, heavy assembly work on a non-continuous basis, intermittent heavy lifting with pushing or pulling (e.g., pick-and shovel work).
Very heavy	Shovelling wet sand (very intense activity, at fast to max), climbing stairs rapidly.

Table 15.2: Permissible heat stress criteria for acclimatized workers (WBGT °C)*.

Work/Rest Regimen	Light	Moderate	Heavy
Continuous 100% work	29.5 °C	27.7 °C	26.0 °C
75% Work, 25% rest, each hour	30.5 °C	28.5 °C	27.5 °C
50% Work, 50% rest, each hour	31.5 °C	29.5 °C	28.5 °C
25% Work, 75% rest, each hour	32.2 °C	31.1 °C	30.0 °C

*Source: American Conference of Governmental Industrial Hygienists (ACGIH - 2001).

15.4. Permissible Limits for Exposure with respect to Heat Stress in India

The maximum wet-bulb temperature of air should not be more than 30 °C (86°F) at 1.5 meters (5 feet) above the floor in any work station as shown in Table 15.3 and a minimum of 30 meters per minute (100 feet per minute) air movement, ventilation through water sprays and evaporative air coolers shall be provided.

Table 15.3: Permissible limits for exposure with respect to heat stress as per the Indian Factory Act, 1947.

Dry Bulb °C Temperature	Wet Bulb °C Temperature	Dry Bulb °C Temperature	Wet Bulb °C Temperature
30	29.0	39	28.1
31	28.9	40	28.0
32	28.2	41	27.9
33	28.7	42	27.8
34	28.6	43	27.7
35	28.5	44	27.6
36	28.4	45	27.5
37	28.3	46	27.4
38	28.2	47	27.3

15.5. Case Study

The study includes 350 workers selected randomly from small and medium scale steel firms engaged in casting and forging processes in the state of Punjab. General information, atmospheric conditions, type of activity and protective aids were recorded by using a validated questionnaire translated into Hindi and Punjabi due to the reason that most workers were not able to understand English. The ambient temperature, WBGT$_{in}$ index has been measured at different locations of each unit for at least 4–5 times during extreme temperature (humid) season from months May to August.

15.5.1. Results

The results of case study revealed that high level of heat stress was observed at the site of melting and heating area including cupola, induction and oil fired furnaces and also the heat treatment units. The

WBGT index in these sites was varied 32.17–37.31 °C and 33.47–38.03 °C in casting and forging units, respectively. The results, when compared with the standards, were found to be very high as shown in Tables 15.4 and 15.5 respectively. The range of WBGT level at various areas such as melting furnaces, moulding drop hammer, grinding and barrelling sections was found significantly higher than 28.5 °C for medium work and 27.5 °C for heavy work i.e. the permissible limits (ACGIH, 2001). Humidex in all sections was also found significantly higher than the prescribed values (Table 15.6), however the only the comparison of WBGT w.r.t. ACGIH, 2001 is shown in Figures 15.3 and 15.4. The higher humidex values are indicators of higher probability of heat strokes. On the other hand study also revealed that standardized working hours given by OSHA were not followed by Indian SMEs as the exposure time found 60–72 hours per week as the workers were working about ten to twelve hours per day and six days in a week. This leads to high level of exposure time per week which is more than 150 per cent higher than the exposure limit in the USA or the European countries. Although the present case study focuses upon the indoor heat stress assessment, but the calculation of WBGT for indoor and out door ares are based upon the equations given below.

For indoor or shaded area like; casting/forging shop under the roof, WBGT is determined as follows::

$$WBGT = 70\% \ T_{wb} + 30\% \ T_g$$

and, for outdoor area, WBGT is calculated as follows:

$$WBGT = 70\% \ T_{wb} + 20\% \ T_g + 10\% \ T_{db}$$

Where, T_{wb} is the wet-bulb temperature (70%), T_g is globe temperature (20%) and T_{db} is dry bulb temperature (10%).

Table 15.4: Average WBGT at various sections in forging industry with respect to ACGIH-2001 standards.

Section	WBGT in Temp (Avg °C)	Recommended WBGT (Avg °C) for Moderate work	Recommended WBGT (Avg °C) for Heavy work
1. Forge section	35.4	28.5	27.5
2. Hot trimming	36.5	28.5	27.5
3. Punching	33.8	28.5	27.5
4. Blank heating	37.9	28.5	27.5
5. Heat treatment	34.9	28.5	27.5
6. Grinding	33.8	28.5	27.5

Table 15.5: Permissible heat stress criteria for acclimatized workers (WBGT values in °C).

Work/Rest Regimen	Light	Moderate	Heavy
Continuous 100 % work	29.5°C	27.7°C	26.0°C
75% Work, 25% rest, each hour	30.5°C	28.5°C	27.5°C
50% Work, 50% rest, each hour	31.5°C	29.5°C	28.5°C
25% Work, 75% rest, each hour	32.2°C	31.1°C	30.0°C

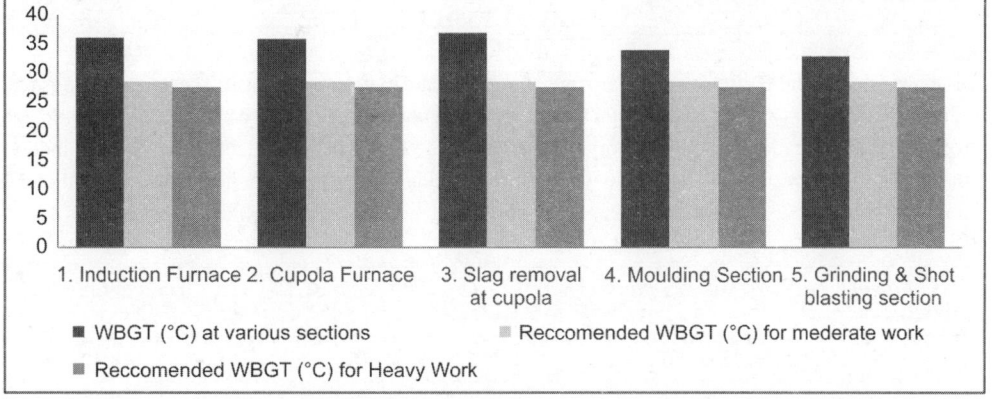

Fig. 15.3: Comparison of average WBGT (°C) recorded at various sections in casting industry with respect to recommended WGBT criteria.

Table 15.6: Degree of comfort for humidex range.

Humidex	Degree of Comfort
20–29	Comfort
30–39	Some Discomfort
40–45	Great Discomfort, Avoid Exertion
46 and over	Dangerous; Probable Heat Stroke

$$\text{Humidex} = T + \frac{5 * \left(\left(6.112 * 10 \left(7.5 * T / (237.7 + T) \right) * H \right) - 10 \right)}{9}$$

where, H is the percentage of humidity in the air and T is the temperature in degrees Celsius.

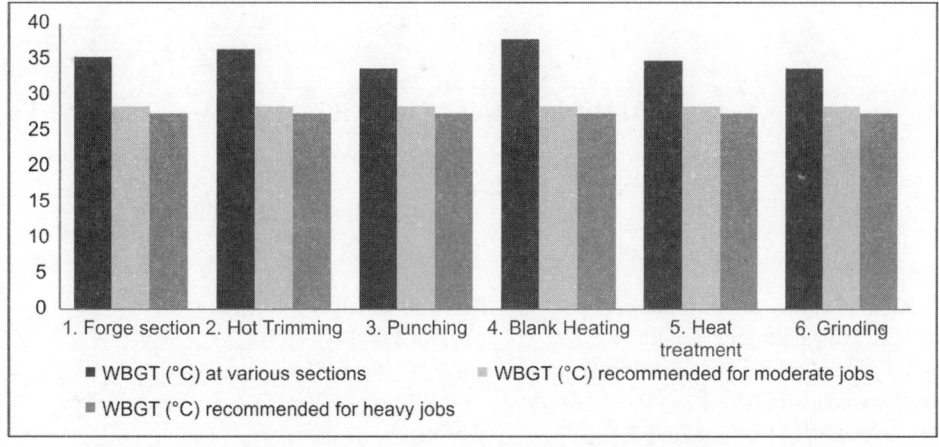

Fig. 15.4: Comparison of average WBGT (°C) recorded at various sections in forging industry with respect to recommended WGBT criteria.

Unsolved Problems

1 Two workers (A and B) are engaged in manual moulding and metal pouring task. They are working at 50 per cent work. Another two workers (C and D) are engaged in hot forging (75 per cent works). The temperatures recorded at four locations by wet bulb, dry bulb and globe are shown below. Calculate the $WBGT_{in}$ indexes for these locations. Investigate whether the workers are working safe or are under heat stress as per the ACGIH norms.

Location of Workers	Wet Bulb Temp °C	Dry Bulb Temp °C	Globe Temp °C
A	30.50	39.45	48.32
B	31.54	38.32	47.87
C	30.75	39.05	47.54
D	31.76	38.64	47.43

2. Two subjects (A and B) are engaged in data entry task. They are working at 100 per cent work. Another two workers (C and D) are engaged in hot forging (50 per cent works). The temperatures recorded at four locations by wet bulb, dry bulb and globe are shown as follows. Calculate the $WBGT_{out}$ indexes for these locations. Investigate whether the workers are working safe or are under heat stress as per the ACGIH norms.

Location of Workers	Wet Bulb Temp °C	Dry Bulb Temp °C	Globe Temp °C
A	30.50	39.45	48.32
B	31.54	38.32	47.87
C	30.75	39.05	47.54
D	31.76	38.64	47.43

Multiple Choice Questions

1. Induced Heat Stress may be reflected as:
    (a) Change in SBP only
    (b) Shifting Hearing Threshold only
    (c) Increased HR only
    (d) All of these
2. The term WBGT stands for
    (a) Wet bulb globe temperature
    (b) Wet bulb glare temperature
    (c) Watt bulb globe temperature
    (d) None of these
3. For un-acclimatized people the WBGT lowers as 22.5 °C then it may lead to restrictions of
    (a) Work ability only
    (b) Time allowed for work only
    (c) Both work ability or time allowed for work
    (d) None of these

4. The restrictions on work time of individuals acclimatized to hot environments begin at about
   (a) 22.5 °C   (b) 24.5 °C
   (c) 26 °C   (d) 28.5 °C
5. Identify the model of equipment used to measure heat stress at workplace?
   (a) Model Quest Temp 35/6°   (b) Model Globe Temp 36/6°
   (c) Model Quest Temp 36/6°   (d) Model Quest Temp 34/6°
6. The maximum wet-bulb temperature of air should not be more than
   (a) 86 °F   (b) 76 °F
   (c) 66 °F   (d) 83 °F
7. The term DBT referred as
   (a) Dry bulb temperature   (b) Dry base temperature
   (c) Dry basic temperature   (d) All of these
8. The PHS stands for
   (a) Pre-determined heat stress   (b) Permissible heat stress
   (c) Permissible high stress   (d) All of these
9. SBP is
   (a) Systolic blood pressure   (b) Systematic blood pressure
   (c) Systolic base pressure   (d) None of these
10. The SBP changes with
   (a) Heat stress   (b) Work hours
   (c) Safety measures   (d) All of these

Answers

1. (d)   **2.** (a)   **3.** (c)   **4.** (c)   **5.** (c)   **6.** (a)   **7.** (a)   **8.** (b)   **9.** (a)   **10.** (a)

Dust Fumes and Respiratory System

16.1. Introduction

Dust is as an aerosol composed of solid inanimate (non-living) particles, as defined by the ILO. The term aerosol implies that airborne particles are carried in or contained in air which may be inhaled. An aerosol can embrace liquid droplets as well as solid particles.

Some dusts are fibro-genic that causes fibrotic changes to lung tissue or toxic, they eventually poison the body systems. Examples of fibro-genic dust are silica, cement dust and certain metals, whereas toxic dust may include arsenic, mercury, beryllium, phosphorous and lead. Some toxic dust, such as arsenic, has an acute effect. Others, such as mercury, have a chronic effect. A number of dusts, although not harmful to health, can have a nuisance effect, for example, dust from combustion of solid fuels.

16.2. Definitions Relevant to Dust and Fumes

Particulate–It is a collection of solid particles, each of which is an aggregation of many molecules.

Mist–It is airborne liquid droplets, e.g., oil mist.

Fumes–These are airborne fine solid particulates formed from the gaseous state usually by vaporization or oxidation of metals, e.g., lead fume.

Vapour–These are airborne liquid droplets given off from the surface of a volatile liquid, e.g., trichloroethylene.

16.3. The Behavior of Dusts

All dusts are potential aerosols and the behavior of particles is influenced by

- (a) The rate of air movement,
- (b) Brownian motion, that is the 'joggling' movement or effect imparted to submicron particles by molecular bombardment and
- (c) The size, density and shape of the particle.

The unit of particle size is the micron, which equals one thousandth of a millimetre, designated μm.

When a particle falls in air it does not accelerate indefinitely. Eventually it reaches a speed at which air resistance equals its weight, and thereafter it falls at constant speed, which is called as 'terminal velocity'. That depends to a great extent upon its size and density.

16.4. Physiology of the Human Lung

The lungs are enclosed in the thoracic cavity and have a sponge-like elastic texture. There are expanded or compressed by movements of the thorax in such a way that air is repeatedly taken in and expelled. They communicate with the atmosphere through the trachea or windpipe, which opens into the pharynx. In the lungs, gaseous exchange takes place. Oxygen from the atmosphere is taken in and carbon dioxide from the blood is released into the lung cavities and eventually to the atmosphere. The trachea divides into two bronchi which enter the lungs and divide into smaller branches. These divide further into bronchioles which terminate in a mass of minute thin-walled, pouch-like air sacs or alveoli. The structure of human respiratory system and process of oxygen exchange is shown in Figure 16.1 (A) and (B) respectively.

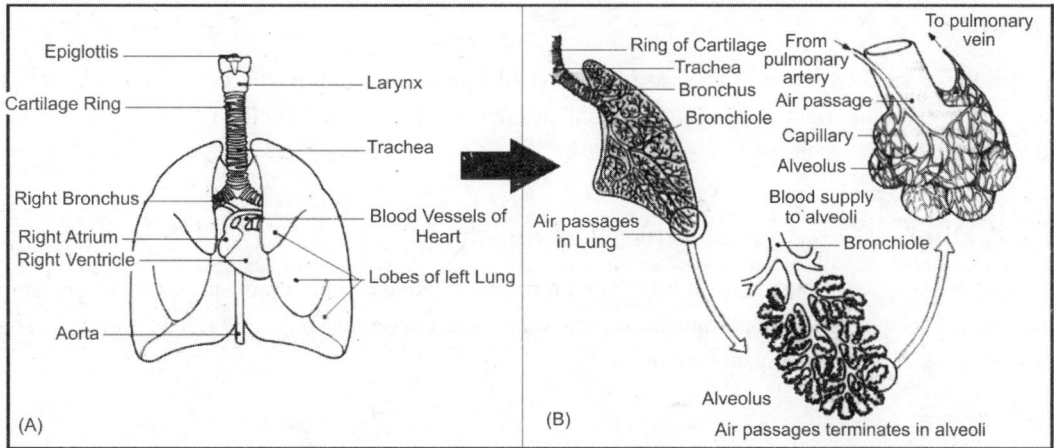

Fig. 16.1: Structure of human respiratory system.

16.5. Physiological Mechanisms of Dust Movement

The mechanisms which induce a particle to move through a particular air pathway into the lung and subsequently be deposited in the lung tissue are the following.

Sedimentation

Dust particles settle under the influence of gravity. The terminal velocity of the sedimenting aerosol is related to the density of the aerosol and to the square of the diameter of the aerosol for those aerosols in the diameter range of 20 μm. Many industrial aerosols are not spheres of uniform shape. However, they may be clumps (aggregates) of particles. The terminal velocity of aggregated aerosols cannot be determined by the above relationship; instead, the aerodynamic diameter must be considered. This is the diameter of a uniform sphere which has the same terminal velocity as the aggregate of other irregular particles. Where the uniform sphere has unit density, the aerodynamic diameter is expressed as the diameter of an equivalent unit density sphere, but where the uniform sphere has the same density as the irregular sphere the diameter is expressed as the stokes diameter, both are measured in microns.

Interception

This is the process whereby irregular particles such as asbestos fibres become caught on the walls of small airways. The length and size of the fibres in relation to the dimension of the airway are important.

Impaction

Impaction takes place through curving in airstream. Suspended aerosols continue under momentum and collide with the wall of the airway. Impaction is related to the velocity of aerosol movement and angular change of direction.

Diffusion

This is the process whereby small aerosols behave like molecules and move freely throughout an air space. It is brought about by the random bombardment of the aerosols by the molecules of the gas in which they are suspended.

16.6. Human Body's Protective Mechanisms

There are a number of mechanisms by which human body endeavours to prevent dust from entering the lungs. The operation of a particular mechanism depends upon the shape and size of particles. The principal protective mechanisms are the following.

The nose

The very coarse hairs lining the nostrils have a filtering effect to trap the larger particles. The cyclonic effect caused by the sudden changes in direction in the nasal passages also causes dust to impinge on the mucous membrane of the nose. In many cases the particle may be expelled by sneezing or blowing the nose.

Ciliary escalator

The surface of the respiratory tract (trachea and bronchi) is lined with special cells, each of which has a cilium growing from its head. The mucous membrane contains mucous glands which secrete a tacky fluid. This forms a sticky film bound up with the cilia. The cilia exhibit a wave-like motion and a particle falling on to the cilia is carried by this motion back to the pharynx, after which it may be expectorated. This mechanism is assisted by mucus which causes particles to adhere to the cilia. Dust deposition locations in the respiratory tract can be broadly classified as given in Table 16.1.

Respirable range particles are, therefore, those particles in the size range 0.43–7.0 microns which enter the various parts of the respiratory tract, those entering the respiratory bronchioles, terminal bronchi and alveoli being the most significant. Consequently, control over particles in this size range is important. Fibres have different deposition characteristics from uniform density spheres. Broadly, asbestos and man-made mineral fibres having a diameter less than 3.5 microns may be regarded as aerodynamically respirable.

Table 16.1: Classification of dust deposition locations in the respiratory tract.

Sr. No.	Size of Dust Particle (μm)	Deposition Location in Respiratory Tract
1.	Above 7	Mouth and throat only
2.	4.7–7	Pharynx
3.	3.3–4.7	Trachea and bronchi
4.	2.1–3.3	Bronchioles
5.	1.1–2.2	Terminal bronchi
6.	0.43–1.1	Alveoli
7.	less than 0.43	Particles less than 0.43 microns tend to remain airborne and are exhaled

Macrophages (phagocytes)

These are wandering scavenger cells with a large nucleus and irregular outline. They move freely through tissue, engulfing bacteria and dust particles in the process. They secrete hydrolytic enzymes which attacks the foreign body, neutralizing its activity to some extent. They are found in alveoli where, after carrying out this scavenging action, they migrate back along the respiratory path way. At the terminal bronchioles they meet the lowest reaches of the ciliary escalator on which they are carried, ultimately to be swallowed or expectorated. In this way macrophage actions supplement respiratory filtration processes.

Lymphatic system

The lymphatic system acts as a form of drainage system throughout the body for further removal of foreign bodies. Lymphatic glands or nodes at specific points in the lymphatic system act as selective filters preventing infection from blood stream. In many cases a localized inflammation occurs in the node.

16.7. Sources of Dust

There are many industrial sources of dust, and they may be classified as:

(a) Dust produced in the cleaning and preliminary treatment of raw materials. Examples: dust resulting from sand blasting operations in foundries, abrasive treatments for the removal of rust.
(b) Dust produced in processes such as refining, grinding, milling and other size reduction processes.
(c) Manufactured dusts for specific treatments or dressings, e.g. in the dressing of seed corn with powdered mercury based fungicides.
(d) Environmental or background dusts, such as those produced by routine sweeping of factory floors, combustion of fuels, the use of packaging materials or road dusts.

16.8. Dust Control Measures

The following aspects are important in the selection of dust control measures.

(a) The type of dust in terms of particle size, weight, density, air velocity and toxicity.

(b) The source of dust in a particular process.

(c) Number of people exposed, duration of exposure (continuous or intermittent) per day, and the number of days per week this emission takes place.

(d) Methods of monitoring emissions, e.g., static sampling, personal dust samplers, and the results of past monitoring activities.

(e) The efficiency of cleaning procedures, manual methods should be replaced by the use of industrial vacuum cleaners.

(f) The efficiency of dust arresting plant, including the system for the maintenance of and testing the efficiency of such plant.

Emphasis should always be placed on control at source by means of dust arresting plant, in preference to the provision and use of respiratory and other protection.

16.9. Dust Control Strategies

Replacement or substitution

Replacement or substitution of hazardous dust producing processes or materials by a suitable alternative should always be considered first. For instance, the use of mercury-based dressing for seed corn instead of the powdered form, or the replacement of toxic dust producing materials by non-toxic materials, is an effective control strategy.

Suppression

In many cases, the use of a wet process, as opposed to a dry process, will be sufficient to reduce the dust hazard. A typical example is in the pottery industry where flint is ground under water due to the danger of fibrogenic dust emission in a dry process.

Isolation

Isolation entails enclosure of the complete process or the actual point of dust production, for instance the total enclosures of large grinding processes or of tipping points for certain dust producing materials, such as coal, to the exclusion of the work force. Tipping points should be provided with efficient dust arresting plant to prevent dust nuisance to people living in the immediate vicinity.

Local exhaust ventilation (LEV)

Local exhaust ventilation points linked to collection and filtration point must be considered. In most cases, it is necessary to install a system of total or partial enclosure in conjunction with cyclone arrestors. It is vital that factors such as particle size, weight and density, together with efflux velocity, are evaluated prior to the selection of a particular form of dust arrest.

Cleaning and house keeping

High standard of cleaning and housekeeping should be maintained where workers are exposed to a dusty process. Failure to do so can lead to an action for bleach of statutory duty and/or common law duty. At the same time as dust suppression plant, depending on its sufficiency will remove the majority of dust from the working environment. Small quantities may escape as a result of handling large defects of plant malfunction. Hand sweeping, using brushes or brooms should be replaced by mechanical vacuum-cleaning equipment. Operator should be trained in the correct use of equipment. Equipment should be serviced and maintained on a regular basis. Such activity should form part of a general cleaning schedule for the area. For example *in situ* vacuum system (ring mains) has been introduced to facilitate the removal of dust from process and storage area. With this system dust is removed to a central collection point through fixed pipe work connected via hosing to hand-held suction devices.

Personal protection

This aspect subdivides into the following areas:

(a) Medical supervision of exposed personnel for early detection of respiratory conditions, supported by annual health screening by occupational health nurses with referral to the occupational physician where appropriate.

(b) The supply, safeguarding and utilization of personal protective equipments like masks etc, which imply the condition of correct type of respiratory protection according to the dust hazard involved, and which the operators should use all the time they may be exposed to dust; also they should wear a dungaree suite, helmet and gloves.

(c) The provision of high standard of welfare amenities, in particular showering and separate work-wear and personal clothing storage facilities.

(d) The frequent training of management and operators in these procedures.

16.10. Dust Explosion

Dust explosion is a mixture of certain particular organic materials with air which form an explosive mixture. These are normally risk-free in their usual condition, when grinded, sanded or refined they may result into very volatile. In fact, during the processing of tea, sugar, starch and potato as well as metals such as zinc and aluminium have resulted into major dust explosions. Other material such as coal, wood, cork, grain and many plastics can form explosive dust clouds.

A mixture of combustible dust and air may burn with dangerous violence, but all mixtures will not do such things. Dust and air having definite concentration can explode but variation in this range may not result into explosion. Low explosive limit is the range of minimum concentration of dust that may be capable of exploding, and the range above which an explosion will not take place is known as upper explosive limit. Furthermore, a range of the volatile dust concentration is not solely a role of chemical concentration of dust.

For an outburst to take place there must be some form of ignition source available. This can be a hot surface, electrical spark, frictional spark or direct flame. The ignition temperature for sugar is $350\,^{\circ}C$, coal $610\,^{\circ}C$, wood $430\,^{\circ}C$, zinc $600\,^{\circ}C$, polystyrene $490\,^{\circ}C$ and magnesium $520\,^{\circ}C$. The lowest explosive concentration for sugar is $350\ mg/m^2$, coal $550\ mg/m^2$, wood $400\ mg/m^2$, zinc $4,800\ mg/m^2$, polystyrene $150\ mg/m^2$ and magnesium $200\ mg/m^2$.

There are several clearly defined stages of a typical factory dust explosion. The preliminary stage, similar to the situation where fine coal dust is thrown onto an open fire, is the typical 'flare up' where there is a sudden release of flame for an instant. This can, however be sufficient to raise locally deposited dust into suspension in air and cause a localised explosion. This primary explosion stage may not result in great degree of damage but is sufficient to send pressure waves in all directions causing further liberation into air of deposited dust. The secondary explosion stage which is much more devastating than the primary stage follows quickly, resulting in extensive damage and often loss of life. Depending upon the layout of the premises and the presence of walls, which may act as temporary baffles, the secondary stage may take place as one great explosion or a series of lesser explosions in different parts of the premises. Most dust explosions take place, however, in specific items of plant such as spray driers, cyclones, settling chambers, powder silos, pneumatic conveying equipment, grinding plant, disintegrators, milling plant and dust collection systems.

16.10.1. Precautions against dust explosions

The frequent removal of deposited dust by industrial vacuum cleaners is one of the most important strategies in preventing dust explosions. Moreover, dust-producing plant should be checked frequently for leakages. Items of plant such as evaporator driers, storage silos and bins, grain elevators, fluid beds and cyclones should be fitted with explosion reliefs, which minimize the devastation by relieving the explosive pressure to a safe area or to atmosphere. Explosion reliefs (vents) may take the form of lightweight panels installed at the top of evaporator driers, elevators and silos. The size of the explosion relief is related to the volume of the installation and its mechanical strength. There are several methods for calculating the size of explosion relief according to the type of installation and particulate under consideration.

In general, any explosion of a flammable mixture, whether dust or gaseous, which, when ignited in a confined space, reaches its maximum pressure in not less than 40 milliseconds, can be brought under control by methods which include suppression, venting, advance inheriting, isolation and automatic plant shutdown.

As a dust explosion is not an instantaneous occurrence but requires a definite time for the development of maximum pressure, it is possible, by the introduction of a suppressant, to arrest the rise of pressure before it reaches dangerous levels. The explosion suppression system in its simplest form consists of a detector, an electrical power unit and a number of suppressors.

An explosion detector and the associated electrical equipment may also be used to open detonator-operated bursting discs, to close high-speed isolation valves, to inert automatically parts of the plant remote from the seat of the explosion and to shut down the plant immediately an explosion occurs. These methods may be used individually but more often are used in combination, depending upon the type and construction of plant and its operating conditions.

Although the fitting of explosion reliefs may prevent devastation of plant by an explosion, this may not be sufficient to stop flame or smouldering material from spreading elsewhere through rotary valves, worms, conveyors or other inlets or outlets for the plant. The use of explosion detector to initiate inserting and isolating arrangements coupled with automatic plant shutdown, therefore, offers an important additional degree of safety which it is often difficult, if not impossible, to achieve in any other way, a typical type of explosion suppression using an explosion detector, electrical power unit and a hemispherical suppressor is shown in Figure 16.2.

Other precautions include the installation of baffle walls in processing areas to prevent the spread of explosion, regular damping down of dusty areas, enclosure of processes and the use of dust arrest plant appropriate to the type of dust produced.

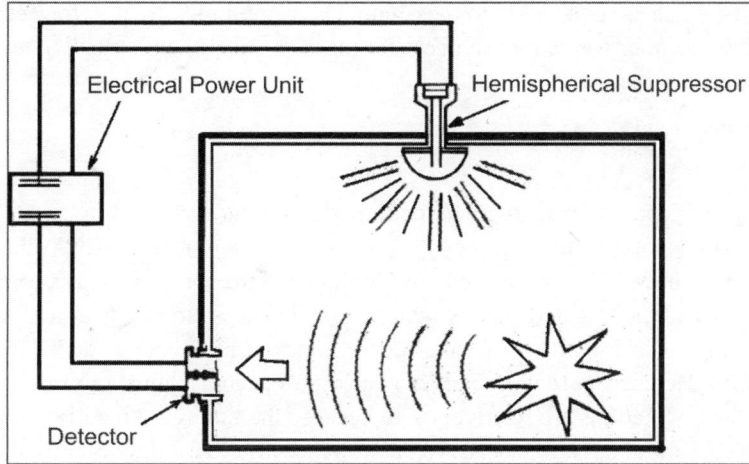

Fig. 16.2: Explosion suppression using an explosion detector, electrical power unit and a hemispherical suppressor.

16.11. Fumes

Fume is formed by the vaporization or oxidation of metals. Typical metallic fumes encountered in industry are lead fume and welding fume, each of which creates ill-effects following inhalation, these are described as follows:

Lead fumes

Environmental control of fume should include damping of process and raw materials, the control molten lead well below 500 °C – the temperature at which fume is produced – and the use of dust and fume control equipment. Any lead process which emits dust and fume should be enclosed and maintained under negative pressure by an enclosing hood, or a hood fitted as close as possible to the source of emission with a capture velocity not less than 1.0 m/sec. Fume must be treated before discharge to atmosphere. Dust and fume arrest plant should incorporate cyclone dust arrestors for the removal of coarse particles and fabric filters or high efficiency wet scrubbers for fine dust and fume particles.

Control measures should be supported by meticulous levels of cleaning and housekeeping, environment and biological monitoring, strict control over personal hygiene and welfare amenities and personal protection measures to prevent the contamination of the body and clothing worn by process workers.

The Work Regulation Act 1980 has an aim to protect people at work exposed to lead. These regulations apply to any work which exposes people to lead, as a mixture of any matter liable to inhale, ingest or immerse into. Duties on the employer include assessing exposure to lead and the provision of adequate controls, including adequate washing arrangements, through to the provision of respiratory protective equipment and medical surveillance. Control measures, respiratory protection and protective clothing provided must be well maintained in working order; also having a strict instruction to all employees to use the measures properly.

Control limits which are equivalent to maximum exposure limits (MELs) for Tetraethyl lead (as Pb) is 0.10 mg m^{-1} (8 hour weighted average) and for lead (other than tetraethyl lead) 0.15 mg m^{-1} (8 hour time weighted average)

Welding fumes

During welding a wide range of airborne particulates are produced according to the base metal and electrodes used. Inhalation of these fumes, gases and dusts may result into 'Welder's lung'. The main constituent of the welding fume are metallic fumes in the form of oxides, together with dust and fumes from flux coatings and metals being welded. The action of heat and ultraviolet light during the welding may result into evolution of ozone, carbon monoxide and nitrogen oxide. These gases are harmful. Heavier particulate matter is also produced as smoke and metal spatter. Most spasmodic welding operations are relatively safe because the fumes are readily diluted by fresh air in the workshop. A serious situation can develop, however, where welding is carried out in confined or unventilated areas.

The following control measures are recommended where welding is to be undertaken regularly. The mechanical ventilation is capable of achieving six to ten air changed per hour. Local exhaust ventilation should be provided at the point of fume production to supplement general ventilation. Portable extraction and filtration units should be used when welding is undertaken *in situ* on production plant. Where welding is carried out in confined spaces, a permit to work system should be operated, together with a system for environmental monitoring. Welders should know the composition of different welding materials in use and any new materials introduced, together with fumes, dust and gases which could be evolved during the welding process. They must also understand the need for different forms of respiratory protection.

16.12. Procedure for Assessment of Respirable Suspended Particulate Matter (RSPM) at Workplace

The procedure involved in assessment of RSPM level at work place includes the use of a dust sampler and pump. There are different types of sampler available, however this text book is confined to use of SKC Leland Legacy pump as shown in Figure 16.3. The procedural steps involved in assessment of RSPM level are as follows:

1. The sampler is attached to the collar of the worker and pump is attached to belt/pant/trouser at the back of the worker or the sampler is hanged on stand within 1 m distance from the nose of the workers (Figure 16.4). The pump is generally set at a flow rate of 9.0 litre/minute.
2. The workers should be instructed not to indulge with the pump and only do their jobs.
3. Filter of 2.0 μm pore size FTFE (Teflon®) is generally used for collecting the samples.
4. Before collecting the sample initial weight of filter is measured, and then final weight after sampling is measured. The difference of these two gives the weight of the sample collected.
5. The initial readings of air sucked and time displayed at the screen of the pump is noted and then pump is started.
6. The pump is run till it gives back flow or gets automatically switched off.
8. Then the final reading of air sampled and time is noted, after that the filter is taken out of the sampler to measure the final weight.

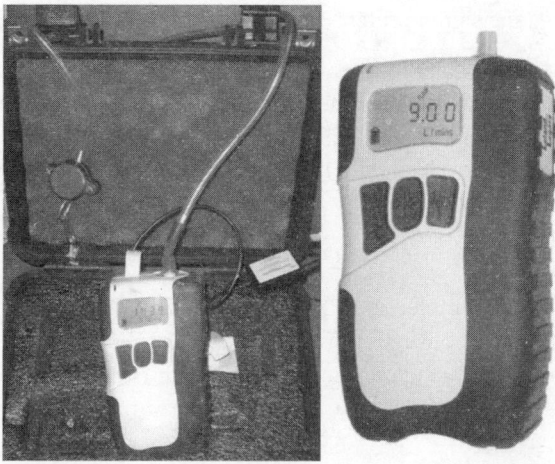

Fig. 16.3: Leland Legacy pumps with sampler for RSPM.

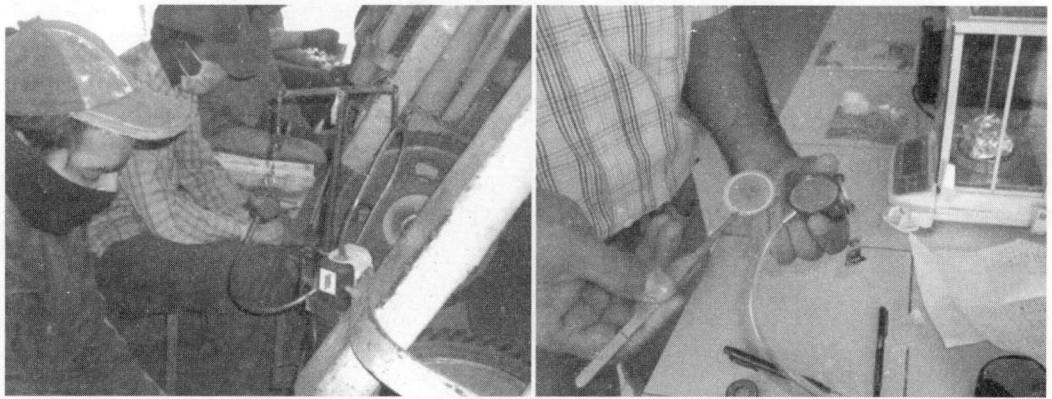

Fig. 16.4: Respirable suspended particulate matter assessment.

For example the RSPM level at various locations in a small scale casting industry is shown in Table 16.2. The following calculations have been used for the same.

Table 16.2: Prevailing RSPM level in various sections of a casting industry.

Name of Section	Initial Weight of Filter (M_1)	Final Weight of Filter (M_2)	Volume of Air Sampled (litre)	V_S = Volume of Air (m3)	Net Wt of Dust (mg)	C = M/V (mg/m^3)	Pump Running Time ('T' hours)	RSPM at 8 hours TWA (mg/m^3)
Moulding/Casting Section								
Location-1	58.6	61.9	550	0.55	3.3	6.00	1.018	47.12
Location-2	57.7	60.2	532	0.532	2.5	4.7	0.985	38.15
Location-3	59.2	61.4	534	0.534	2.2	4.12	0.99	33.33
Location-4	58.9	61.3	580	0.580	2.4	4.14	1.074	30.82
Location-5	59.2	62.2	590	0.590	3	5.08	1.092	37.23

Mass of particles found on the sample filter 'Ms' $= (M_2 - M_1)$

Where, Ms = mass found on the sample filter,
M_1 = the weight (mg) of the clean filter before sampling,
M_2 = the weight (mg) of the sample containing filter,
Note: The blank filters must be subjected to the same equilibrium conditions.
The sampled volume is:

$$Vs = Q \times T/1000$$

Where, Vs = volume of the air sampled in m^3, Q = average flow rate of air sampled, L/minute,
T = sampling time, min, 1000 = conversion from L to m^3
The concentration of the particulate matter in the sampled air is expressed in milligrams/m^3.

$$C = Ms/Vs$$

Where, C = mass concentration of particulate matter, mg/m^3,
Ms = mass found on the sample filter in mg,
Vs = volume of air sampled in m^3, RSPM level at 8 hours TWA (mg/m^3) = $(C/T) \times 8$.
Where, T is running time in hours, C = mass concentration of particulate matter, mg/m^3

16.13. Dust Formation

Air currents carry minute firm particles formed by various breakdown processes, called dust. Dust is a very fine separated solid particle act as airborne without any chemical or physical change other than fracture as defined by the Mine Safety and Health Administration (MSHA). The particles remain airborne settle having large size and others in the air indefinitely.

A very minute particle having size enough to inhale is known as respirable dust and enter deep into the respiratory system beyond the body's natural clearance mechanisms of cilia and mucous likely to be retained.

Dust that can be inhaled, but trapped in the nose, throat, and upper respiratory tract is inhalable dust as described by the Environmental Protection Agency, USA (EPA). Inhalable dust has median aerodynamic diameter about 10 μm, as total dust include airborne particles apart from of their size. Industrial and health related issues are growing as dust emissions are increasing such as respiration problems, irritation to eyes, skin, nose, and damaging of machines/equipment. The American Conference of Governmental Industrial Hygienists (ACGIH) has adopted various standards to assess the severity of health related risks in a working environment act as threshold limit values (TLVs) about which all workers may work for 8 hours per day over extended periods of time without adverse effects.

The severity of health problems depends on the concentration, mineralogical, composition on weight basis (mg/m^3) and quantity basis million particles per cubic foot (mppcf) of air, of dust particles that may cause respiratory related diseases, known as pneumoconiosis, if exposed for a long time period.

16.14. Case Study

The study contains casting and forging units of small and medium scale (SMEs) located in northern India.

In India, industries are on growth at a very fast rate due to globalization and liberalization, particularly the small and medium enterprises (SMEs) industries such as foundries, mining industries and food processing industries. Consequently, the employment rate is also increasing and more and more people are absorbed in jobs including white as well as blue collar labour jobs. However, there remains a gap in terms of use of advanced technologies by SMEs. The shop floor workers are exposed to dust, fumes, gases and other hazardous chemicals, and consequently they are suffering from different respiratory disorders and lungs diseases. This scenario is more in industry and workplaces where, dust and fumes are inevitable in the work environment. The small and medium scale casting firms where the consideration to workers' safety is quite less and moreover the workers involved are uneducated and unaware of vulnerability of the health due to respirable dust and hence they are insensitive towards the use of personal protective/safety equipment (PPEs) at the workplace.

There are several advantages associated with Small Scale Industries like as excess quantity of employees in their financial system and helps to promote the economy balanced development across all the regions. The various industrial sectors as casting and forging make up a significant amount of service (Singh et al., 2010, www.indiaprwire.com). Occupational health and safety practices are ignored sometimes due to the reason of production targets (Singh et al., 2010).

The main objective of the case study was to investigate the level of occupational exposure to respirable suspended particulate matter (RSPM) and assess the level of various respiratory symptoms.

16.15. Materials and Methods

Industrial growth is the backbone of any country and in India's case small scale and medium scale industries are its backbone. While encouraging its growth the occupational hazards of workers are still ignored. So it had to be found whether the respiratory dust exceeding the limit or no use of protective equipment was the real problem. Qualitative and quantitative data were collected in various sections of casting firms.

16.15.1. Qualitative data

A self-designed comprehensive questionnaire is designed which provided us with the qualitative data such as the age, weight, smoking habits, drinking habits, the amount of work exposure, the level of dust exposure, overtime they had to put in, diseases they are suffering from, etc. Around 224 workers were personally interviewed from different sections of an industry. Qualitative data was collected through the questionnaire in various sections like machining, grinding, and moulding section, 224 workers were interviewed personally.

16.15.2. Quantitative data w.r.t RSPM level.

SKC Leland Legacy pump was used for sampling of dust on 2.0 μm pore size, PTFE (Teflon®) filter at flow rate of 9.0 litre/minute. The grinding section was found with the highest dust concentration among all the sections. The machine was put on the stand near the worker on whom the sample had to be collected, eight samples were taken in grinding section and five samples were taken in machining section, eleven samples were taken in moulding section. The data was statistically analyzed to get to the conclusion. Following are the ways in which machine was installed on various sites.

16.15.3. Results and discussion

It is evident from Figure 16.5 that the RSPM level was highest in moulding/casting and grinding section as compared to the machining (Lathe, drilling , broaching) section. Whereas, in other sections RSPM level was found below the prescribed limit of 4 mg/m^3 by the Indian factory Act. But still a high number of persons were suffering from many respiratory diseases as it will be shown by the qualitative data because they were not wearing protective equipment.

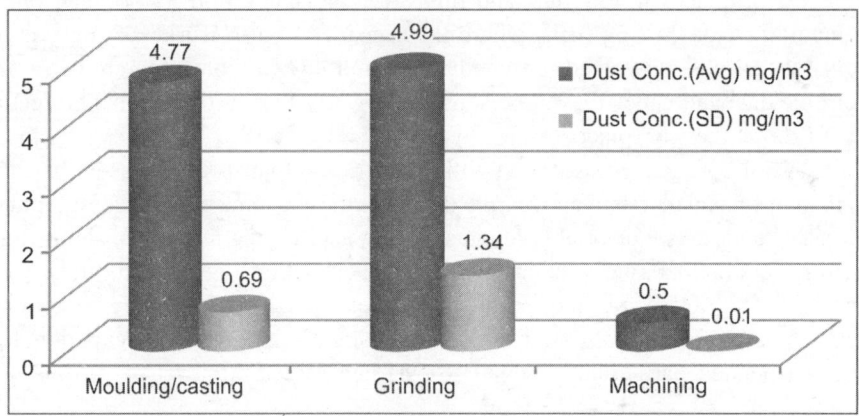

Fig. 16.5: Dust concentration in various sections.

The majority of (53 per cent) workers were unaware as management does not explain the benefits of using PPEs and also in addition to this around 54 per cent workers reported that management does not enforce to wear PPEs at workplace, whereas in only 28 per cent workers reported that they are using proper PPEs. Only 25 per cent workers admitted that they use nose/mouth mask properly. These figures reveal that occupational safety and protection is being ignored in small scale casting units. Therefore, respiratory health hazards risk along with the other health risks is quite obvious. It was also reported by majority of the workers that they opt for an overtime of 3–4 hours per day. Therefore, these workers were exposed to hazardous conditions for 60–72 hours without proper PPEs.

Most of the workers were engaged in grinding, forging, moulding and painting sections out of which approximately 28 per cent of workers were smokers, 37 per cent consume tobacco, 17 per cent 'gutkha[2]'and 'chutki[3]', 9 per cent beetle leaf chewing. On the other hand, 43 per cent workers consume 100–750 ml of alcohol per month. If we talk about the respiratory issues, then 43 per cent have cough, 45 per cent chronic cough, 42 per cent phlegm, 41 per cent chronic phlegm, 35 per cent wheezing, 39 per cent breathlessness, running nose by 28 per cent and throat irritation by 30 per cent workers. On the other hand 10 per cent have asthma as shown in Figure 16.6.

Overall it is observed that workers working in grinding, moulding and casting sections were exposed to high amount of dust fumes and gases as compared to machining section and still workers in these sections were suffering from diseases at the same level as in other sections. This is mainly because workers working even in sections like grinding and machining were unaware to wear any protective equipment. Workers were having smoking and drinking habits. They consumed high amounts of low quality liquor and also consumed tobacco and cigarettes.

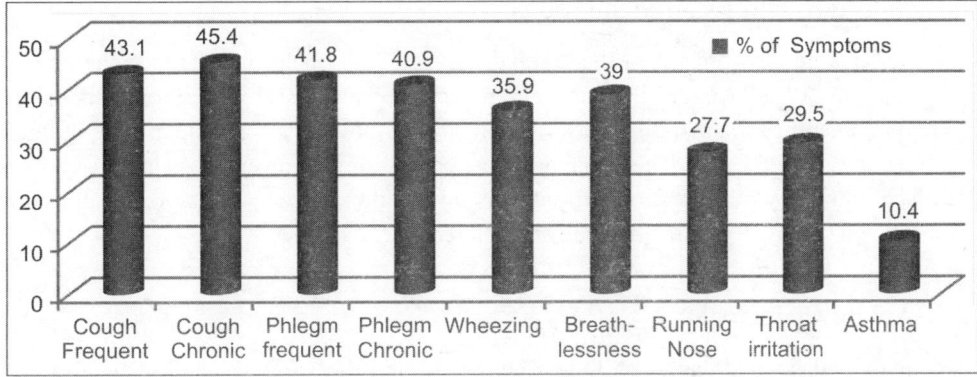

Fig. 16.6: Respiratory symptoms in various sections.

16.15.4. Conclusions

The small scale and medium scale industries have lacked in wearing proper protective equipment (Figure 16.7). This is either due to their lack of education or due to negligence of the company. As they are not wearing any protective equipment health problems facing them are acute. So they must be educated in protective equipment and be made to wear them. This study concluded that workers are exposed to RSPM higher than that of the recommended limits thus they reported with considerable proportion of respiratory symptoms. The management of these casting units takes notice of this problem and they agreed that this was hampering the efficiency of the workers and in turn was having an effect on their production.

Fig. 16.7: Sampling on shop floor in grinding section, workers without complete PPE.

16.15.5. Solutions and recommendations

Some solutions were also offered to the company which could help them reduce the amount of respirable dust. Dust can be controlled through prevention, control systems, dilution and isolation. During bulk manual material handling tasks, the avoidance of dust is very unfeasible but that may minimize by designing the task. The remaining dust particles may prevent by using dust collection systems. The other way of protection is by means of isolation in which workers are placed in an enclosed cab and supplied with fresh, clean and filtered air.

16.15.6. Acknowledgement

The assistance and help extended by the management and the workers of the casting units for conducting personal interviews and collecting dust samples in each section very thoroughly is acknowledged by the authors.

16.16. Lung Functions Measurement using Spirometry

Spirometry is a dominant instrument used to notice, track and manage patients having lung disorders. Spirometry becomes very consistent and easy to include into a routine office visit. It measures the rate during forced breathing manoeuvres by which lung changes volume. Spirometry begins with a full inhalation, followed by a forced expiration that rapidly empties the lungs. These efforts are recorded and graphed. Normal lungs generally can empty more than 80 per cent of their volume in six seconds or less.

16.16.1. Method for spirometry

Spirometry requires considerable patient effort and cooperation. Insufficient attempt may lead to misdiagnosis and inappropriate treatment. Normal spirometry values may vary, and interpretation of results relies on the parameters used. The normal ranges for spirometry values vary depending on the subject's height, weight, age, sex, and racial or ethnic background. American Thoracic Society (ATS) Standardization of Spirometry has recommended the following parameters for pulmonary functions test (PFT).

  (i) Forced Vital Capacity (FVC): The volume of air powerfully and maximally exhale out of the lungs awaiting no more can be expired after deepest possible breath. FVC is usually expressed in litres.

 (ii) Forced Expiratory Volume per Second (FEV_1): The volume of air forcibly exhaled from the lungs in the first second of a forced expiratory (FVC) manoeuvre. It is expressed as litres. This PFT value is critically important in the diagnosis of obstructive and restrictive diseases.

(iii) Ratio of Forced Expiratory Volume (at 1sec) to Forced Vital Capacity, expressed as a percentage (FEV_1/FVC) ratio: The percentage of the total FVC was expelled from the lungs during the first second of forced exhalation. This PFT value is critically important in the diagnosis of obstructive and restrictive diseases.

(iv) FEF_{25-75}: It is a measure of the flow rate in litres per second of the middle half of a FVC test. This test is a sensitive test for the presence of obstructive airway disease. The value of looking at the middle half becomes clear when you realize that the first quarter of the FVC test is in part affected by the subject's effort in overcoming the inertial forces which resist thoracic wall expansion.

(v) Peak Expiratory Flow Rate (PEFR): Peak Expiratory Flow Rate (PEFR) is a measure of the highest expiratory flow rate during the PFT test. It is measured in litres of air expired per second or minute (L/sec or L/minute). Since it is a measure of the peak or maximum flow of expired air, it becomes a sensitive test for the presence of obstructive disease. Patients with a low PEFR would have to be further evaluated for obstructive pathologies. This is a useful measure to see if the treatment is improving obstructive diseases like broncho-constriction, that is secondary to asthma.

(vi) Forced Inspiratory Vital Capacity (FIVC): A forced inspiratory vital capacity is the maximal volume of air inspired with a maximally forced effort from a position of maximal expiration. FIVC indicate expiratory airway obstruction.

16.16.2. Interpretation of spirometry

Pulmonary function abnormalities can be grouped into two main categories: obstructive and restrictive defects. This grouping of defects is based on the fact that routine spirogram measures two basic components–air flow and volume of air out of the lungs. Generally the idea is that if flow is impeded, the defect is obstructive and if volume is reduced, a restrictive defect may be the reason for the pulmonary disorder.

16.16.2.1. *Obstructed airflow*

The patency (dilatation or openness) is estimated by measuring the flow of air as the patient exhales as hard and as fast as possible. Flow through the tubular passageways of the lung can be reduced for a number of reasons:

- Narrowing of the airways due to bronchial smooth muscle contraction as is the case in asthma.
- Narrowing of the airways due to inflammation and swelling of bronchial mucosa and the hypertrophy and hyperplasia of bronchial glands as is the case in bronchitis.
- Material inside the bronchial passageways physically obstructing the flow of air as is the case in excessive mucus plugging, inhalation of foreign objects or the presence of pushing and invasive tumour.
- Destruction of lung tissue with the loss of elasticity and hence the loss of the external support of the airways as is the case in emphysema.
- External compression of the airways by tumours and trauma.

16.16.2.2. *Restricted airflow*

'Restriction' in lung disorders always means a decrease in lung volumes. This term can be applied with confidence to patients whose total lung capacity has been measured and found to be significantly reduced. Total lung capacity (TLC) is the volume of air in the lungs when the patient has taken a full inspiration. The TLC cannot be measured by spirometry because air remains in the lungs at the end of a maximal exhalation i.e. the residual volume or RV. The TLC is therefore the sum of FVC + RV.

16.16.3. Equipment for spirometry test

To perform the FVC manoeuvre, the patient must first breathe in deeply to his full extent. The patient (subject) then places the mouthpiece fixed over the transducer in the mouth and expels the air in his lungs as quickly as possible. Once all the air in the lungs has been expelled, the subject must breathe in (inhale) as quickly as possible, with the mouthpiece still in his mouth, until the lungs are full. There are numerous spirometers available for measuring the lung functions, however spirometer model RMS Helios 401 with reusable mouth piece and nose clip is shown in Figure 16.8.

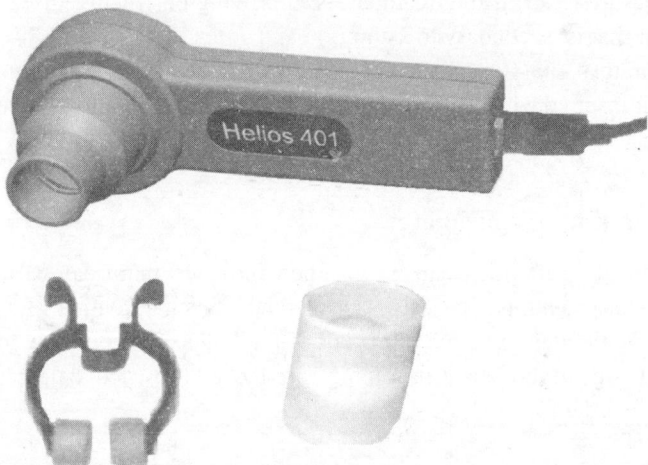

Fig. 16.8: RMS Helios 401 Spirometer, reusable mouth piece and nose clip.

16.16.4. Protocol and procedure of spirometry test

The spirometry test is normally conducted as per the following procedure:

1. Fix the transducer in the transducer housing of the hand piece.
2. The transducer has an arrow on the surface. Insert the transducer in the direction indicated by the arrow.
3. The transducer will click into position when correctly connected. Proper fixation is very important.
4. Fix the mouthpiece (disposable or reusable) over this assembly.
5. The mouthpiece will click into position when correctly connected.
6. Connect the transducer assembly to the computer using the USB-Serial cable.
7. Train the subject (person to be tested) in test performance. The patient's collaboration is essential to carry out the manoeuvre correctly. Use a nose clip to allow air to flow only through the subject's mouth. The patient must sit or stand upright holding the hand piece to his mouth and throughout the manoeuvre the subject should try to keep his back straight as much as possible.
8. The subject should hold hand piece unit in such a way that the air passage is completely unobstructed. Ensure the subject's hands, fingers or clothing, etc., do not obstruct the air

flow. The area in front of the subject should also be kept clear to avoid back-draft of air entering the turbine and affecting the readings.

9. Click Start icon and select either the pre-medication or post medication in the dialog box which appears to specify which type of test is to perform the manoeuvre.

10. Click Stop icon in the main menu. Alternatively, keep the transducer in place until the device detects the end of the expiratory manoeuvre according to ATS (American Thoracic society) criteria.

11. This criterion is satisfied when the volume accumulated during the last second is lower than 0.03 litres.

12. After the manoeuvre is performed, the display screen will show the graph obtained. The manoeuvre number, will be shown either in the center column of the screen (FVC) or along the bottom of the screen (SVC and MVV).

13. After each successive manoeuvre is performed, a dialog box, specific to the type of test, appears (Figure 16.9). The observed values of the current manoeuvre and the previous manoeuvre are compared. Based upon these readings, and the action suggested by the software, the user can *accept* or *reject* the manoeuvre. If the manoeuvre is rejected then the observations are lost. If the manoeuvre is accepted then the readings get stored as a numbered manoeuvre. Only accepted manoeuvres are added to the list of manoeuvres performed.

14. Out of the list of accepted manoeuvres, the user can carry out analysis using the parameters as described in section 16.17.1. The spirogram plotted after performing FVC is shown in Figure 16.9.

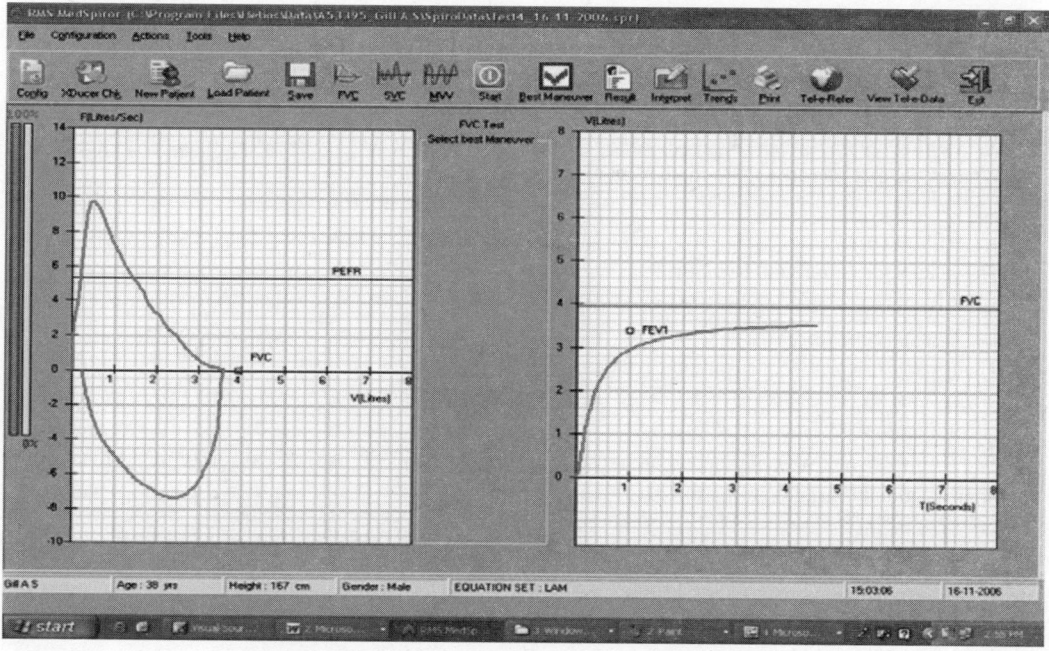

Fig. 16.9: Spirogram plot after performing FVC test.

Multiple Choice Questions

1. Which are the composing particles of dust?
   - (a) Solid
   - (b) Liquid
   - (c) Solid and Liquid
   - (d) Solid and Gas

2. Fibro-genic dust is composed of
   - (a) Silica
   - (b) Certain Metals
   - (c) Cement Dust
   - (d) All of Above

3. All dusts are potential aerosols and the behavior of particles is influenced by
   - (a) air movement
   - (b) 'joggling' movement
   - (c) density and shape of the particle
   - (d) All of Above

4. RSPM stands for
   - (a) Respirable suspended particulate matter
   - (b) Required suspended particulate matter
   - (c) Respirable suspended particle matter
   - (d) None of these

5. A particulate may be
   - (a) Liquid particles
   - (b) Solid particles
   - (c) Gaseous particles
   - (d) All of Above

6. The unit of particle size is measured in
   - (a) One thousandth of millimetre
   - (b) Ten thousandth of millimetre
   - (c) One thousandth of centimetre
   - (d) Ten thousand of centimetre

7. Which of the following gas/gases is released by blood into lung cavities
   - (a) Oxygen
   - (b) Carbon dioxide
   - (c) Oxygen and carbon dioxide
   - (d) None of these

8. In lower respiratory tract diaphragm refers as
   - (a) Skeletal muscle of respiration
   - (b) Organ of gas exchange
   - (c) Chest cavity
   - (d) All of above

9. Upper respiratory tract nasal cavity functions as
   - (a) Filters, warms and moistens air
   - (b) Passage of air, food and liquid
   - (c) Production of sound
   - (d) Air sacs for gas exchange

10. What is the size of particle at which remains airborne and exhaled?
    - (a) 0.43 microns
    - (b) 0.43–0.66 microns
    - (c) 0.66 microns
    - (d) Less than 0.43 microns

11. The phagocytes are found to be in
    - (a) Alveoli
    - (b) Bronchi
    - (c) Diaphragm
    - (d) Lungs

12. The lower explosive concentration of sugar is
    - (a) 350 mg/m^3
    - (b) 450 mg/m^3
    - (c) 550 mg/m^3
    - (d) 650 mg/m^3

13. The Control limits which are equivalent to maximum exposure limits (MELs) for Tetraethyl lead for (8 hour weighted average) is
    - (a) 0.10 mg m^1
    - (b) 0.20 mg m^1
    - (c) 0.14 mg m^1
    - (d) 0.12 mg m^1

14. During welding a wide range of airborne particulates are produced according to
    - (a) Electrode
    - (b) Base metal
    - (c) Electrode and base metal
    - (d) None of these

15. What is the pore size of filter used for Respirable Suspended Particulate Matter (RSPM) at workplace?
    (a) 2.0 μm
    (b) 3.0 μm
    (c) 4.0 μm
    (d) 5.0 μm
16. What should be the flow rate of pump while measuring time weighted (RSPM) level at a location.
    (a) 9.0 litres/minute
    (b) 9.5 litres/minute
    (c) 8.0 litres/minute
    (d) 10.0 litres/minute

Answers

1. (c) **2.** (d) **3.** (d) **4.** (a) **5.** (b) **6.** (a) **7.** (b) **8.** (a) **9.** (a) **10.** (d)
11. (a) **12.** (a) **13.** (a) **14.** (c) **15.** (a) **16.** (a)

Occupational Health and Safety

17.1. Introduction

This chapter comprises of three parts; first part of the chapter briefly describe the environmental law w.r.t. legal control of hazardous substances and processes, followed by basics of safety management. The second part describes about scenario of occupational health and safety in India and abroad. The third part of the chapter represents the results of exploratory case study to bring out the occupational health and safety practices in SMEs and addresses some important issues such as status of health and safety practices in Casting & Forging SMEs, level of use of PPEs, exertions, fatigue and anxiety due to work exposure, work schedule, overtime, alcohol/tobacco/smoking etc intake, work posture and musculoskeletal disorders, health symptoms like respiratory symptoms, BP, hearing disability, etc. and at the end the conclusions of the case study.

17.2. Environmental Law – Legal Control of Hazardous Substances and Processes

There is a huge number of chemicals along with extremely toxic substances increasing at a fast pace and their control and effective management is becoming extremely difficult.

About 5 million chemicals have been synthesized in the past 4-5 decades at the rate of 120,000 every year; about 50–70 thousand chemicals are being used extensively in millions of commercial products. There has been a tremendous increase in the chemical and alkali industry since independence. A total of 515 units spread over 20 states in India have been identified as hazardous installations by a nationwide survey undertaken by ILO.

Today India is the leading producer of pesticides with more production in South Asia. There are some reports available which reveal that many factories near Udaipur are engaged in making acids, secret trade name, and highly toxic chemical for exports. It is a banned item in the western countries because of toxicity and poisonous gases and liquid effluents that emanate from it during its manufacture.

Under the confidentiality clause for protection of trade secrets, the chemical formula has been kept a secret from public and officials and consequently the regulatory agencies have failed to exercise control over them. Therefore in this way some laws are itself hindrances against the legal control of hazardous substances in the country.

The idea of management of hazardous substances is relatively new to the Indian industrial, administrative and legal backgrounds. However, the problem of toxic hazards has already touched dangerous proportions. Therefore, there is a great need of emphasis on these issues.

In fact, the magnitude of the problem was realized by all only after the Bhopal Gas Tragedy. The disaster was the greatest failure of the present time to prevent Environmental Pollution and Human Health Hazards, caused by the accidental release of chemicals into the air.

Similarly the leakage of Oleum Gas from the Shriram Food and Fertilizers Industries in Delhi was another example of such an accident. These disasters have forced State Government to measure the adequacy and safety and pollution control norms in hazardous industries and also to evaluate whether they a pose a danger to the nearby community, some of the laws are listed as follows:

(a) Environmental Act 1986
(b) Indian Factory Act 1948
(c) Insecticides Act
(d) Explosives Act
(e) Petroleum Act
(f) Indian Boiler Act

A survey of the past legislations reflects that perhaps it was the Indian Penal Code, which first dealt directly with negligent conduct in relation to poisonous substances, it mentioned that whosoever does act with poisonous substances in a manner so rash or negligent as to endanger human life or to likely to cause injury or harm to other knowingly; such as acid attacks, neglects to take order with any such substance in his/her possession as is sufficient to guard against danger from such substance, shall be punished with imprisonment of 6 months or a fine of Rs 1000 or both.

The Explosives Act was framed to regulate the manufacture, stocking (holding), sales, conveyance and import of explosive materials. These provisions do not meet the present day emergency requirements. An amendment in this act in 1978, included some modern explosives and provided for variations in the existing conditions of licenses. The Indian Arms Act of 1978 is also inadequate with respect to penalties and scope of its provisions for dealing properly with bombs and other explosives. No sentence of imprisonment can be enforced, only a fine of Rs 3000 can be imposed.

The Insecticides Act deals with the manufacture, import, transfer, sales and distribution of insecticides with a view to preventing risk to human being and other related matters. A review committee of Tiwari and Group observed that the Act did not discourage the use of organo-chlorine pesticides which were in disfavour all over the world for their proven detrimental effects on living natural resources and environment. Thus this act is totally inadequate as a high level of residues is recorded in food stuff, animal tissues and even human fat.

In addition we also have the Water Act (Prevention and Control of Pollution) and the Air Act (Prevention and Control of Pollution). The Factory Act deals with the specific type of pollution concerned with specific category of hazardous substances keeping in view the objectives of regulating the discharge of environmental pollutants and handling of hazardous substances.

The Environment Act was enacted as a comprehensive law to deal with environmental protection as well as for combating pollution and related matters. The Environment Act places the responsibility on the Central Government for laying down the procedure and safeguards for handling of hazardous substances for the prevention of accident, which may cause environmental pollution. This Act imposes duty on persons handling hazardous substances to comply with the prescribed procedural safeguards and every person carrying or running any industry, operation or process, shall be bound to render all assistance to the person empowered by the central government for carrying out the statutory functions, if it fails to do so without any reason he will be punished.

The Factory Act 1948 is another measure, which deals with Environmental Pollution, caused by hazardous processes which is restricted to only inside premises.

The specific law to control water pollution was enacted only in 1974, only after the Stockholm Conference on Environmental Pollution held in 1972. It was followed by another enactment to control Air Pollution in 1981. The delay in enacting such environmental laws has been due to slow industrial progress in India, as well as the bureaucratic mentality coupled with the low level of public awareness of hazardous.

17.3. Indian Penal Code

The need for curbing acts of environmental pollution has been felt for a long time and hence the Indian Penal Code has specified punishment for polluting the atmosphere, affecting the health of person. It states that whoever voluntarily mitigates the health of atmosphere in any place so as to make it noxious for general health of public or passing along a public way, may be imposed a fine which may extend to Rs 500.

17.3.1. The Orient Gas Company Act 1857

It is one of the oldest statute that contains provisions for the regulations of Orient Gas Company that may be discharged during its operations and eventually results in the pollution of air or water. It states that "whenever any gas shall escape from any pipe laid down or setup or belonging to the said company, they shall immediately prevent such gas from escaping after receiving the written notice, and in case the company doesn't remove the cause of complaint within 24 hours even after receiving the notice, then it shall be penalized Rs 50 each day, during which the gas should be suffered to escape". This Act is limited to gas as leakage from pipes and didn't provide for any other forms of air pollution.

17.3.2. The Boilers Act 1923

It deals with the standards of construction of boilers. The Act provides for certified boiler attendants and a central boiler board. The board is empowered to determine the maximum pressure up to which the boiler maybe used.

17.4. Management of Safety at Enterprise Level

A good safety performance is always the result of well-planned and coordinated efforts on the part of the enterprise. It is not achieved by mere compliance with the statutes, campaigns or scattered promotional activities.

Safety is directly linked with quality of work, decisions associated with work, working conditions, place of work, method of work, procedures employed, persons involved with the work as depicted in Figure 17.1. In any organization, there are two important components to assure the safety at a required level of efficiency; (a) machine, equipment and plants, (b) manpower.

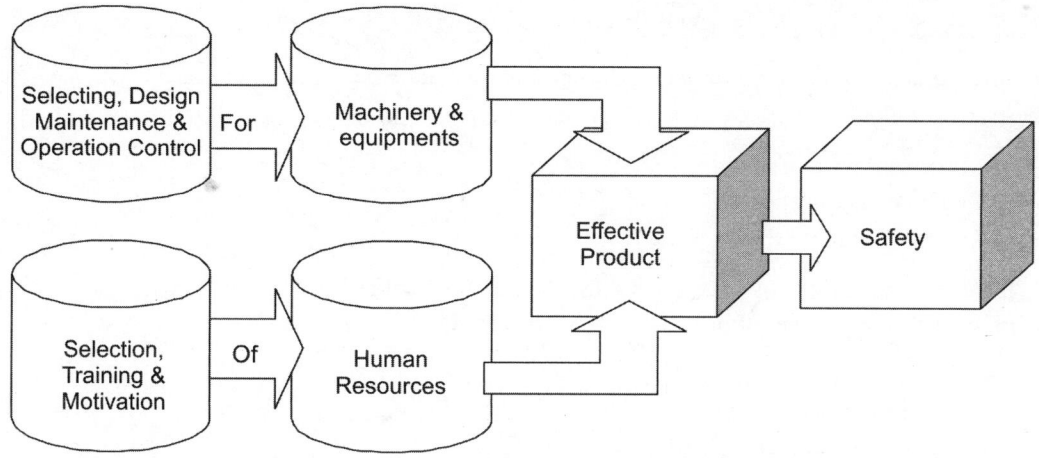

Fig. 17.1: Safety management at an enterprise level.

17.4.1. Safety management is closely related to production/operations management

Safety management mainly involves the following issues;

1. The function of safety management is complicated by some factors like changing technology, and practical difficulties in evaluating and anticipating, the problems of the future.
2. Changes in social values and increasing demand from societies and governments.
3. The vulnerability of human component, which closely interacts with all of the activities and decides results.

On the basis of above cited issues, safety management formulates the following main objectives;

1. To improve the overall standard of safety and health at work.
2. To integrate the concept of safety and health in the overall management of enterprise and organization.
3. To develop good manufacturing practices with employees on efficiency and safety.

Once the issues are identifies, objectives are formulated, the next most important step is to take the necessary actions to accomplish those objectives, therefore safety management need to take the following actions;

1. Identification and evaluation of risks.
2. Providing safe conditions of work.
3. Controlling and directing the actions of persons so as to ensure safe performance.

   - Circumstances, conditions and materials which may cause harm or injury to humans working in the form of unguarded machines, poor illumination, exposure to chemicals and defective wire loops.
   - Unsafe work methods-exceeding the safe speed limit, unauthorized work positions, unsafe work positions and failure to use protective equipment.
   - Maladjustment between machine and man like physical strength and intellectual demands.

17.5. Factors Relating to Work Environment

There are mainly three factors related to work environment which affects the safety management, these are as follows:

1. Poor supervision
2. Poor communication
3. Poor relations on the shop floor

Identification of the root cause of problems is the most important thing, therefore occurrence of any accident fire a number of questions, such as:

- If the purchase was proper?
- Responsibility of upkeep?
- Inspection at proper intervals?
- Environment of storage?
- Any similar accidents in the past?
- Whether instructions for safe use handed out/instructed?
- Whether standard operating procedures are in place for safe use?

Satisfactory answer to these questions will lead to the tenable answer to the root cause.

- Absence of inbuilt checks and defective management policies on procurement as well on maintenance and safety maybe leading to problems.
- A manager has a number of tools, which can help him identify the causes correctly and timely and help in the choice of control measures.
- For general accident prevention and identification of frequent causes of accidents, the most useful tools are:

    1. Judgement and knowledge of managers
    2. Safety inspections
    3. Investigation into accidents

Thus through these tools the managers are in a position to consider the possible causes and evaluate the consequences in a given set of circumstances, provided they are necessary to safety and health issues.

- The regular inspections carried out on the shop floor cannot escape the notice of the workmen. This is in effect an indirect way of demonstrating the management's trust in safety for employees. Thus, contributing to better shop floor relations.
- Safety inspections bring out the area where wastes can be decreased, processes and productivity improved, resulting in the better management of services.
- Safety inspections in context with employees and discussion with them on their difficulties in ensuring security will result in useful suggestions, better understanding and mutual benefits to increase in the safety programme.
- Safety inspections should be objective systematic, regular and should be backed with sincerity of purpose.

17.6. Occupational Health and Safety

According to the International Labour Organization (ILO), *Occupational Health and Safety includes the societal, psychological and physical welfare of workforce; that is the 'complete human being'.* Occupational Health and Safety is a discipline with a broader scope which aims at:

- Encouragement and preservation of the highest degree of physical, psychological and societal welfare and safety of employees within all professions.
- Avoidance of harmful effects on the health of workers and employees due to their work environment.
- Safeguarding of workers during their service from hazardous work conditions.
- Providing and maintaining an occupational environment as per the physical/psychological capabilities and limitations of workers.
- Adjustment and redesigning of work to make compatible with individual's capabilities.

Occupational health not concerned merely with the occupational syndromes, rather it also includes all the factors affecting health of workers. With development of new technology, there is a requirement to recognize the risk factors of modern occupational hazards. India immediately needs up to date occupational health safety (OHS) legislation with sufficient enforcement mechanism; besides it also requires establishment of centres of excellence in occupational medicine to grab up with the rest of the world (Pandve and Bhuyar, 2008). For the successful implementation of occupational health and safety practice, it is necessary that both employers and workers associate and participation in health and safety programmes. It is also mandatory to consider the issues related to occupational medicine, industrial hygiene, toxicology, education, engineering safety, ergonomics, psychology, etc.

It is normally observed that occupational health issues are relatively less preferred than occupational safety issues due to the difficulty to deal with. On the other hand, when health is addressed, the safety is considered inherently, since a healthy workplace is by description also a safe workplace, although the contrary of the same still may not be true, i.e., a safe workplace may not be necessarily a healthy workplace. The most significant point is that both health and safety must be addressed and considered simultaneously in every workplace. On the whole, it is uttered that the definition of *occupational health and safety* covers both health and safety the broadest perspectives.

17.7. Deprived Work Environment

Wherever the poor working conditions exist, there are the potentials of hazard to workers' health as well as their safety. Insanitary and unsafe work environment can originate anywhere, wherever the workplace may be indoor or outdoor. For example in Indian casting and forging industry, the workplace hazardous conditions are inevitable and can impose many health and safety hazards. A typical illustration may be the use of heavy machinery inside forging units like drop hammers, presses, barrels, heating furnaces and moulding machines, induction/cupola furnaces in casting industry. Workforce engaged in various activities get occupational exposure of noise, vibrations, heat stress, dust, fumes, gasses and other toxic chemicals like in spray painting, nickel plating etc. There is a need to put efforts and to implement occupational health and safety at work places to prevent industrial accidents and diseases, and concurrently there is strong need to recognize the association between workers' health and safety, the workplace, and the environment inside as well outside the workplace.

17.8. Significance of Occupational Health and Safety

Work in industry plays a vital role in people's lives, since most workers spend at least eight hours a day in the workplace. Therefore, work environments should be safe and healthy. The industrial developmental potential for small and medium enterprises (SMEs) and the growth prospects of the national economy of a country cannot be achieved without health and safety practices in the work place (Ahasan, 2002). Unfortunately, a few employers realise and assume less accountability for the safeguard issues related to workers' health and safety. Indeed, a number of employers are still unaware that they have the moral as well as legal liability for protecting the workers. Consequently, the job-related accident and diseases are recurrent throughout the globe. For example; in casting and forging industry every day workers deal with substantial health hazards such as: noise and vibration, heat stress, dusts/fumes/gases, heavy load lifting and work posture.

17.9. Indian Scenario of Steel (Casting and Forging) Industry

The Indian Institute of Foundry Men (IIFM) represents the foundry industry and has more than three thousand five hundred members. The foundries support automobile, machine building, sanitary needs, household needs, and related industries. Around forty per cent of the outputs of the foundries are for the automobile sector. The foundry industry has four thousand five hundred foundries in India producing around 7 million tons of castings; many of them are small or medium scale. The industry gives employment to more than five million people directly and three times that number indirectly (www.indiaprwire.com).

According to the Association of Indian Forging Industry (AIFI) a major portion of this industry consists of small and medium enterprises (SMEs). About 200 organised and 1000 unorganised forging units in the country spread across Pune, Chennai, Delhi, Ludhiana and Jalandhar. Steel forgings are an integral part of auto industry. The total capacity at present is estimated to be about 1.5 million tons per annum. As per estimates the industry provides direct employment to about 2 lakhs people, contributing directly to the livelihood of more than three quarter of a million people.

Traditional labour oriented markets are changing towards automation and mechanization. However, at the same time general awareness about the occupational health and safety is not being spread to the society (Jaiswal et al., 2006). Under the current scenario of unemployment workers are scared of loss of their jobs; hence people are ready to work in even hostile conditions and without health and safety consideration. This has made them more susceptible to the hazards of technology than their counterparts in developed countries. Industrialization in India is primarily focused on production, whereas health and safety have a very low priority (Jaiswal et al., 2006). Except a few major reputed public and private industries, other industrialists are insensitive towards the importance of occupational health and safety. The employer of small scale units are totally lagging behind in providing occupational health and safety to the workers. Therefore the manpower employed in small scale casting and forging units are exposed more to occupational noise, heat stress, musculoskeletal strain, dust, etc. (Singh. et al. 2010).

17.10. Present Scenario of Occupational Health and Safety in Developing Countries

A number of studies have been reported on occupational health hazards, most of them are carried out in developed countries. However the developing countries like India is still far behind in this consideration. Mizoue et al. (1999) characterized the Japanese OHSs (occupational health services) for SMEs through comparison with the Finnish services. In Japan OHSs were mostly directed at workers' health management, whereas in Finland they were used for assessment of risk at work. Both countries had prepared OHSs specific to SSEs; the Finnish approach is to integrate various services at the workplace, whereas the Japanese one was to establish a new organization specific to SSEs. To identify the healthcare status in SSEs in Korea, 5,080 factories participated in the government-funded subsidy program in 1997, surveyed by Hyesook et al. (2002). The researchers reported that, overall morbidity of the workers in these SSEs was higher than the national average for both general and occupational diseases. Authors recognized the desirability for a program targeted to occupational health services specifically in Korea. It also concluded that program could be superior replica model for other fast developing nations. The other researchers, Kim and Paek (2000) had recommended that the companies run by better management with better health and safety record would experience less bankruptcy during economic downfall.

Rongo et al. (2004) assessed work-related hazard and health problems in SSI in Dar es Salaam in Tanzania. Researchers reported that workers were highly exposed to multiple health hazards and use of protective equipment was poor. In China, Zhi et al. (2000) reported that 83 per cent of the SSI surveyed in county towns had at least one type of occupational hazard and noise-induced hearing loss was one of the seven types of occupational diseases cited. Gomes et al. (2002) conducted a cross-sectional study to assess the exposure to noise and heat and studied the level of occupational hygiene practices at a foundry and a bottling plant in a rapidly developing country (Dubai, UAE). Thermal stress, relative humidity, ventilation, illumination and noise levels were measured at different work units at foundry and soft drink bottling factory. Thermal stress and noise levels were high (exceeding 90 dB), while relative humidity, ventilation, illumination were low at foundry as compared to bottling plant. But the same study did not report the chronic effects of hazards on health parameters.

In Indian scenario, the statistics for the overall incidence/prevalence of occupational disease and injuries for the country is not adequately compiled in an easily accessible format. However, Leigh et al. (1999) have estimated an annual incidence of occupational disease between 924,700 and 1,902,300 and 121,000 occupational disease caused deaths in India. Based on the survey of injury incidence in agriculture, a study by Mohan and Patel (1992) in Northern India, an annual incidence of 17 million injuries per year (2 million moderate to serious) and 53,000 deaths per year in agriculture sector alone was estimated. A report by National Institute of Occupational Health (1999) records more than 3 million people working in various type of mines, ceramics, potteries, foundries, metal grinding, stone crushing, agate grinding, slate-pencil industry, etc. These workers are occupationally exposed to free silica dust and are at potential risk of developing silicosis. In India there is a huge gap in epidemiological evidence from different industry specific as well as exposure specific areas (Jaiswal et al., 2006). Global investors import both hazardous industries as well as new technologies. Hence, traditional labour oriented markets are changing towards automation and mechanization. However, at the same time general awareness about the occupational health and safety is not being spread to the society (Agnihotram, 2005). Within this situation, the workers have to join workplaces with low resources settings including health and safety consideration. This has made them more susceptible to the hazards of technology than their counterparts in developed countries. Industrialization in India is primarily focused on production, whereas health and safety have a very low priority (Jaiswal et al., 2006). Except a few major, reputed public and private industries, other industrialists are insensitive towards the importance and need of occupational health and safety.

In another study reported by Atmaca et al. (2005) at concrete traverse, iron and steel, cement and textile factories around Sivas (Turkey). 73.83 per cent of workers in the industries were disturbed from the noise in their workplaces. Noise causes the problem of nervousness on workers at a rate of 60.96 per cent. Whereas 30.86 per cent of the workers have ailments such as ringing in the ear and hearing losses. At the same time 85.94 per cent of the workers reported that they do not have periodical hearing tests. The same study reported that ear protection accessories were being used in the industries by a rate of 32.94 per cent. The rate of using ear protection accessories at the cement factory at which the noise level is at the highest rate of 7.69 per cent. Hence, the industries at Sivas have the problem of noise.

Dembe (2008) reviewed a variety of studies on history of concerns about long working hours and the current scientific evidence regarding their effects on workers' health, reported that long working hours increases the risk for hypertension and cardiovascular disease along with relationships between long hours at work and musculoskeletal injuries, diabetes, and chronic infections. Evidence also suggests that overtime and extended work schedules can lead to depression and other psychological conditions. The ethical considerations regarding long working hours are to be thought of as questions about the type of society we want to create. There are number of studies reported on occupational health hazards like noise, temperature, dust, etc.

17.11. Scarcity of Occupational Health Specialist in India

The World Health Organization, WHO, estimates occupational health risks as the tenth leading cause of morbidity and mortality (Pingle, 2011). In India, occupational health is not integrated with primary healthcare, and it is the mandate of the Ministry of Labour, not the Ministry of Health (Kesavachandran and Rastogi, 2005). The statistics of the occupational hazards in the country are very alarming as the rate of mortality due to occupational diseases is high (Kesavachandran and Rastogi, 2005). There is an urgent requirement of contemporary OHS legislation through sufficient enforcement mechanism and also institute the centres of excellence to grasp up with the rest part of the world (Agnihotram, 2005). There is a huge gap of human resource in occupational health; hence, there is a strong need of improvement in the current training capacity. There are twenty-one institutes across the country. The existing capacity for training is about 460 specialists. This number is inadequate considering the population of India's working class. If we look at the statistics with respect to occupational hygienists in India, there are merely 100 qualified occupational hygienists in the country, while the Factories Act reveals that there is a requirement to monitor the environment for a large number of harmful substances (Zodpey et al., 2009). At the same time we have simply 1125 qualified occupational health professionals in our country against a requirement of more than 8000 qualified occupational health doctors (www.dgfasli.nic.in/info1.htm). In the 60th World Health Assembly, the World Health Organization has spoken over major gaps among and inside countries with respect to exposure of workers as well as local communities to occupational hazards, and also their access to occupational health services (www.dgfasli.nic.in/info1.htm). The occupational health training is carried out in a few medical colleges for graduate and postgraduate diplomas and degrees. Jaiswal et al. (2006) have stressed upon strengthening the skills and broadening our knowledge in occupational health and safety. Therefore, recommended that academic community should take initiative to demonstrate its leadership to address the long neglected concern of occupational health and safety. There is a need of epidemiological research to determine the exposure and occupational risks for which the public–private partnership is very important (Agnihotram, 2005).

17.12. Need for Research on Occupational Health and Ergonomics in Steel Industry

Manufacturing industry especially the casting and forging (iron and steel) industry comprises a substantial part of the occupation in India. In the poor economic conditions and apprehension of unemployment, the workers have accepted the jobs with low resources settings including health and safety consideration. This has made them more prone to the hazards of technology. Industrialization in India is primarily focused on production, whereas health and safety have a lower priority (Jaiswal et al., 2006). Except a few major, reputed, public and private industries, other industrialists are insensitive towards the importance of occupational health and safety. The employer of small scale units are totally lagging behind in providing occupational health and safety to the workers, therefore the manpower employed in small scale casting and forging SMEs are more exposed to occupational noise, heat stress, musculoskeletal strain and dust, etc. The developmental potential for small and medium enterprises (SMEs) and growth prospects of national economy of a country cannot be achieved without health and safety practices at the work place (Ahasan, 2002). At the same time, a very limited literature is available, which addresses the occupational health and safety issues of these SME's. Therefore, there is a need to assess the occupational health and safety practices, level of occupational hazards at the shop floor, subsequently to investigate the level of health deterioration among the workforce engaged in casting and forging (steel) industry.

17.13. Case Study: An Exploratory Study for Assessment of Awareness with respect to Occupational Health, Safety and Environment in SMEs

Exploratory study was conducted through questionnaire survey. The questionnaire included demographic descriptors, the nature of job/process, exposure of workers (in years) to hazardous conditions like noise, temperature, dust/fumes/gasses, and chemicals, working hours, shifts and overtimes, personal information of the workers regarding work posture, physical load, reaction to noise, temperature and dust, sleep disturbances, water intake, diseases, alcohol and tobacco consumption, use of PPEs, work time injuries, auditory, visual and overall health and job satisfaction. Subjective observations were taken regarding the use of PPEs at the shop floor and the responses of the workers regarding the same were asked in the questionnaire. Most of the workers of these companies were very less educated and unable to understand English as such, and hence statements of the questionnaire were translated in both languages (Punjabi and Hindi). The responses of the workers were recorded. The height and weight of each worker was measured manually using a measuring tape and weighing balance (Figure 17.2).

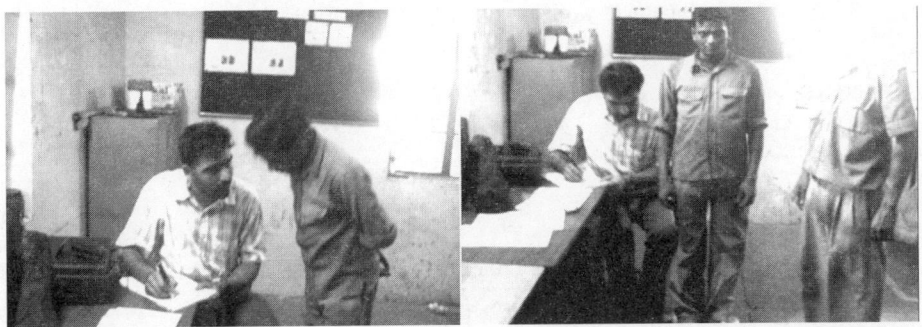

Fig. 17.2: Questionnaire survey and interview conducted with the industry workers.

17.13.1. Subjects for questionnaire survey

The present exploratory study included randomly selected 572 shop floor male workers of total twelve casting and forging SME. Non-probability convenience sampling was used to select SMEs and subjects. Hence, the approachability and availability of workers in the industry was the criteria for selection. Once, the industry management permitted the workers to volunteer they were taken for the study. Oral consent was received from these subjects who volunteered for the study. These workers were performing different jobs in various sections such as moulding molten metal pouring, grinding, forging, punching, blanking, welding, gas cutting, electroplating and painting. Around 95 per cent of these workers were performing their jobs manually.

17.13.2. Demographic data

The demographic parameters are shown in the Tables 17.1 to 17.6, it is evident that more than sixty per cent of workers are of the age group of 26 – 42 years. With the increase in age, the number of workers steadily decreases (Table 17.1). Majority of the workers were either at primary level education (28.50 per cent) or uneducated (25.53 per cent) as shown in Table 17.2. They have no formal training and they were unaware of the safety and protection at the workplace. In addition, they were unaware of the hazardous effects of work conditions. The majority of the workers were married (75.17 per cent). The married subjects were working for more number of years than the unmarried subjects, thus they have more exposure to noise, temperature, dust, physical workload, etc. (Table 17.3). The distribution of workers interviewed from various sections is shown in Table 17.4.

Table 17.1: Distribution of industry workers with respect to age, N (%).

Age Range (Years)	18–22	22–26	26–30	30–34	34–38	38–42	42–48
Workers (%)	62 (10.84%)	91 (15.91%)	92 (16.08%)	108 (18.88%)	105 (18.36%)	62 (10.84%)	52 (9.09%)

Table 17.2: Distribution of industrial workers with respect to education level, N (%).

Level of Education	Uneducated	Primary	Middle	Matric	Senior Secondary	Above Senior Secondary
Workers	146 (25.53)	163 (28.50)	130 (22.73)	81 (14.16)	40 (6.99)	12 (2.10)

Table 17.3: Distribution of workers with respect to marital status and exposures to working conditions environment (years), N (%).

Exposures to Working Conditions Environment (Years)			Number of Workers (%)		
			Noise	Temperature	Dust/Fumes
1–4 years	Worker's marital status	Married	68 (11.89)	54 (9.441)	50 (8.741)
		Unmarried	102 (17.83)	107 (18.71)	110 (19.23)
5–9 years		Married	120 (20.98)	128 (22.38)	130 (22.73)
		Unmarried	28 (4.89)	25 (4.371)	28 (4.89)

Contd.

Contd.

Exposures to Working Conditions Environment (Years)			Number of Workers (%)		
			Noise	Temperature	Dust/Fumes
10–14 years	Worker's marital status	Married	90 (15.73)	110 (19.23)	104 (18.18)
		Unmarried	12 (2.1)	10 (1.75)	4 (0.7)
15–19 years		Married	96 (16.8)	84 (14.7)	96 (16.8)
		Unmarried	0 (0)	0 (0)	0 (0)
≥ 20 years		Married	56 (9.79)	54 (9.44)	50 (8.74)
		Unmarried	0 (0)	0 (0)	0 (0)
Total		Married	430 (75.17%)	430 (75.17%)	430 (75.17%)
		Unmarried	142 (24.83%)	142 (24.83%)	142 (24.83%)

Table 17.4: Distribution of workers interviewed in each section of SMEs, N (%).

Total No. of Units (12)	G. Total of Workers	Number of Workers in Casting	Number of Workers in Forging
Workers Interviewed in Sections	572 (100)	224 (39.16)	348 (60.84)
Molding/Casting	102 (17.83)	102 (17.83)	NA
Gas Cutting/Welding	34 (5.94)	34 (5.94)	NA
Blank cut/Trim/Punching	56 (9.79)	NA	56 (9.79)
Drop Forging	84 (14.68)	NA	84 (14.69)
Broaching/Machining	51 (8.92)	NA	51 (8.92)
Grinding	126 (22.03)	43 (7.52)	83 (14.12)
Barrelling	23 (4.02)	NA	23 (4.02)
Quality Check Inspection	34 (5.94)	16 (2.79)	18 (3.15)
Nickel Plating/Painting	32 (5.59)	18 (3.15)	14 (2.45)
Tool Room/Maintenance	30 (5.24)	11 (1.92)	19 (3.32)
Total Workers	572 (100)	224 (39.16)	348 (60.84)
Total Employees in Firms	1500	600	900
Response Rate (%)	38.13	37.33	38.13

*Overall response rate = $\frac{572}{1500} \times 100$ = 38.13, NA: not applicable.

17.13.3. Fatigue, noise-induced syndromes, N (%)

As far as the reporting with respect to noise induced syndromes was concerned, around 94 per cent of the workers experienced speech interference, out of which 42 per cent workers reported 'always' speech interference. These workers were engaged in forging, blanking, punching, trimming, barrelling, and grinding sections. Twenty seven per cent (27 per cent) of the workers, who were working in machine moulding, shot blasting, sizing gauging reported 'often' speech interference and rest of the workers in the tool room, nickel plating, quality and cupola/induction furnace operators reported 'low' speech interference. The overall speech interference level considered at five point scale was on higher side (2.89). This speech interference hinders the day-to-day necessary communication and has resulted in noise annoyance amongst the workers. At the same time about 7 per cent of the workers reported 'always' annoyed, 10 per cent workers reported often, and 18 per cent workers reported sometime, 23 per cent of the workers felt seldom annoyed and rest of the 41 per cent workers never felt annoyed due to high noise levels. Overall noise annoyance level at five point scale was 1.18. Headache and disturbed sleep/anxiety was reported by very few subjects. The level of headache and anxiety were 0.83 and 0.57, respectively (Figures 17.3 and 17.4). These are also on lower side and this could be attributed to the basic two reasons. First, the workers have accepted noise as a part of their job. Secondly, the hearing threshold was shifted, and the workers have been adapted to high noise levels. The workers who reported noise annoyance were having less than five years of work exposure. The prevailing noise annoyance could be due to speech interference, as the workers were unable to communicate in the noisy environment. Concurrent to the increased adaptability, the workers also suffered from hearing loss; thus they got adapted to the high noise levels.

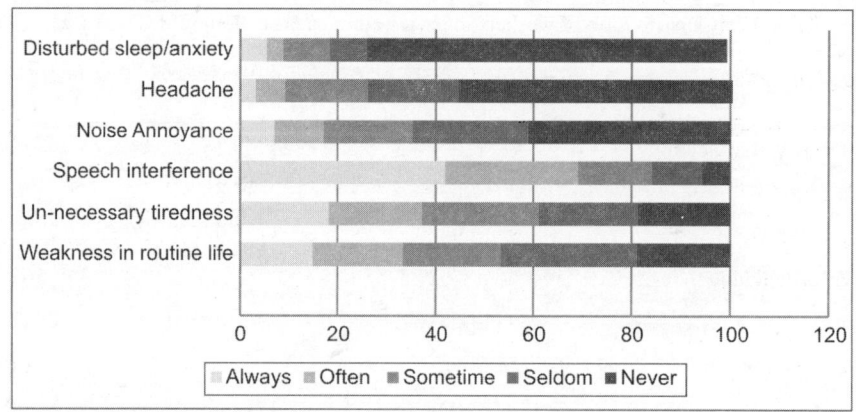

Fig. 17.3: Percentage distribution of workers with respect to fatigue/noise induced syndromes.

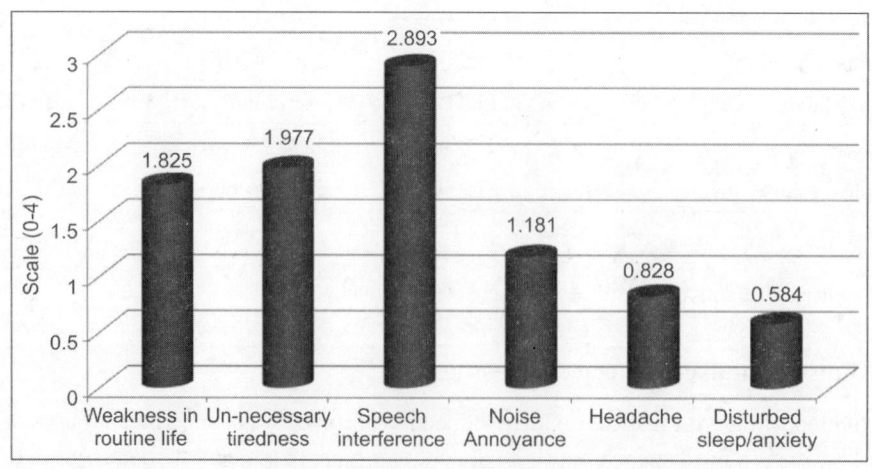

Fig. 17.4: Fatigue, noise induced syndromes at a 0–4 point scale.

Scale ranges: always = 4, Often = 3, Sometime = 2, Seldom = 1, Never = 0

$$\text{Speech Interference} = \frac{41.96 \times 4 + 27.10 \times 3 + 15.03 \times 2 + 10.14 \times 05.77 \times 0}{100} = 2.893$$

17.3.4. Work schedule and shifts of industrial workers

As far as the work schedule was concerned, only 28 per cent of the workers were working in shifts. These workers were mainly from the forging, grinding and moulding sections. Rest of the 72 per cent workers were working day time. Around 84 per cent of the workers were working more than 8 hours/day. Out of these 85 per cent of the workers reported additional overtime of 12–24 hours/week. The distribution of workers according to the work schedule is shown in Figure 17.5. OSHA work hour's norms are not being followed in Indian SMEs because in most of these units, workers work for 10–12 hour/day for six days/week, i.e., the total working hours were 60–72 hours per week. These working hours were significantly higher than the prescribed limits of 40 hours/week in the USA or the European countries and as well as the prescribed limit of 46 hours/week as per the Indian Factory Act. Most of the workers were doing 2–4 hours/day overtime, i.e., 12–24 hours/week. Hence they were under total exposure of 20–32 hours, i.e., 50–80 per cent per week higher than exposure time per week in the USA or the European countries. The distribution of workers with respect to overtime per week is shown in Figure 17.6. Such long working hours may perhaps result in increased risk of cardiovascular, respiratory, and hearing impairments along with musculoskeletal disorders and injuries (Dembe, 2008). This provides serious cause of concern for long working hours and challenges the ethical implication on unconventional shift work and long workhour schedules.

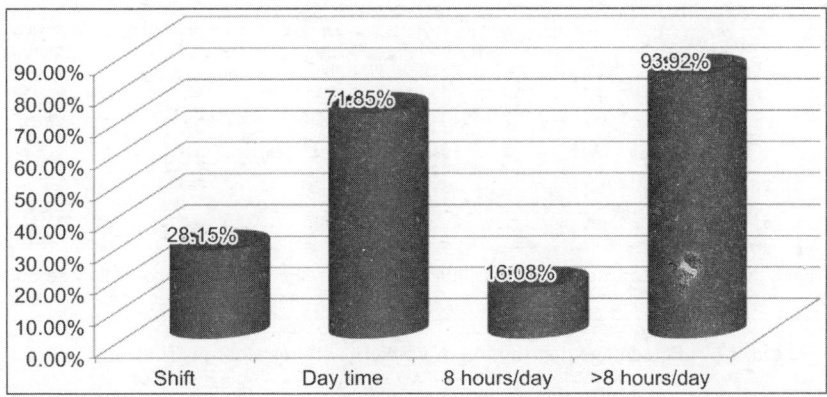

Fig. 17.5: Percentage distribution of workers with respect to shift and work schedule.

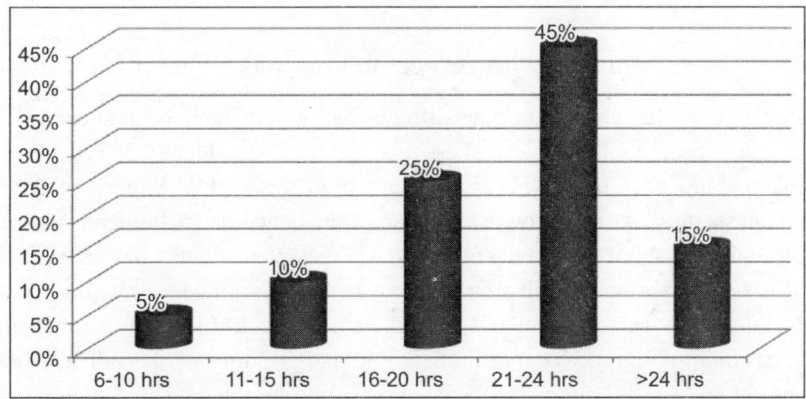

Fig. 17.6: Distribution of workers with respect to overtime per week, *N* (%).

The casting and forging industry workers work for overtime not because they worship the occupation but due to their poor economic conditions Hence they work more to earn more. The prevailing performance standards in SMEs generally did not include sufficient rest allowances. Most of the firms hire contractual labours, and the targets given to the workers were not based upon a time and motion study (work study) analysis, however there must be a scientific base for production and wage rates.

17.3.5. Distribution of workers w.r.t. various personal habits (intakes)

Responses with respect to different types of intakes by the workers is shown in Figure 17.7. Around 25 per cent of the workers admitted that they smoke (1–3 cigarettes or biddies[1] /day), and 30 per cent workers reported, that they were habitual tobacco consumers. Eleven per cent (11%) of workers was habitual of having 'gutkha[2]' and 'chutki[3]', six per cent of workers reported the habit of beetle leaf chewing. Forty seven per cent (47%) of the workers reported about 40–750 ml per month intake of alcohol. Consumption of glucose/salt water during the summer season was reported by only 18 per cent of workers.

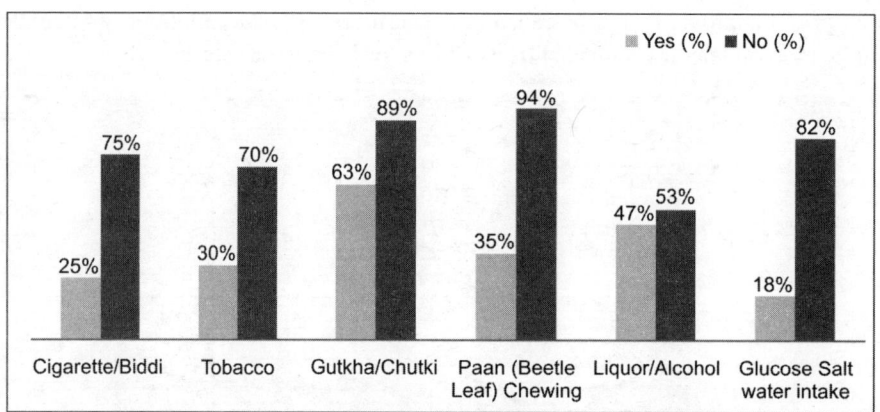

Fig. 17.7: Percentage distribution of workers with respect to various intake.

[1] Biddi is a locally manufactured substitute of cigarettes, which is smoked by very low income group population in India.

[2,3] Gutkha and Chutki are chewable flavoured products made from tobacco.

17.3.6. Distribution of workers with respect to musculoskeletal disorders (MSD)

The MSD complaints of the workers in their different body parts have been shown in Figure 17.8. About fifty seven per cent (57%) of the workers reported neck/shoulder stiffness, 58 per cent of the workers complained low back pain, whereas 56 per cent of the workers experienced wrist stiffness and forearms pain/ stiffness. As far as lower body parts were concerned, a large proportion of workers (42 per cent) complained about pain in leg muscles, whereas 25 per cent workers were experiencing knee/ankle stiffness. The reason for such problems may be long working hours without any appropriate work–rest schedule. Moreover, reporting of low back pain and neck stiffness was much higher among the moulding, casting/pouring, heavy martial handling and grinding workers. Therefore, the heavy load lifting and the excessive bending work postures could be the cause of such musculoskeletal problems, which can be corrected by the ergonomic interventions, as described in chapter-10.

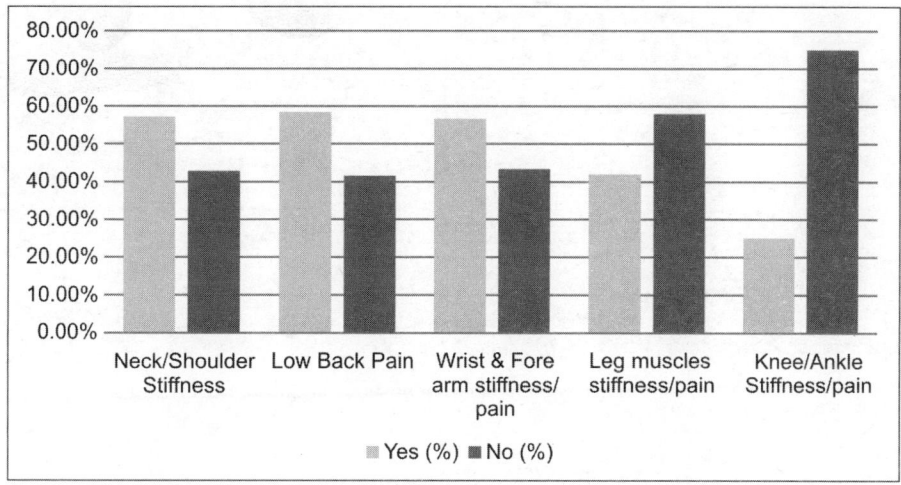

Fig. 17.8: Distribution of workers with respect to musculoskeletal complaints, *N* (%).

17.3.7. Distribution of workers with respect to health-related symptoms

Apart from the MSDs, the workers also experienced other health problems such as; respiratory, hearing and cardiovascular symptoms, which are shown in Table 17.5.

Table 17.5: Distribution of workers with respect to various health symptoms.

Symptom	*Total*	*Yes, N (%)*	*No, N (%)*
Phlegm	572	254 (44.41%)	318 (55.59%)
Wheezing	572	242 (38.46%)	330 (61.54%)
Breathlessness	572	240 (40.21%)	332 (61.54%)
Asthma	572	50 (7.34%)	522 (92.66%)
Hearing disability	572	380 (70.00%)	192 (30.00%)
Blood Pressure	572	101 (17.66%)	471 (82.34%)
High BP	572	20 (3.50%)	--
Low BP	572	81 (14.16%)	--

In case of the respiratory problems, around, 44 per cent reported frequent phlegm, 38 per cent reported wheezing, and 40 per cent reported breathlessness, whereas asthma was reported by 7 per cent of the workers only. These workers were from casting/moulding, gas cutting and welding, forging, grinding, and painting/nickel plating sections. As far as the hearing disability was concerned, 70 per cent workers reported hearing disability due to workplace noise; these workers were from moulding, forging, punching, blanking and barrelling sections. Problem of blood pressure (BP) was reported by 17.66 per cent workers only and out of which only 20 (3.5 per cent) workers reported high BP and rest (14.16 per cent) workers reported low BP. This may be due to the higher sweat losses from excessive heat stress, which needs further investigation. BP of exposed group v/s control groups is further investigated in subsequent chapter 18.

Fig. 17. 9: Workers working in bending postures at different workplaces.

17.3.8. Workers' subjective responses with respect to work posture and sweat losses at workplace

The distribution of workers working under various work postures is shown in Table 17.6. The overall score at five point scale of working postures in sitting and standing position were 2.25 and 2.56, respectively. About 68 per cent workers reported heavy lifting with a score of 2.27. The frequency of load lifting was much higher in workers engaged in moulding, casting, grinding and gas cutting sections. Seventy per cent of workers reported working in bending posture with score of 3.08 at five point scales, which could be the reason for higher prevalence of low back and neck pain. Since the proportion of workers working under unsafe bending posture is considerably higher, therefore investigation cases have been described in chapter 10. Regarding the work posture analysis and lifting tasks, respectively. The workers in bending postures at different work places are shown in Figure 17.9. About 90 per cent of workers reported heavy sweating and almost the same proportion of workers reported 3–9 litre/day of water intake. As shown in Figure 17.7, eighty two (82%) per cent of workers revealed that, they were taking water without mixing anything, whereas only 18 per cent workers were taking lemon/salt with water. Nearly the same proportion of workers has reported tiredness and weakness in daily routine life. For the same reason it is very obvious that merely simple (plane) water intake is not sufficient to compensate the sweat losses. Some firms also revealed that, they had offered lemon/glucose/salt water as a substitute for tea, but the workers had preferred to take tea instead of lemon–salt water. This could be due to the reason that workers were not aware of sweat losses as its long term consequence.

17.3.9. Distribution of workers with respect to auditory, eye sight and overall health

In case of auditory, visual and overall health, the subjective responses on a five point scale, given by the workers is shown in Table 17.7. About 55 per cent of the workers reported weak eyesight at score of 2.49, around 70 per cent workers reported weak hearing capacity with lowest score of 1.94, whereas overall weak health was reported by 53 per cent workers with score of 2.50 at four point scales. The level of hearing capacity was lowest at five point scale, which could be due to the non-use of ear protection and high level of noise exposure. Hence, there is a need of further investigation into auditory functions of the workers, the same has been analyzed and discussed in chapter 14.

17.3.10. Distribution of workers with respect to awareness of health and safety, use of personal protective equipment (PPEs) and workplace injuries

The safety measures and the workplace injuries are shown in Table 17.8. In the case of injuries at the workplace, about 15 per cent workers reported major injuries at the workplace, whereas, minor injuries were reported by 59 per cent workers (this includes; hand or foot burns due to hot metal, crushing of fingers and foot under heavy jobs, small cuts, etc.). The reasons for injury could be lack of concentration or negligence of the workers. But, at the same time, it was also observed that the all sizes of PPEs were not provided to the workers, thus they wear improper PPEs, which further increases the chances of injury. As far as the use of occupational protective equipment was concerned, about 51 per cent workers were not aware of the benefits of using the personal protective equipment. Around 55 per cent workers revealed that management did not explain the benefits of using PPEs and 56 per cent workers reported that management does not enforce wearing PPEs. In the case of the use of PPEs at work place, 78 per cent workers were not wearing PPEs while working.

Table 17.6: Distribution of workers with respect to work posture N (%).

Working Posture	Total	Always	Often	Some time	Seldom	Never	Scale (0–4)
Sitting (stool/desk, squatting & cross legged)	572	152 (26.57)	120 (20.98)	113 (19.76)	95 (16.61)	92 (16.08)	2.253
Standing	572	183 (31.99)	140 (24.47)	125 (21.85)	61 (10.66)	63 (11.01)	2.557
Walking	572	159 (27.80)	132 (23.08)	123 (21.50)	86 (15.03)	72 (12.59)	2.384
Lifting heavy load	572	135 (23.60)	155 (27.10)	103 (18.01)	92 (16.08)	87 (15.21)	2.277
Bending	572	306 (53.50)	102 (17.83)	95 (16.61)	45 (7.87)	24 (4.19)	3.085[a]
Sweating at work place	572	324 (56.64)	188 (32.87)	60 (10.49)	Nil	Nil	3.462

Scale ranges: always = 4, Often = 3, Sometime = 2, Seldom = 1, Never = 0

Bending Posture[a] = $\dfrac{(4 \times 53.50 + 3 \times 17.83 + 2 \times 16.61 + 1 \times 7.87 + 0 \times 4.19)}{100} = 3.085$

Table 17.7: Distribution of workers with respect to auditory, visual and overall health, N (%).

Parameter	Total	V Good	Good	Little Weak	Very Weak	Scale (1–4)
Eye Sight	572	105 (18.36)	150 (26.23)	237 (41.44)	80 (13.98)	2.489
Hearing Capacity	572	72 (12.58)	100 (17.48)	120 (20.98)	280 (48.95)	1.936[a]
Orall Health	572	87 (15.21)	185 (32.34)	225 (39.34)	75 (13.12)	2.496

Hearing Capacity [a] = $\dfrac{12.58 \times 4 + 17.48 \times 3 + 20.98 \times 2 + 48.95 \times 1}{100} = 1.936$

Scale Ranges: Very Good = 4, Good = 3, Little Weak = 2, Very Weak = 1,

Table 17.8. Distribution of workers with respect to awareness of health and safety, use of PPEs and work place injuries, *N* (%).

Parameter	Total	Yes, N (%)	No, N (%)
Awareness of benefits of PPE	572	281 (49.13)	291 (50.87)
Does management explain the benefits of PPE?	572	260 (45.45)	312 (54.55)
Does management force you to wear PPE?	572	250 (43.71)	322 (56.29)
Use of PPE at the work place	572	123 (21.51)	449 (78.49)
Dungaree	572	141 (24.65)	431 (75.35)
Gloves	572	180 (31.50)	392 (68.50)
Goggles	572	222 (38.80)	350 (61.20)
Gum Shoes/boots	572	211 (36.90)	361 (63.10)
Nose/mouth mask	572	199 (34.80)	373 (65.20)
Ear plugs/muffs	573	153 (26.80)	419 (73.20)
Helmet	572	184 (32.20)	388 (67.80)
Others	572	103 (18.01)	469 (81.99)
Major injury at workplace	572	86 (15.04)	486 (84.96)
Minor injury at workplace	572	337 (58.92)	235 (41.08)
Does company conduct regular medical checkups	572	96 (16.78)	476 (83.22)
Does company provide medicine or medical facility like ESI?	572	332 (58.04)	240 (41.96)
Satisfaction with health care/medical facility	332	104 (31.33)	228 (68.67)

Table 17.9: Details of score at five point scale with respect to use of PPEs by the workers (%).

Use of PPE at Work Place	(i) Always	(ii) Often	(iii) Sometime	(iv) Seldom	(v) Never	Score on Scale (0–4)
Dungaree	7%	10%	8%	5%	70%	0.79 [a]
Gloves	9%	12%	10%	9%	60%	1.01
Goggles	12%	15%	13%	10%	50%	1.29
Gum Shoes/Boots	10%	12%	15%	10%	53%	1.16
Nose/Mouth Mask	9%	14%	12%	9%	56%	1.11
Ear Plugs/Muffs	5%	12%	10%	9%	64%	0.85
Helmet	7%	15%	11%	5%	62%	1.00
Others (Turban/Safa/Cap)	20%	35%	15%	5%	25%	2.20 [a]

Scale ranges: always = 4, Often = 3, Sometime = 2, Seldom = 1, Never = 0

[a] Dungaree $= \dfrac{7 \times 4 + 10 \times 3 + 8 \times 2 + 5 \times 1 + 70 \times 0}{100} = 0.70$

[a] Others (turban/Safa/cap) $= \dfrac{20 \times 4 + 35 \times 3 + 15 \times 1 + 25 \times 0}{100} = 2.20$

Data also revealed that only 25 per cent of the workers were wearing dungarees, 32 per cent of the workers were using gloves and 39 per cent of the workers were using eye protection (goggles). As far as the use of gum shoes/boots is concerned, only 37 per cent workers were using them, at the same time nose masks were used by 34 per cent workers. The ear protection was found be used by only 26 per cent workers. Ear protection and dungaree were the lowest used PPEs at five point scale with score of 0.85 and 0.79 respectively. The detail of level of use of PPEs at a five point scale is shown in Table 17.9.

17.3.11. Distribution of workers with respect to reasons for non-use of PPEs at workplace

The different reasons reported by the workers for non-use of PPEs are shown in Figure 17.10. About 32 per cent of workers reported that they felt uncomfortable on wearing PPEs. About 12 per cent of workers reported, they were not habitual of using PPE, whereas 22 per cent worker admitted their own negligence. About 34 per cent of the worker revealed that management do not even provide the PPEs.

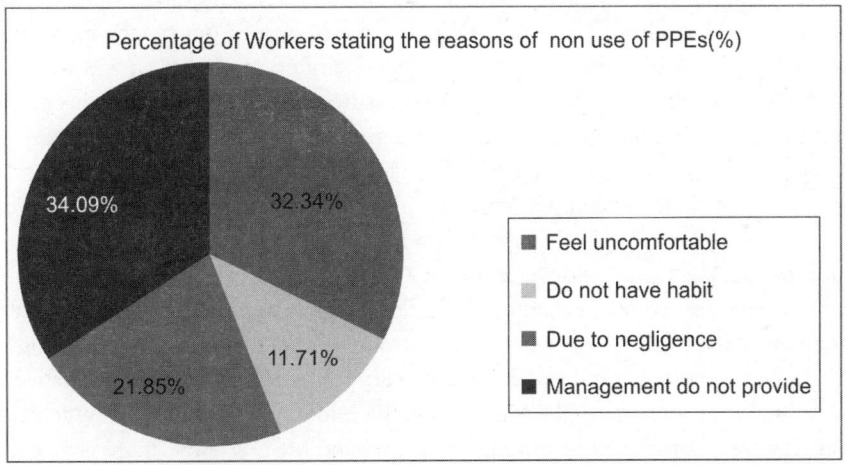

Fig. 17.10: Distribution of workers with respect to various reasons for non-use of PPEs.

Majority of workers were not using PPEs at work place and around 66 per cent workers were not wearing PPEs either due to bad design as they feel uncomfortable after wearing the available PPEs or were not being provided by the managements. In both the cases the management was responsible; however the workers may be negligent in this concern. Therefore, there is a strong need that management should provide the ergonomically designed PPEs to the workers.

As already described, the industry lag in enforcing the use of PPE and mostly the managements of small and medium scale units do not even bother about it. Another significant factor could also be that workers work under sub-contractors, who literally ignore these aspects. It is also true that workers do not expect much of health and safety care from the management; rather they have accepted hazardous conditions as a part of their job. Consequently, the workers have to work under low resources settings with respect to health and safety.

It is very pertinent to mention that the Casting and Forging industry of the region is primarily focused on executing the production targets and getting more orders from their customers, whereas health and safety concern was at lowest priority. This is in accordance with previous study; Jaiswal et al., 2006, that industrialization in India is primarily focused on production, whereas health and safety have a very low priority. Certainly, it is their responsibility to implement the health and safety measures and provide counselling to the workers to educate them regarding the benefits of wearing PPEs. The study validates findings of the international labour office (ILO) which estimate that in year 2001, there were 2.2 million deaths due to work-related injuries and diseases (www.ilo.org/safework). Occupational work is indeed an essential component of human society. It provides basics needs (food shelter and clothes) satisfaction to families and communities. But it is very important to safeguard the interests of industrial workers in terms of their health and safe environment and let us not forget that health is non-negotiable.

17.3.12. Conclusions

The exploratory study was focused to reveal occupational health and safety practices in SMEs and addresses some important issues such as: status of health and safety practices in casting and forging SMEs. From the exploratory study, it is concluded that workers were unaware about the health hazards resulting from the work environment. The workers were suffering from speech interference due to occupational noise, which hinders the day-to-day necessary communication and resulted in noise annoyance amongst the workers. The lower blood pressure due to work exposure is contradictory to the literature; thus investigation into cardiovascular autonomic control (BP & HRV) has further been carried out on a sample of workers in chapter 18. The respiratory problems due to work exposure were reported by around 40 per cent workers, therefore respiratory health assessment (qualitative) has already been carried out in this chapter. Around 16.70 per cent of the workers reported hearing disability due to the work place noise, auditory functions (hearing threshold and latency time) investigation has been carried out in chapter 14. The complaints of musculoskeletal disorders (MSD) were higher amongst the workers. Hence, there is a strong need of ergonomic interventions to analyse the prevailing working postures and make the necessary improvements. The workers were found to work for additional over time of 12–24 hours/week. Such long working hours may increase the risk for cardiovascular, respiratory and hearing impairments and MSD. Hence, there is a need to strictly enforce the Labour Laws keeping in view the health concern of workers. Workers were not wearing PPEs due to bad design as they felt uncomfortable after wearing the available PPEs. However, in most of cases PPEs were not being provided to the workers. The casting and forging industry of the region is primarily focused on the production, whereas health and safety of workers was at the lowest priority, this is in accordance with Jaiswal et al. (2006). In essence, the health and safety practices seldom exist in casting and forging SMEs of the region.

Multiple Choice Questions

1. What is the spread form of ILO?
   (a) International labour organization
   (b) Indian labour office
   (c) International labour office
   (d) None of the above

2. Bhopal gas tragedy occurred in?
   (a) 3rd December 1984
   (b) 15rd Dec 1985
   (c) 1995
   (d) None of these

3. In Bhopal gas tragedy about 5 Lakhs people were exposed to gas and about:
   (a) 20,000 died
   (b) 50,000 died
   (c) 30,000 died
   (d) None of these

4. The Indian Penal Code has specified punishment for polluting the atmosphere, states that whoever voluntarily mitigates the health of atmosphere in any place so as to make it noxious for general health of public or passing along a public way, may be imposed a fine extended up to
   (a) Rs 1500
   (b) Rs 5000
   (c) Rs 500
   (d) All of the above

5. Occupational health not concerned merely with the occupational syndromes, rather it also includes
   (a) all the factors affecting health of workers
   (b) the factors affecting only physical health of workers
   (c) the factors affecting only mental health of workers
   (d) none of the above

6. AIFI stands for
   (a) All India Forging Industry
   (b) Association of Indian Forging Industry
   (c) Association of International forging Industry
   (d) None of the above

7. The full form of SMEs is
   (a) Steel Manufacturing Enterprises
   (b) Small and Medium Enterprises
   (c) Small Manufacturing Enterprises
   (d) None of above

8. Noise Induced syndromes include
   (a) Noise annoyance and head ache
   (b) Disturbed sleep and anxiety
   (c) Speech interference only
   (d) All of these

9. Respiratory symptoms consist of
   (a) Cough and phlegm
   (b) Wheezing and breathlessness
   (c) Asthma
   (d) All of the above

10. The spread form of PPE is
    (a) Personal practice equipment
    (b) Personal protective equipment
    (c) Personal protective environment
    (d) Any one of the above

11. Which of the following is the reason due to which workers do not wear PPE at work place?
    (a) Workers feeling uncomfortable
    (b) Worker's negligence
    (c) Management do not provide and enforce them
    (d) All of the above

12. The risk of MSD among workers can be reduced by
    (a) Correcting work posture
    (b) correcting work schedule
    (c) Ergonomic interventions
    (d) All of the above

Answers

1. (c)   **2.** (a)   **3.** (a)   **4.** (c)   **5.** (a)   **6.** (b)   **7.** (b)   **8.** (d)   **9.** (d)   **10.** (b)
11. (d)   **12.** (d)

Cardiovascular Health of Steel Workers

18.1. Introduction

In chapter 17 work related physical conditions for the protection of workers has been investigated in small and medium enterprises, which explored that managing of various enterprises lacking within this concern. Before that, it is also observed in chapters 14, 15, and 16 that employees of steel industry were highly exposed to noise, dust, and heat stress especially during the extreme weather conditions in summer season. Thus, workers working in low resource settings were found highly exposed to onsite hazardous work conditions. The present chapter exhibits a case study which investigated the effect of onsite work conditions and physical activities on cardiovascular parameters of industry workers. This chapter comprises two parts: first part explains general background of cardiovascular health followed by methods to assess the cardiovascular functions. Second part is dedicated to a case study of comparison between exposed and control group, process wise variation in the parameters and effect of duration of work experience on the cardiovascular functions (parameters). Subsequently, it describes about the influence of alcohol, smoking and tobacco consumption on the cardiovascular parameters.

18.2. Statistical Tools used for Analysis

As mentioned above in the introduction that, for each type of parameter two groups of population were included: one group of male subjects (workers) from casting and forging firms, i.e., exposed group and another group of male subjects from the randomly chosen non-exposed population, i.e., control group. All the parameters were statistically investigated using the *t*-test and ANOVA for multiple comparisons among various sub-groups with a confidence level of 95 per cent. The results were also analyzed for association of physiological parameters with habits like alcohol, smoking and tobacco ingestion. The conclusions were drawn on the basis of results and analysis.

18.3. Cardiovascular Autonomic Control (HRV) and Work Conditions

The working environment of steel SMEs is very risky and unavoidable. The work-related physical condition and safety measures are ignored at the workplaces (Singh et al., 2010). It has been established by Togo and Takahashi (2009) that the occupational exposure to hazardous settings along with work related stressors, and work timing had been associations with low HF component of power spectral density. The workers are exposed to noise, heat stress, dust and fumes, etc. which are known factors leading to decreased parasympathetic activity (Shannon et al., 2001 and Fujino et al., 2007). The noise annoyance perceived by an individual is critical factor towards the cardiovascular effects of noise hazards (Lang et al., 1992 and Peter et al., 1993). There is a need that investigations into parasympathetic nervous system activity be focussed to protect occupational cardiovascular health (Togo and Takahashi, 2009).

There are a lot of explorative research studies based on the health effects of noise exposure in various occupations. In various occupations the exposure to high noise levels is strongly related to the possibility for perceptive myocardial infarction death (Hugh et al., 2005). There has been a significant effect of noise annoyance on increasing diastolic blood pressure (BP) (Peter et al., 1993). Prolonged experience of sound level above 85 dB (A) could result into the most significant factor in rising BP with in perceptive subjects (Lang et al., 1992).

Occupational hazards like noise are the known factors leading to decreased parasympathetic activity (Shannon et al., 2001 and Yoshihisa et al., 2007). The workers who had perceived noise exposure at workplace were found with a higher prevalence of myocardial infarction (Rai et al., 1981). The noise level of high range as 88–107 dB (A) may affect the cardiovascular system as a worker engaged in occupation for 6–8 hours in a day (Yoshihisa et al., 2007). It had been established that there may be a risk of high blood pressure among noise sensitive individuals if there is prolonged contact with the intensity of noise over 85 dB (A). Thus the noise annoyance perceived by an individual is critical factor towards the cardiovascular effects of noise hazards (Lang et al., 1992 and Peter et al., 1993).

Beside the hazardous effects of hostile work conditions and occupational stresses, a plenty of studies have been conducted by various authors on the positive affects of physical work out, on the cardiovascular heath parameters. On the same track some of the studies have established that, the individuals engaged in various sports activities were found to have lower resting heart rates (Shin et al., 1997, Fagard et al., 1999 and Loimaala et al., 2000). It has been approved that lesser heart rates are at least partially the result of increased parasympathetic tone (Yataco et al., 1997, Shin et al., 1997, Seals et al., 1989 and Levy et al., 1998). Antelmi et al. (2004) also reported an inverse correlation of HRV with heart rate. Some other studied like; Sunkaria et al. (2010) reported increased level of heart rate variability among the individuals practicing the yoga activities.

Some other studies by Shin et al. (1997), Molgaard et al. (1994), Urstad et al. (1998), Goldsmith et al. (1997), Rossy et al. (1998) and Dixon et al. (1992), reported that increased parasympathetic tone by vigorous activity (Monahan et al., 2000). Hamer et al. (2007) also reported that routine workout and physical fitness could help provide defence from the injurious effects of mental stress exposure.

As far as the effect of age on HRV is concerned, it is well recognized that HRV is pessimistically associated with the age (Agelink et al., 2001). However, most of the studies by Liao et al. (1995), Antelmi et al. (2004), Shannon et al. (1987) and Korkushkio et al. (1991) had found the decline levels of HRV after an age of around 40 years. The interventions like aerobic exercises that affect HRV had shown greater effects among younger than older adults (Hautala et al., 2003). Other authors also reported the large increase in heart rate variability components among the youth subjects (Paul et al., 2006). Peterson et al. (1988) observed an inverse relationship of obesity with the heart rate variability. On the other hand, Kageyama et al. (1997) also reported inverse relationship between body mass index with cardiac parasympathetic activity reflected by RSA amplitude by considering the influence of age, body fat, smoking condition and alcohol drinking on RSA.

Some researchers reported the positive effects of alcohol or neutral effect on cardiac health and HRV. Researchers of Harvard reported that a modest intake of alcohol reduced the risk of coronary heart disease (Manson et al., 1992). An extensive survey of all main studies on heart diseases by Porte et al. (1985) found that 'Alcohol ingestion is associated to total death volume in a U-shaped style, i.e. the modest drinkers (consumer) have a least total mortality in comparison with total abstainers and heavy consumers of alcohol. There has been a very low influence of occasionally short alcohol consumption on the cardio risks (Razay et al., 1992). Another investigation had also reported a decline of 49 per cent in coronary heart disease (CHD) among male population consuming moderate level of alcohol (Blackwelder et al., 1980).

Cigarette smoking is also well known risk for cardiovascular ailments. Routine smokers have been found with increased sympathetic activity in resting condition, and acute exposure to cigarette smoke is also a powerful stimulation effects on sympathetic activity. The majority of the effects of smoking on neuro-cardiovascular regulation have been attributed to the major ingredient of cigarette, i.e., nicotine. The authors like Barutcu et al. (2005), have reported loss of cardiac vagal tone among the intense smokers which became more evident predominantly in parasympathetic manoeuvre like controlled respiration. Dietrich et al. (2007) also reported an association between occupational exposure to environmental tobacco smoke (ETS) and lower HRV, higher HR among aged people. On the other hand authors like Kageyama et al. (1997), Murata et al. (1994) had failed to find a chronic effect of smoking on HRV.

18.4. Effect of Work Conditions, Alcohol and Smoking on Blood Pressure

The past researchers (Lang et al., 1992, Peter et al., 1993) have studied the effect of annoyance due to workplace noise jointly with workplace group support, nightshift, and work satisfaction on blood pressure. The study reported a significant effect of noise annoyance on increasing the diastolic blood pressure (DBP). An elongated exposure to occupational noise level at more than 85 dB (A) can result in high BP among sensitive persons (Lang et al., 1992). A significant association has also been observed between occupational noise induced hearing loss and high BP, for example, Ni Chun-Hui et al. (2007) reported an increased systolic blood pressure (SBP) and diastolic blood pressure (DBP) among the subjects having hearing loss at higher frequency at significant value of $P \leq 0.05$. Kathleen et al. (1992) investigated the influence of work stress on the blood pressure of 129 healthy individuals including males and females. The study concluded that average BP increased as the work load increases. Melamed et al. (1999) observed high levels of SBP and DBP among workers performing complex jobs under high occupational noise exposure over a period of 2–4 years. Another study by Narlawar et al. (2006) investigated the occurrence of hypertension and hearing loss in iron and steel industry workers in Nagpur (India). Prevalence of hypertension among continuously exposed (CEG) workers (25.51 per cent) was appreciably high than intermittently exposed (IEG) workers (14.05 per cent). Overall the study stated prevalence of hypertension and hearing loss are common among workers having constant occupational noise exposure.

As far as the effect of age on BP is concerned, high level of; age, BMI and alcohol consumption increases blood pressure as reported by Shaper et al. (1981). Other researchers investigated and found that BP is associated with the BMI. Wannamethee et al. (1994) reported a significant influence of BMI on blood pressure rather than on heart rate. Similar observation was reported by Talbott et al. (1990), i.e., BMI and alcohol drinking raised the SBP and DBP with significant association.

A number of researchers have investigated the cardiovascular effects of alcohol intake. Excessive alcohol intake has been found associated with hypertension. Makoto et al. (2000) reported that excessive consumption of alcohol may result in the formation of high rate of stress on the human body. Thierry et al. (1987) observed an optimistic relationship among arterial hypertension and alcohol intake among workers of small and medium-sized companies in the Paris region. Other authors like Yasushi et al. (2001) concluded that regular alcohol intake increased BP in normotensive workers aged 45–54 years, but not a mild intake. Thadhani et al. (2002) reported that risk of constant hypertension follows a J shaped curve among the youth female subjects, demonstrating a low risk with mild intake and high risk among heavy drinkers. MacMahon (1987) reported that moderate alcohol consumption reduces the blood pressure. O'Callaghan et al. (1995) also reported that there is no significant change in BP by occasionally alcohol consumption. Serge et al. (2004) had also found that, occasionally alcohol consumers (60 g alcohol/day) suffer considerably very less risk of mortality.

The acute effect of cigarette smoking has been observed to raise blood pressure by vasoconstriction and accelerated heart rate. However studies failed to report the chronic effects of regular smoking as observed by Benowitz et al. (1984) and Aronow et al. (1971). Some other researchers like Kagamimori et al. (1980), Arkwright et al. (1982), Lang et al. (1983) reported the non-existence of association between smoking habit and blood pressure. But other authors, Elliott (1980) and Dyer et al. (1982) reported that smoking raised blood pressure. While the others like Green et al. (1986), Gofin et al. (1982) and Istvan et al. (1999) failed to confirm this finding. But to the contrary, Seltzer (1974), Green et al. (1986), Savdie et al. (1984) and Agner (1983) reported low blood pressure of smokers in contrast to non-smokers. Lang et al. (1983) also reported a significant but negative relationship of systolic blood pressure (SBP) with cigarette smoking. However, in India the non-smoking male subjects were found with high risk of blood pressure with significant positive association (Gupta et al., 2007, Khuraibet et al., 2000, Hazarika et al., 2002), which is similar to the Western population as reported by these researchers (Bolinder et al., 1992, Bolinder et al., 1998, Accortt et al., 2002, Schroeder et al., 1985, Westman et al., 1995, Hergens et al., 2008).

18.5. Heart Rate Variability (HRV)

Heart rate variability refers to the regulation of the sino-atrial node (the natural pacemaker of the heart) by the sympathetic and parasympathetic branches of the autonomic nervous system. Heart rate variability (HRV) is beat-to-beat variation in heart rate (i.e., in R–R intervals) under resting conditions. These beat-to-beat variations occur due to continuous changes in the sympathetic and parasympathetic outflow to the heart (Malik, 1998 and Malik, 1995). HRV has been shown to be a good tool to quantify the tone of autonomic nervous system to the myocardium. Though, in principle, the heart rate variability can be measured over any length of recorded ECG, in reference to the guidelines of Task Force (1996) at least 5 minutes of ECG must be recorded to measure sympathetic and parasympathetic tone. The HF element is normally considered as a sign of vagal modulation (parasympathetic). The LF element is modulated by both the sympathetic and parasympathetic nervous systems. In this sense, its understanding is more controversial. A few researchers consider LF power, particularly when expressed in normalised units, as a measure of sympathetic modulations; whereas others understand it as a combination of sympathetic and parasympathetic activity (Task Force, 1996, Malliani et al., 1991, Ori et al., 1992, Togo and Takahashi 2009).

18.6. Method for HRV Assessment

Heart rate variability (HRV) is evaluated by time domain and frequency domain methods. The time domain method is based upon the number of time units (in milliseconds) within the beat-to-beat (R-R) intervals of the heart, and calculated from the variations between the conventional normal beat-to-beat intervals. In this method, statistical tools are applied to quantify the variations in RR intervals and then its parameters are computed. Most of the conventional time domain parameters (i.e. SDNN, SDSD, RMSSD, NN50 and pNN50) are marker of parasympathetic activity. Frequency domain measures of HRV offer data on the distribution of the elements of HRV with the aid of power spectral density analysis. Spectral analysis of HRV is characterized by three main elements: the high frequency (HF) element (0.15–0.40 Hz) measures the effect of the vagus nerve in modulating the sino–atrial node; the small frequency (LF) element (0.04–0.15 Hz) gives an index of sympathetic influence on the heart, especially when these are calculated in standardized units; the smallest frequency (VLF) element (0.003–0.04 Hz) reflects the effect of various issues on the heart, including chemoreceptors, thermo-receptors, the renin–angiotensin system, and other non-regular issues. More or less, all of the unpredictability from a short term spectral analysis of HRV is confined in these three elements (Malik, 1998, Malik, 1995, Task Force, 1996).

18.7. Protocol and Procedure for HRV Recording

Demographic profile (age, height, weight, experience) should be noted in the prescribe form. The inside room temperature of the cabin needs to be maintained at 24–26 °C. It is very important to give instructions to the subjects about the test requirements, and following are some of the main instructions given to the subjects:

1. To abstain from intake of alcohol, tobacco/smoking and coffee before the test, to have two hour of fasting before the test, to have a sound sleep in the previous night before reporting to the laboratory, etc.
2. To avoid drugs known to affect cardiac autonomic functions like cough cold medicines, etc., and to wear comfortable clothing for the testing.
3. Each of the subjects should be trained to close their eyes and relax while breathe normally, and also as far as feasible not to cough, sigh, move and very important not sleep during data acquisition.
4. A short term ECG 5 minutes is normally recorded at an acquisition rate of 512 Hz as per the recommendation of Task Force (1996). The process of ECG recording of a subject in spine rest position is shown in Figure 18.1 and followed by ECG showing PQRS waves in Figure 18.2.

Fig. 18.1: ECG recording of a subject in spine rest position.

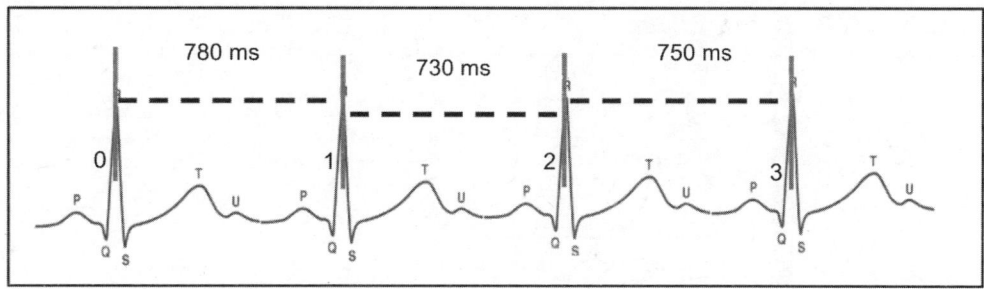

Fig. 18.2: Normal ECG recording with P, QRS and T waves.

A time-based tachogram is generated for each 5 minutes recording (Figure 18.3). All recordings are visually examined and manually over-read to verify beat classification. Non-sinus beats are eliminated along with R–R interval after the ectopic interval. The time domain parameters; mean RR, mean HR, standard deviation of the R–R intervals (SDNN), the root mean square of the successive R–R interval differences (RMSSD), and the percentage of R–R intervals that differed by 50 ms (pNN50) are calculated for each 5-min recording. Although researcher may use different algorithms, however, Fast Fourier Transformation (FFT) algorithm can be utilized to generate heart rate spectra for each 5-minute recording as described in Figure 18.4. Analysis can be performed on time domain elements of HRV like mean HR, SDNN, RMSSD, NN50, pNN50, and frequency domain components of HRV like low-frequency component (LF 0.04–0.15 Hz) for sympathetic, high-frequency component (HF 0.15–0.40 Hz) for parasympathetic activities and (LF/HF) ratio for autonomic tone.

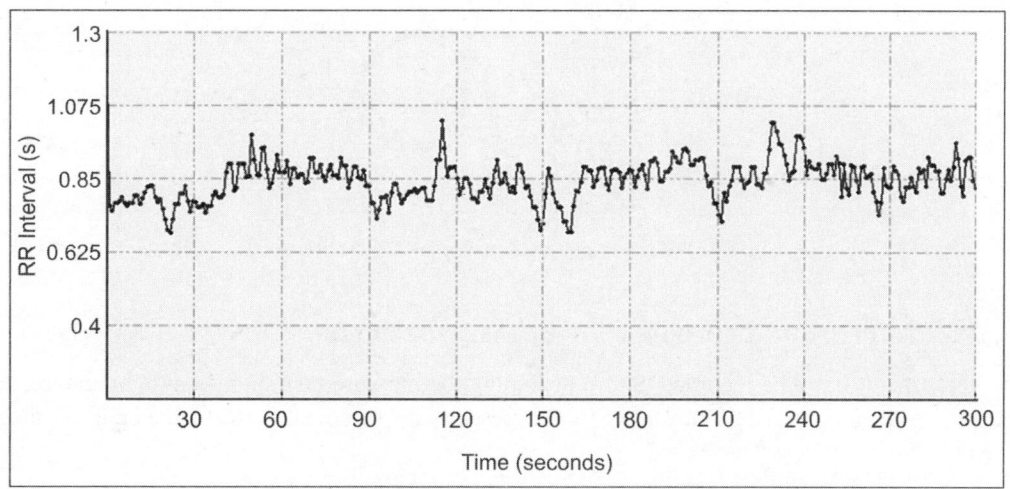

Fig. 18.3: Tachogram plotted for 300 sec recoding of ECG.

18.8. Blood Pressure (BP) Measurement

The pressure level of blood varies frequently and effected by various physiological and environmental issues. So, it is the most essential to manage these issues up to high extent or else it will not be possible to record in the blood pressure data forms.

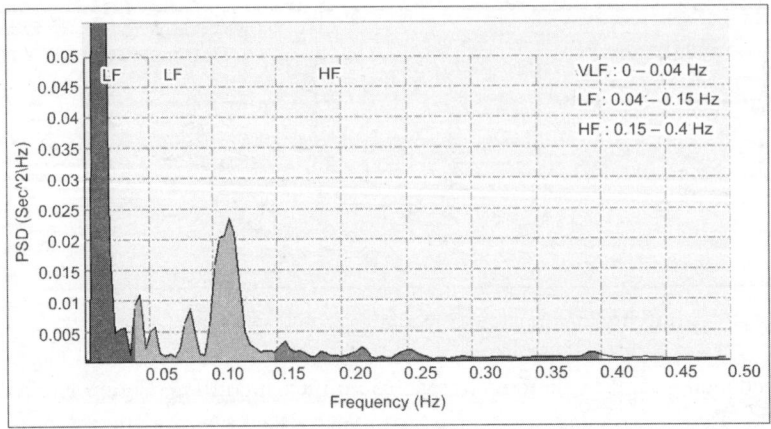

Fig. 18.4: FFT Power Spectral Density (PSD) spectrum.

18.9. Equipment for BP Measurement

The simple mercury sphygmomanometer is suggested, and in terms of accuracy it should be preferred over automated apparatus (Figure 18.5).

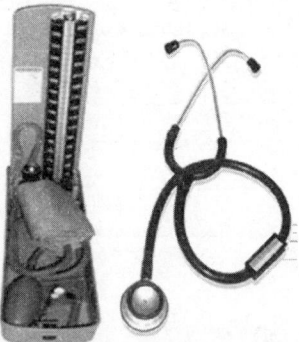

Fig. 18.5: Mercury sphygmomanometer and stethoscope.

18.10. Protocol and Procedure for BP Measurement

The items that need to be recorded before beginning the measurement are: room temperature, arm circumference and cuff used. Before the blood pressure measurement is made, the subjects should be instructed as follows:

1. To give up intake (anything else than water), which influence blood pressure sixty minutes earlier than recording. Also to avoided painful procedures and exercise before one hour.
2. The subjects should be asked to pass urine before the measurements.
3. The subjects should wear comfortable clothes preferable with half sleeve shirts, etc., so that the upper right arm is bare and, he/she must sit on a chair quietly for about 5 minutes before measure the BP.
4. A comfortable temperature (24–26 °C) should be maintained and the BP should be recorded on a suitable form. Three measurements can be taken one minute apart and averaged.

Case study: Comparison of cardiovascular functions (HRV & BP) of industry workers v/s controls

This case study is focused upon comparison of cardiovascular functions (HRV & BP) of industry workers v/s controls. This study included a sample of 138 male worker engaged in various activities of steel small and medium scale steel firms (casting and forging SMEs), and working under hostile conditions. Cardiovascular parameters (HRV and BP) were measured as per the protocols described in Sections 18.7 and 18.10 respectively. The findings of the study are described and discussed in the following 18.11 and onward sections. Based upon the literature review and personal visit of the industry, a hypothesis was proposed as follows:

- Hypothesis (H2): The Cardiovascular function parameters (HRV & BP) of exposed group are significantly lower than the controls.

18.11. Comparison of Exposed Group v/s Control Group for Cardiac Autonomic Control (HRV)

The exposed group and the control group were statistically comparable for mean age (SD) (at 'p' > 0.05). However the BMI of industry subjects was found lower than the control subjects. The 't'-test results for mean age (SD), mean BMI (SD) and HRV parameters of both the groups are also shown in Table 18.1. The industry workers showed significantly lower resting heart rate with mean HR (SD) of 68.27 (10.23) beats per minute as compared to the control group with mean HR (SD) of 74.24 (10.26) at 'p' value less than 0.01. They also showed considerably improved parasympathetic activity as shown in NN50, pNN50 and high frequency component (HF ms^2) of power spectrum (at 'p' value of less than 0.05), which revealed better cardiac autonomic control amongst the industry workers.

Table 18.1: The 't'-test results of exposed v/s control group for HRV (cardiac autonomic control).

Parameter	Exposed Group (N = 138)	Control Group (N=50)	Difference of Means	'p' Value
	Mean (SD)			
Age	29.65 (7.628)	31.30 (8.591)	-1.645	0.211
BMI	21.385 (3.686)	23.383 (2.630)	-1.998	0.001
Mean RR	0.898 (0.120)	0.822 (0.102)	.0767	0.000
Mean HR	68.27 (10.23)	74.240 (10.261)	-5.969	0.001
NN50	61.620 (54.011)	44.180 (42.645)	17.440	0.041
pNN50	22.931 (19.614)	15.584 (16.046)	7.347	0.019
LF (ms^2)	183.45 (171.650)	149.41 (58.967)	34.041	0.223
HF (ms^2)	108.21 (115.695)	67.894 (67.086)	40.236	0.024
LF/HF Ratio	2.216 (0.962)	2.309 (0.668)	-0.094	0.526

p value ≤ 0.05; Significant at 95 per cent confidence interval and p value ≤ 0.01; Significant at 99 per cent confidence interval.

18.12. Comparison of Exposed Group v/s Control Group for Blood Pressure (BP)

In case of blood pressure (BP), the industry subjects were found to have significantly lower systolic blood pressure (SBP) and mean blood pressure (MBP) parameters in comparison with the control group at p <0.05. Body Mass Index (BMI) of exposed group subjects was also lower than the control group at ≤ 0.05. The 't'-test results of age, BMI, SBP, DBP and MBP of both the groups are shown in Table 18.2.

Table 18.2: The 't'-test results of exposed v/s control group for BP.

Parameter	Exposed Group (N = 132)	Control Group (N = 51)	Difference of Means	'p' Value
	Mean (SD)			
Age (Years)	29.65 (8.14)	31.98 (8.87)	-2.329	0.093
BMI	21.48 (3.78)	23.43 (2.62)	-1.94735	0.001
Systolic BP	116.72 (9.62)	121.01 (6.15)	-4.2868	0.004
Diastolic BP	79.35 (7.82)	81.48 (4.93)	-2.1261	0.073
MBP	91.87 (7.98)	94.65 (5.04)	-2.7832	0.022

p value ≤ 0.05 Significant at 95 per cent confidence interval and p value ≤ 0.01 Significant at 99 per cent confidence interval.

Since the industry subjects, i.e., exposed group showed significantly better cardiovascular autonomic control as reflected in NN50, pNN50 and HF component of power spectrum along with lower resting HR and BP (at 'p' 0.05), therefore:

Hypothesis (H_2): The Cardiovascular function parameters (HRV & BP) of exposed group are significantly lower than the controls is rejected

18.13. Effect of Workplace Conditions on Cardiovascular Autonomic Control (HRV and BP)

The effect of occupation on HRV components (cardiac autonomic control) was found be insignificant, except the resting mean HR (SD). This difference exists due to significantly reduced mean HR of workers engaged in moulding/casting section as compared to control group (Table 18.3). The post hoc Tukey's analysis for multiple comparison is shown in Table 18.4, which reveals that, the workers engaged in moulding/casting section showed significantly lower mean HR (SD) of 65.60 (9.84) in comparison with the controls with mean HR (SD) 74.24 (10.3) at $p ≤ 0.05$.

18.14. Effect of Workplace Conditions on BP Parameters

The type of work condition was found to be associated with the BP as shown in Table 18.5. The industry workers who transfer the hot work piece from the oil fired furnace to the forger were found with the lowest BP parameters. These workers were doing their jobs at a higher level of WBGT and therefore they suffered from sweat losses. Post hoc Tukey's analysis for multiple comparisons demonstrates that, the workers engaged in transferring the hot work piece from furnace to forger showed significantly lower SBP as compare to the tool room workers and control subjects. At the same time the workers engaged in moulding/casting tasks were having significantly lower SBP than the tool room workers (Table 18.5). The tool room workers were doing relatively lighter tasks at lower WBGT. The workers who were engaged in transferring the hot work piece from furnace to forger showed significantly lower mean BP (MBP) as compared to tool room workers (Table 18.6).

Table 18.3: ANOVA results for effect of work conditions on HRV (autonomic control).

HRV Parameter	Punching Blanking (n=19)	Hot Forging (n=18)	Hot Job Handling (n=12)	Hammer Operator (n=13)	Grinding (n=43)	Moulding /Casting (n=20)	Tool Room/ Welding (n=13)	Control (n=50)	'p' Value
	HRV Parameters Mean (SD)								
Age	25.63 (7.47)	34.06 (8.53)	26.42 (4.66)	26.62 (2.98)	28.91 (6.05)	32.00 (8.43)	32.50 (10.12)	31.94 (8.96)	0.003
BMI	20.09 (3.11)	22.22 (6.04)	19.01 (2.33)	21.90 (2.31)	21.74 (3.19)	20.64 (1.88)	23.95 (4.51)	23.38 (2.63)	0.000
Mean RR	0.86 (0.11)	0.88 (0.12)	0.92 (0.16)	0.93 (0.11)	0.89 (0.14)	0.93 (0.13)	0.85 (0.11)	0.82 (0.10)	0.006
Mean HR	70.68 (9.58)	69.11 (9.42)	66.92 (11.95)	65.38 (7.90)	68.46 (11.52)	65.60 (9.84)	71.42 (8.62)	74.24 (10.3)	0.019
NN50	70.42 (66.05)	61.61 (60.74)	65.58 (28.01)	50.00 (28.62)	60.90 (54.44)	68.25 (62.84)	47.83 (52.99)	44.18 (42.64)	0.472
pNN50	27.95 (23.46)	18.94 (20.14)	27.89 (9.68)	20.28 (12.23)	22.88 (20.39)	23.83 (22.38)	17.55 (19.50)	15.58 (16.05)	0.204
LF	179.95 (164.146)	(196.78) (172.95)	168.40 (97.3)	128.58 (65.52)	196.02 (189.82)	252.12 (233.32)	84.00 (64.01)	149.41 (158.96)	0.144
HF	179.95 (164.146)	94.64 (86.11)	93.17 (67.27)	73.53 (54.89)	111.35 (133.96)	142.14 (152.03)	92.72 (104.59)	67.89 (67.08)	0.228
LF/HF Ratio	1.99 (1.11)	2.36 (0.78)	2.35 (1.09)	1.99 (0.52)	2.26 (0.92)	(2.43) (1.08)	1.92 (1.13)	2.31 (0.67)	0.592

p value ≤ 0.05 Significant at 95 per cent confidence interval and p value ≤ 0.01 Significant at 99 per cent confidence interval.

1: Punching blanking, 2: Forgers, 3: Hot job handling furnace to forger, 4: Hammer operator, 5: Moulding/casting, 6: Grinding, 7: Tool room/welding, 8: Control group.

Table 18.4: Tukey's analysis multiple comparison for mean HR (SD) for group 2 with others.

Parameter	Sub Group (I)	Sub Group (J)	Difference of Means	'p' Value
Mean HR (SD)	6	1	-5.0842	0.782
		2	-3.5111	0.965
		3	.21538	1.000
		4	-1.3167	1.000
		5	-2.8651	0.969
		7	-5.8167	0.778
		8	-8.6400	0.036

Table 18.5: Effect of occupation on BP parameters using ANOVAs results.

Para meter	Punching Blanking (n = 18)	Forgers (n = 19)	Furnace to Hammer Hot Job Handler (n = 10)	Hammer Operator (n = 9)	Grinding (n = 44)	Moulding/ Casting (n = 19)	Tool Room/ Welding (n = 13)	Control Group (n = 51)	'p' Value
	Mean (SD)								
Age (Yrs.)	25.50 (7.56)	33.95 (10.18)	25.50 (4.55)	25.78 (2.73)	29.11 (6.19)	31.37 (8.12)	34.31 (11.00)	31.98 (8.88)	0.002
BMI	20.07 (3.14)	22.29 (6.14)	19.65 (2.59)	22.07 (2.57)	21.65 (3.21)	20.78 (2.07)	23.69 (4.35)	23.43 (2.62)	0.001
Systolic BP	118.51 (9.24)	117.68 (11.99)	109.86 (11.93)	114.37 (6.33)	116.97 (9.12)	114.28 (6.65)	122.41 (8.99)	121.01 (6.15)	0.002
Diastolic BP	78.59 (9.46)	82.81 (6.16)	75.86 (8.78)	78.67 (5.39)	79.53 (8.01)	76.42 (6.95)	82.18 (7.165)	81.47 (4.93)	0.068
Mean BP	91.90 (8.35)	94.43 (7.46)	87.20 (9.49)	90.56 (5.37)	92.01 (8.11)	89.04 (6.10)	96.23 (8.41)	94.65 (5.04)	0.006

p value ≤ 0.05 Significant at 95 per cent confidence interval and p value ≤ 0.01 Significant at 99 per cent confidence interval.

1: Punching blanking, 2: Forgers, 3: Hot job handling furnace to forger, 4: Hammer operator, 5: Grinding, 6: Moulding/casting, 7: Welding/tool room, 8: Control group.

Table 18.6: Tukey's analysis multiple comparison for SBP, DBP and MBP.

Group (I)	Group (J)	'p' Value		Group (I)	Group (J)	'p' Value
		Systolic BP	Mean BP			Systolic BP
3	1	0.179	0.722	6	1	0.827
	2	0.282	0.180		2	0.934
	4	0.946	0.972		3	0.905
	5	0.265	0.998		4	1.000
	6	0.891	0.555		5	0.953
	7	0.015	0.050		7	0.050
	8	0.006	0.050		8	0.093

p value ≤ 0.05 Significant at 95 per cent confidence interval and p value ≤ 0.01 Significant at 99 per cent confidence interval.

1: Punching blanking, 2: Forging, 3: Hot job handling furnace to forger 4: Hammer operator, 5: Grinding, 6: Moulding/casting, 7: Welding/tool room, 8: Control group.

18.15. Effect of Work Exposure on Cardiovascular Autonomic Control (HRV and BP) Within Exposed Group

The results with respect to effect of work exposure on cardiovascular autonomic control are explained as follows.

As far as the effect of work exposure on HRV components was concerned, only pNN50 was found be affected (Table 18.7). Post hoc Tukey's analysis for multiple comparisons shown in Table 18.8, revealed that the workers with work exposure of 10–15 years showed significantly lower value of pNN50 as compared to the workers with work exposure up to 5 years.

Table 18.7: Effect of work exposure on autonomic control in exposed group.

Parameter	Work Exposure				'p' Value
	Up to 5 Years (n = 59)	5 to 10 Years (n = 40)	10 to 15 Years (n = 22)	Above 15 Years (n = 17)	
	Mean (SD)				
Age	25.42 (6.11)	28.53 (4.21)	34.09 (5.55)	39.59 (7.97)	0.000
BMI	19.94 (2.21)	22.33 (3.03)	22.99 (4.28)	22.07 (6.12)	0.001
Mean RR	0.92 (0.13)	0.86 (0.10)	0.89 (0.15)	0.91 (0.15)	0.124
Mean HR	66.54 (10.13)	70.73 (7.99)	69.32 (11.74)	67.71 (12.61)	0.235
NN50	73.10 (54.75)	59.03 (52.23)	43.18 (48.98)	51.35 (55.18)	0.110
pNN50	28.72 (20.07)	19.93 (18.29)	16.25 (17.72)	17.92 (18.85)	0.019
LF	192.64 (165.47)	179.71 (145.79)	191.34 (242.20)	149.32 (144.01)	0.825
HF	123.32 (127.02)	110.20 (112.92)	91.33 (135.22)	70.28 (57.08)	0.369
LF/HF Ratio	2.0816 (1.03)	2.17 (0.80)	2.58 (0.96)	2.33 (0.99)	0.197

Table 18.8: Post hoc Tukey's analysis for multiple comparisons for HRV components among various subgroups.

Parameter	Group (I)	Group (J)	Difference of Means	'p' Value
Age	1	2	-3.10127*	0.049
		3	-8.66718*	0.000
		4	-14.16451*	0.000
BMI	1	2	-2.38767*	0.006
		3	-3.05048*	0.003
		4	-2.12427	0.124
pNN50	1	2	8.78953	0.115
		3	12.47203*	0.048
		4	10.80439	0.172

p value ≤ 0.05 Significant at 95 per cent confidence interval and p value ≤ 0.01 Significant at 99 per cent confidence interval.

Sub-group 1: Exp >5 years, Sub-group 2: 5–10 years, Sub-group 3: 10–15 years, Sub-group 4: Exp >15 years.

18.16. Effect of Work Exposure on BP Parameters

Amongst the industry subjects BP parameters were found to be significantly associated with the work exposure. The industry subjects with work exposure less than 5 years showed significantly lower

SBP, DBP and MBP as compared to the subjects with work exposure of 10–15 years as well as the subjects with work exposure more than 15 years. However, the same workers showed insignificant difference from the subjects with 5–10 years of work exposure (Tables 18.9 and 18.10).

Table 18.9: Effect of work exposure on BP among industry workers.

Parameter	Work Exposure				'p' Value
	Up to 5 Years (n = 55)	5–10 Years (n = 35)	10– 5 Years (n = 25)	Above 15 Years (n=17)	
	Mean (SD)				
Age (Years)	24.92 (5.81)	28.70 (4.42)	34.36 (5.89)	40.27 (8.26)	0.000
BMI	20.09 (2.32)	22.16 (3.22)	22.89 (4.24)	22.05 (5.93)	0.004
Systolic BP	113.68 (8.61)	117.70 (7.68)	120.50 (14.57)	120.53 (14.57)	0.008
Diastolic BP	76.03 (7.53)	80.19 (6.76)	83.53 (7.49)	83.53 (7.49)	0.000
Mean BP	88.58 (7.71)	92.69 (6.75)	95.87 (9.28)	95.78 (9.28)	0.000

Table 18.10: Post hoc Tukey's analysis for multiple comparison of SBP, DBP and MBP w.r.t. work exposure.

Group (1)	Group (J)	'p' Value		
		SBP	DBP	MBP
1	2	0.235	0.058	0.001
	3	0.022	0.001	0.004
	4	0.051	0.002	0.454
2	3	0.694	0.405	0.492
	4	0.750	0.424	1.000
3	4	1.000	1.000	0.072

p value ≤ 0.05 Significant at 95 per cent confidence interval and p value ≤ 0.01 Significant at 99 per cent confidence interval.

Group 1: Subjects with Work Exp (1–5), Group 2: subjects with Work Exp (5–10), Group 3: Subjects with Work Exp (10–15), Group 4 Subjects with Work Exp (>15).

18.17. Discussion

18.17.1. Cardiovascular autonomic control (HRV)

The results of present study revealed that the exposed group subjects (industry workers) showed adaptive response to the physical work load. The exposed group subjects revealed considerably less resting mean heart rate (HR) as compared to controlled group. Concurrently the industry workers demonstrated significantly improved parasympathetic activity as reflected in NN50, pNN50, HF component of power spectrum signifying that there was a positive response to the physical work load. It has been agreed that lower heart rates are at least partially the result of increased parasympathetic tone (Yataco et al., 1997, Shin et al., 1997, Seals et al., 1989 and Levy et al., 1998). Antelmi et al. (2004) also reported an inverse correlation of HRV with heart rate ($p < 0.001$). Rennie et al. (2003) observed a significant association with high rate of HF capacity in males during high-intensity activity, such as lifting and carrying of heavy jobs, hot material handling (Shin et al., 1997, Molgaard et al., 1994, Urstad et al., 1998, Goldsmith et al., 1997, Rossy et al., 1998, Dixon et al., 1992), but not all vigorous activities (Schuit et al., 1999, Loimaala et al., 2000). Hamer et al. (2007) also reported that work out and physical robustness may possibly act as a defence against the harmful effects of job related psychological stress (Noise annoyance) exposure.

On the other hand control group subjects were physically less active and live a sedentary life style. Therefore, they showed significantly higher resting mean HR (SD) along with significantly lower parasympathetic activity as compared to the industry workers. The modern sedentary life style without regular physical activities (exercise) could be credited to the lower cardiac autonomic control amongst control group.

This, body idleness may results into crucial cardiovascular risks by means of obesity, reduced cardiovascular fitness (Blair and Brodney, 1993), raised blood pressure (Duncan et al., 1985). Thus it is realized that a specific level of physical exercise/work is mandatory for the control group to maintain/ improve the autonomic tone or sympatho-vegal balance. This alteration can be corrected by non-pharmacologic interventions like physical exercise, yogic exercises (Sunkaria et al., 2010) and alike.

18.17.2. Comparison of exposed group vs control group for blood pressure

The past researchers (Lang et al., 1992, Peter et al., 1993) have studied the effect of work-related noise annoyance along with the combined effect of group support at work, working in night shift, and job contentment on blood pressure. Noise annoyance significantly increased diastolic blood pressure (DBP) (Peter et al., 1993). Ni Chun-Hui et al. (2007) found that both SBP and DBP of subjects with hearing loss at high frequency were significantly higher than subjects with normal hearing at $P \leq 0.05$. Kathleen et al. (1992) investigated the effect of high job strain on blood pressure and reported that men with high job strain showed higher mean blood pressure.

But in the present study, blood pressure of exposed group subjects was observed appreciably lesser as compared to the control group subjects with $p \leq 0.05$. Also exposed group subjects had significantly lower body mass index (BMI) as compared to control group with $p \leq 0.05$. In the present study, the industry workforce was engaged in many different daily moderate to heavy as described in chapters 14 and 17, work tasks like manual material handling tasks, forging, moulding, pouring, grinding, etc. The workers generally work more than 8 hours/day, i.e., 10–12 hours per day.

In the present study, industry workers showed significant hearing loss due to occupational noise exposure. Moreover the workers do not feel much of noise annoyance unlike reported by Peter et al. (1993) and Lang et al. (1992). This could be possibly due to adaptability to the workplace noise level and occupational hearing loss. Therefore, our study reported insignificant relationship between hearing loss and blood pressure in workers uncommon with Wu et al. (1987), Peter et al. (1993) and Lang et al. (1992), who observed significant outcome of noise irritation on increasing diastolic blood pressure (DBP) and major increase of BP among sensitive individuals.

18.17.3. Effect of workplace conditions on cardiovascular autonomic control (HRV and BP)

As far as the effect of workplace condition on HRV and autonomic control was concerned, there was insignificant difference in HRV components of workers engaged in different workplaces. However, there was significant difference of mean HR at 'p' value less than 0.05. The ANOVA with multiple comparison revealed that, the workers engaged in moulding/casting section showed significantly lower mean HR (SD) as compared to the control group subjects at 'p' value less than 0.05. These workers were doing heavy/very heavy activities like shovelling of sand, lifting and carrying of moulds and crucibles containing molten metal and molten metal pouring, etc. Thus physical exercise was inherently performed by the workers. This is in accordance with the type of work activities such as modest and heavy tasks; there occurs a reasonable decrease in heart rates as reported by Rennie et al. (2003). Other researchers like Shin et al. (1997), Fagard et al. (1999) and Loimaala et al. (2000) also reported lower resting heart rates in individuals engaged in some aerobics activities.

18.17.4. Effect of workplace conditions on BP

The analysis of variance ANOVA results demonstrates a significant influence of work place condition on the BP parameters. The multiple comparisons using Tukey's analysis in ANOVA revealed that, the workers engaged carrying of job from furnace to forger were having significantly lower SBP and MBP as compared to the tool room workers and controlled individuals. The subjects engaged in moulding/ casting tasks were having lower SBP than the tool room workers. These workers were doing their jobs at a high WBGT and therefore they sweat a lot. It has been previously discussed in Chapter - 17 that industry workers did not use glucose or salt water to compensate the sweat losses. Thus the chronic swat losses could be attributed to the lower BP parameters similar the past researcher like, Salonen et al. (1983), Khaw and Connor (1988) who reported an association between blood pressure and dietary salt intake. However the BP of workers of other sections did not differ significantly. The tool room workers were doing relatively lighter tasks and at lower WBGT, moreover they were having significantly higher BMI as compared to the workers of hot material handling and moulding/ casting section. Similarly, the control subjects were living a sedentary life style without much of physical activities.

18.17.5. Effect of work experience (Years) on cardiovascular autonomic control

As far as the effect of work experience (years of work exposure) on cardiovascular health parameters (HRV & BP) is concerned, the outcome of the case study is in contrary to the normal belief and the literature. A very less effect of number of years of working has been observed on the HRV and BP, these are described in the following sections.

18.17.5.1. *Effect of work exposure on HRV components in exposed group*

As far as the effect of work exposure on HRV is concerned only one component, i.e., pNN50 was affected. The subjects with work exposure up to 5 years showed significantly high value of pNN50 as compared to the subjects with higher work experience. At the same time the BMI and age is also increasing. Therefore the significant difference of pNN50 could be the confounding effect of age and BMI.

18.17.5.2. *Effect of work exposure on BP in exposed group*

The past studies investigated the relationship between hostile work conditions and BP. Johsson and Hansson (1977) reported significantly higher systolic and diastolic blood-pressure in male industrial workers with prolonged exposure to noise and auditory impairment than the control of same age with normal hearing. However, Tarter and Robins (1990), studied the incidents of hypertension and higher mean blood pressure among subjects exposed to industrial noise of ≥85 dB (A) for at least five years. There was no correlation between MBP and hearing loss at 4000 Hz and the period of work exposure to high-noise areas. Narlawar et al. (2006) also reported an optimistic relationship among time period of exposure and occurrence of high blood pressure. This indicated that hypertension and hearing loss were common among the labourers with consistent exposure to high occupational noise. Whereas, Lang et al. (1986), conducted a study on 249 men working in a foundry of the Paris area and reported that SBP was optimistically connected to the time period of the occupational exposure to a noise level of more than the prescribed limits of 85 dB (A).

The present study also reported a significant influence of work exposure on the BP. The industry subjects with work exposure up to five years, showed significantly low SBP, DBP and MBP as compared to the subjects with work exposure of 10–15 years and the subjects with work experience of more than 15 years. However, there was insignificant difference from the subjects with 5–10 years of work experience. The workers with wok exposure of five years were also having significantly lower age and BMI than the workers with experience of half a decade to one and a half decades. Although our results also showed an increasing trend of BP with work exposure in accordance with past studies by Johsson A and Hansson L (1977), Tarter SK and Robins TG (1990) and Narlawar et al. (2006), but in the present study the industrial workers did not show prevalence of hypertension, as the SBP/DBP and MBP were still under the normal range, i.e., 120/80. This increasing trend of BP, although under normal range, could be attributed to the age factor, since the work experience and age of the worker increases simultaneously. It has been previously discussed in chapters 13 and 14 that, the workers were exposed to noise and are prone to the hearing loss but at the same time they showed significantly lower BP as compared to the control subjects. This might be due to the heavy physical activity and sweat losses at the work place. Thus, our study did not show prevalence of hypertension amongst the workers with hearing loss in common with past study by Hirai et al. (1991) which had also not shown significant relationship between the hearing loss and the prevalence of hypertension. Thus, a long-lasting exposure to industrial noise contributes towards hearing disability. However, elevation of blood pressure was not observed in labours engaged under work conditions with high noise levels. Wu et al. (1987) however, uttered that hear disability cannot be referred as a suitable noise exposure index to indicate association with blood pressure.

18.17.6. Cardiovascular effects of occupational noise

A persistently noise exposed subjects of several workplaces were found associated with high risk of acute myocardial infarction death (Hugh et al., 2005). The influence of work noise and level of satisfaction during various shifts on the BP has been observed in a study and there has been a significant amplification in DBP due to noise irritation (Peter et al., 1993). But in Indian environmental conditions of casting and forging industry, although the workers were highly exposed to high levels of workplace noise (Ref. chapter 13), prolonged work experience under high occupational noise levels may result into noise induced hearing loss (NIHL). As described in chapter 17, that the industry workers also showed minimum level of occupational noise syndromes like noise annoyance and anxiety etc.; rather they established sound at job, an integral occupational element. At the same time the workers were not using hearing protection like hearing plugs and ear muffs. It has been shown that industry workers showed significant hearing loss in comparison with the control subjects. It is understood that throughout the adaptability to noise exposure and NIHL, the autonomic tone might have been affected. But the physical activities at the workplace might have possibly conquered the effect of noise annoyance, and thus the HRV components; NN50, pNN50, HF (ms^2) were significantly improved.

Our results are in contradiction to the results shown by other authors such as Togo et al. (2009), and Hugh et al. (2005). The workforce had occupational exposure to noise, heat, fumes or dusty environment, which are known factors leading to decreased parasympathetic activity (Shannon et al., 2001 and Fujino et al., 2007). The workers who perceived noise exposure at work place were found with a higher prevalence of myocardial infarction (Rai et al., 2007). It had been established that, prolonged experience to noise above 85 dB (A) is critical parameter for high BP among noise sensitive individuals. Although the noise annoyance perceived by an individual is critical factor towards the cardiovascular effects of noise hazards (Lang et al., 1992 and Peter et al., 1993). But, the present study did not confirm these facts as most of the workers were insensitive towards work place noise.

18.18. Summary

The workers were engaged in various activities involving physical efforts. Hence the physical workout was inbuilt in their jobs. The study can be summarised that, the occupational noise exposure leads to the hearing disability and compliance to the noisy environment. However, the risks of cardiovascular problems due to noise exposure appear to be voided by the heavy physical activities at work place. Thus an exposure of less than 10 years to physically demanding and even to the antagonistic environmental conditions (noise etc.) showed improved HRV components. Past studies Rennie et al. (2003) also reported associations between moderate and vigorous activity and increased HRV. High energy demanding task has been indicated as closely linked to rising HRV, indicative of a probable method, so as to reduce coronary heart disease risk.

Multiple Choice Questions

1. What is the extended form of HRV?
   - (a) Heart Rate Variables
   - (b) Hearing Range Variability
   - (c) Heart Rate Variability
   - (d) None of the above
2. Heart rate variability refers to the regulation of the sino-atrial node of the heart by the sympathetic and parasympathetic branches of the
   - (a) Autonomic nervous system
   - (b) Central nervous system
   - (c) conscious
   - (d) None of these
3. The heart rate variability can be measured over any length of recorded ECG, but in reference to the guidelines of Task Force (1996), at least _____ of ECG must be recorded.
   - (a) 5 minutes
   - (b) 5 sec
   - (c) 15 minutes
   - (d) None of these

4. The power spectral density anlysis HRV includes
   (a) HF (0.15–0.40 Hz) only
   (b) HF (0.15–0.40 Hz) and LF (0.04–0.15 Hz)
   (c) VLF (0.003–0.04 Hz) only
   (d) All of the above
5. Before HRV recording the subject is instructed to
   (a) Abstain from intake of; alcohol, tobacco/smoking tea coffee
   (b) Avoid Drugs affecting HRV
   (c) Close their eyes and relax while breathe normally
   (d) All of the above
6. The inside room temperature of the cabin for HRV recording need to be maintained at
   (a) 20 °C
   (b) 24–26 °C
   (c) 25 °C
   (d) All of the above
7. High value of NN50, pNN50 and high frequency component (HF ms^2) of power spectrum is an indicator of;
   (a) Increased Sympathetic activity
   (b) Increased Parasympathetic activity
   (c) Reduced Sympathetic activity
   (d) None of above
8. Sympatho-vegal balance is indicated by which of the HRV components
   (a) HF only
   (b) HF and LF
   (c) LF/HF Ratio
   (d) None of the above
9. Before the blood pressure measurement is made, the subjects should be instructed to
   (a) Give up intake (anything else than water), which influence blood pressure sixty minutes earlier than recording
   (b) Avoided painful procedures and exercise before one hour
   (c) Sit quietly for about 5 minutes
   (d) All of the above
10. The normal value of SBP/DBP for an adult person is
   (a) 130/70
   (b) 120/80
   (c) 130/80
   (d) Any one of the above

Answers

1. (c)   **2.** (a)   **3.** (a)   **4.** (d)   **5.** (d)   **6.** (b)   **7.** (b)   **8.** (c)   **9.** (d)   **10.** (b)

References

Accortt, N. A., J. W. Waterbor, C. Beall, and G. Howard. 2000. 'Chronic Disease Mortality in a Cohort of Smokeless Tobacco Users, *Am J Epidemiol* 156:730–37.

Agate, J. N. 1949. 'An Outbreak of Cases of Raynaud's Phenomenon of Occupational Origin'. *Br J Ind Med.* 6:144–63.

Agelink, M. W., R. Malessa, B. Baumann, T. Majewski, F. Akila, T. Zeit, and D. Ziegler. 2001. 'Standardized Tests of Heart Rate Variability: Normal Ranges Obtained from 309 Healthy Humans, and Effects of Age, Gender, and Heart Rate, *Clin Auton Res* 11:99–108.

Agnihotram, R. V. 2005. 'An Overview of Occupational Health Research in India'. *Indian J Occup Environ Med.* 9(1):10–14.

Ahasan, M. R. 2002. 'Occupational Health, Safety and Ergonomic Issues in Small and Medium-Sized Enterprises in a Developing Country'. Diss., University Of Oulu.

Ahmad, H. O., J. H. Dennis, O. Badran, M. Ismail, S. G. Ballal, A. Ashoor, et al. 2001. 'Occupational Noise Exposure and Hearing Loss of Workers in Two Plants in Eastern Saudi Arabia'. *Ann Occup Hyg.* 45: 371–80.

American Conference of Governmental Industrial Hygienists. 2001. *Documentation of the Threshold Limit Values and Biological Exposure Limits, Notice of Intended Changes.* Cincinnati, American Conference of Governmental Industrial Hygienists.

American Thoracic Society. 1987. 'Standardization of Spirometery 1987 Update'. *Am Rev Respiratory Respir Dis.* 136:1285–98.

———. 1991. 'Lung Function Testing: Selection of Reference Values and Interpretative Strategies'. *Am Rev Respir Dis.* 144:1202–18.

———. 1994. 'Standardization of Spirometery 1994 Update'. *Am J Respir Crit Care Med.* 152:1107–36.

ANSI S3.21-1978 (R-1992). *American National Standard for Manual Audiometry.*

ANSI S3.6 (1996). *Specifications for Audiometers.*

Antelmi, I., R. S. De Paula, A. R. Shinzato, C. A. Peres, A. J. Mansur, and C. J. Grupi. 2004. 'Influence of Age, Gender, Body Mass Index, and Functional Capacity on Heart Rate Variability in a Cohort of Subjects without Heart Disease'. *Am J Cardiol.* 93:381–85.

Astrand, P. O., and L. Rodahl. 1986. *Textbook of Work Physiology* (3rd edition). New York: McGraw-Hill.

Atmaca, E., I. Peker, and A. Altin. 2005. 'Industrial Noise and Its Effects on Humans'. *Pol J Environ Stud.* 14(6):721–26.

Bates, G., and V. Miller. 2005. 'The Effects of the Thermal Environment on Health and Productivity'. Paper presented at the 6th International Scientific Conference of the International Occupational Hygiene Association, Pilanesburg, Bojanala, September 19-23.

Bedi, R., D. K. Shukla, and A. Sachdeva. 2004. 'Effects of Impact Noise on Humans – A Case Study of Drop Forge Hammers'. Paper presented at the 7th International Symposium of the East European Acoustics Association, St. Petersburg, Russia, June 8-10.

Benowitz, N. L., F. Kuyt, and P. Jacob. 1984. 'Influence of Nicotine on Cardiovascular and Hormonal Effects of Cigarette Smoking'. *Clin Pharmacol Ther.* 36:74–81.

Blackwelder, W. C., K. Yano, G. G. Rhoads, A. Kagan, T. Gordon, and Y. Palesch. 1980. 'Alcohol and Mortality. The Honolulu Heart Study'. *Am J Med.* 68(2):164–69.

Bolinder, G. and U. de Faire. 1998. 'Ambulatory 24-h Blood Pressure Monitoring in Healthy, Middle-aged Smokeless Tobacco Users, Smokers, and Nontobacco Users'. *Am J Hypertens.* 11:1153–63.

Bolinder, G. M., B. O. Ahlborg, J. H. Lindell, et al. 1992. 'Use of Smokeless Tobacco: Blood Pressure Elevation and Other Health Hazards Found in a Large-scale Population Survey'. *J Intern Med.* 232:327–34.

Branton, P. and G. Grayson. 1967. 'An Evaluation of Train Seats by an Observation of Sitting Behaviour'. *Ergonomics.* 10:35–51.

Buckhout, R. 1964. 'Effect of Whole Body Vibration on Human Performance'. *Human Factors* 6:157–63.

Cadaveira, F., F. Díaz, and C. Grau. 1990. 'Short- and Middle-latency Auditory Evoked Potentials in Abstinent Chronic Alcoholics: Preliminary Findings'. *Electroencephalogr Clin Neurophysiol/Evoked Potentials Sect.* 77(2):145–50.

Celik, O., S. Yalçin, and A. Oztürk. 1998. 'Hearing Parameters in Noise Exposed Industrial Workers'. *Auris Nasus Larynx.* 25(4):369–75.

Chaffin, D. B. C. Belson, S. D. Ianni, and P. A. Punte. 2001. *Digital Human Monitoring for Vehicle and Workplace Design*. Warrendale, Pennsylvania: Society of Automotive Engineers.

Chaffin, D. B., G. B. J. Andersson, and B. J. Martin. 1999. *Occupational Biomechanics* (3rd edition). New York: Wiley.

Chaffin, D. B. 1997. 'Biomechanical Aspects of Workplace Design'. *Handbook of Human Factors and Ergonomics* (2nd edition)., Edited Gavriel Salvendry. New York: Wiley.

Chaney, F. B., and K. S. Teel. 1967. 'Improving Inspector Performance through Training and Visual Aids'. *J Appl Psychol.* 51:311–15.

Chen, C. J., Y. T. Dai, Y. M. Sun, Y. Chang, and Y. J. Juang. 2007. 'Evaluation of Auditory Fatigue in Combined Noise, Heat and Workload Exposure'. *Ind Health.* 45:527–34.

Christopher, D. W, John, D. L., Yili, L., Sallie, E. G. B. 2004. *An Introduction to Human Factors Engineering*, New Delhi: OHI Learning Pvt Ltd.

Crook, M. A., and F. J. Langdon. 1974. 'The Effects of Aircraft Noise in Schools around London Airport'. *J Sound Vib.* 12:221–32.

Davies, H. W., K. Teschke,, S. M. Kennedy, M. R. Hodgson, C. Hertzman, and P. A. Demers. 2005. 'Occupational Exposure to Noise and Mortality from Acute Myocardial Infarction'. *Epidemiology.* 16:25–32.

Dembe, A. E. 2008. 'Ethical Issues Relating to the Health Effects of Long Working Hours'. *J Bus Ethics.* 84:195–208.

Dennis, J. P. 1965. 'The Effect of Whole Body Vibration on a Visual Performance Task'. *Ergonomics.* 8:193–205.

Directorate General, Factory Advice Service and Labour Institutes. *The Factories Act, 1948 (Act No. 63 of 1948) as amended by the Factories (Amendment) Act, 1987*. Chapter IVA Provisions Relating to Hazardous Processes, Section 41F. http://dgfasli.nic.in/statutes1.htm.

Dixon, E. M., M. V. Kamath, N. McCartney, and E. L. Fallen. 1992. 'Neural Regulation of Heart Rate Variability in Endurance Athletes and Sedentary Controls'. *Cardiovasc Res.* 26:713–19.

Edholm, O. G., and K. F. H. Murrell. 1973. *The Ergonomics Society: A History 1949–1970*. London: Ergonomics Research Society.

Elliott, J. M., and F. O. Simpson. 1980. 'Cigarettes and Accelerated Hypertension'. *NZ Med J.* 91:447–49.

Fagard, R. H., K. Pardaens, and J. A. Staessen. 1999. 'Influence of Demographic, Anthropometric and Lifestyle Characteristics on Heart Rate and Its Variability in the Population'. *J Hypertens.* 17:1589–99.

Floyd, E. A., A. K. Keaton, J. T. Clark, and H. K. Rucker. 1995. 'Chronic Ethanol Ingestion Alters Parameters of Mid-latency Auditory Evoked Potentials in Male Rats'. *Alcohol.* 12(1):15–22.

Floyd, E. A., J. D. Reasor, E. L. Moore, and H. K. Rucker. 1997. 'Effects of Chronic Ethanol Ingestion on Mid-latency Auditory Evoked Potentials Depend on Length of Exposure'. *Alcohol.* 14(3):269–79.

Floyd, W. F., and J. S. Ward. 1968. 'Anthropometric and Physiological Considerations in School, Office and Factory Seating'. *Ergonomics.* 12:132–39.

Forstho, A., P. Hehnert, and H. Neigen. 2001. 'Comparison of Laboratory Studies with Predictions of the Required Sweat Rate Index (ISO 7933) for Climates with Moderate to High Thermal Radiation: A Technical Note'. *Appl Ergon.* 32:299–303.

Fujino, Y., H. Iso, and A. Tamakoshi. 2007. 'A Prospective Cohort Study of Perceived Noise Exposure at Work and Cerebrovascular Diseases among Male Workers in Japan', *J Occup Health.* 49:382–88.

Garg, A., J. S. Moore,, and J. M. Kapellusch. 2007. 'The Strain Index to Analyze Jobs for Risk of Distal Upper Extremity Disorders: Model Validation', Paper presented at the 2007 international conference of the The Institute of Electrical and Electronics Engineers, Singapore, December 2-4.

Goldsmith, R. L., J. T. Bigger Jr, R. C. Steinman, and J. L. Fleiss. 1992. 'Comparison of 24-hour Parasympathetic Activity in Endurance-trained and Untrained Young Men'. *J Am Coll Cardiol.* 20:552–58.

Gomes, J., O. Lloyd, and N. Norman. 2002. 'The Health of the Workers in a Rapidly Developing Country: Effects of Occupational Exposure to Noise and Heat'. *Occup Med.* 52(3):121–28.

Green, M. S., E. Jucha, and Y. Luz. 1986. 'Blood Pressure in Smokers and Nonsmokers: Epidemiologic Findings'. *Am Heart J.* 111:932–40.

Guillemin, V., and P. Wechsberg. 1953. 'Physiological Effects of Long Term Repetitive Exposure to Mechanical Vibration'. *J Aviat Med.* 24:208–21.

Gupta, B. K., A. Kaushik, R. B. Panwar, V. S. Chaddha, K. C. Nayak, V. B. Singh, R. Gupta, and S. Raja. 2007. 'Cardiovascular Risk Factors in Tobacco-chewers: A Controlled Study'. *J Assoc Phys India.* 55: 27–31.

Habibullah, N. Saiyed, and Rajnarayan R. Tiwari. 2004. 'Occupational Health Research in India'. *Ind Health.* 42:141–48.

Hautala, A. J., T. H. Makikallio, A. Kiviniemi, R. T. Laukkanen, S. Nissilä, H. V. Huikuri, and M. P. Tulppo. 2003. 'Cardiovascular Autonomic Function Correlates with the Response to Aerobic Training in Healthy Sedentary Subjects'. *Am J Physiol Heart Circ Physiol.* 285:1747–52.

Hazarika, N. C., D. Biswas, K. Narain, H. C. Kalita, and J. Mahanta. 2002. 'Hypertension and Its Risk Factors in Tea Garden Workers of Assam'. *Natl Med J India.* 15:63–68.

Henderson, D., and R. P. Harnemik. 1986. 'Impulse Noise: Critical Review'. *J Acoust Soc Am.* 80 (2).

Hendrick, D. J. 1996. 'Occupational Lung Disease - 9: Occupation and Chronic Obstructive Pulmonary Disease (COPD)'. *Thorax.* 51:947–55.

Hergens, M. P., M. Lambe, G. Pershagen, A. Terent, and W. Ye. 2008. 'Risk of Hypertension in Swedish Male Snuff Users: A Prospective Study'. *J Intern Med.* 264(2):187–94.

Hergens, M. P., M. Lambe, G. Pershagen, A. Terent, and W. Ye. 2008. 'Smokeless Tobacco and the Risk of Stroke'. *Epidemiology.* 19(6):794–99.

Hignett, S., and L. McAtamney. 2000. 'Rapid Entire Body assessment (REBA)'. *Appl Ergon.* 31: 201–05.

Hirai, A., Takata, M., Mikawa, M., Yasumoto, K., Iida, H., Sasayama, S., Kagamimori, S. 1991. 'Prolonged exposure to industrial noise causes hearing loss but not high blood pressure: a study of 2124 factory laborers in Japan'. J Hypertension. 9(11):1069-73.

Huddleston, J. H. F. 1970. 'Performance on a Visual Display Apparently Vibrating at One to Ten Hertz'. *J Appl Psychol.* 54:401–08.

Ighoroje, D. A., C. Marchie, and E. D. Nwobodo. 2004. 'Noise Induced Hearing Impairment as an Occupational Risk Factor Among Nigerian Traders'. *Niger J Physiol Sci.* 19(1–2):14–19.

International Labour Office (Geneva). 1995. *Introduction to Work Study.* Edited. George Kanawaty. Bombay: Universal Publishing Corporation.

Indian Standard. 1981. *Assessment of Noise With Respect to Community Response.* New Delhi: Bureau of Indian Standards.

———. 1994. *Assessment of Noise During Work for Hearing Conversation Purpose (First Revision).* New Delhi: Bureau of Indian Standards.

Jaiswal, A., B. K. Parto, and C. S. Pandav. 2006. 'Occupational Health and Safety: Role of Academic Institutions'. *Indian J Occup and Environ Med.* 10(3):97–101.

Johnson, A., and Hanson, L. 1977. 'Prolonged Exposure to a Stressful Stimulus (Noise) as a Cause of Raised Blood Pressure in Man'. *The Lancet.* 309(8002): 86-87

Kageyama, T., N. Nishikido, Y. Honda, Y. Kurokawa, H. Imai, T. Kobayashi, T. Kaneko, and M. Kabuto. 1997. 'Effects of Obesity, Current Smoking Status, and Alcohol Consumption on Heart Rate Variability in Male White-collar Workers'. *Int Arch Occup Environ Health.* 69:447–54.

Karlsmose, T. B., M. Lauritzen, A. P. Engberg. 2000. 'A Five-Year Longitudinal Study of Hearing in a Danish Rural Population Aged, 31–50 Years'. *Br J Audiol.* 34:47–55.

Kathleen, C. L., J. R. Turner, and A. L. Hinderliter. 1992. 'Job Strain and Ambulatory Work Blood Pressure in Healthy Young Men and Women'. *Hypertension.* 20:214–18.

Keenan, J. P., P. R. Freeman, and R. Harrell. 1997. 'The Effects of Family History, Sobriety Length and Drinking History in Younger Alcoholics on P300 Auditory Evoked Potentials'. *Alcohol.* 32(3): 233–39.

Khaw, K. T., and E. B. Connor. 1988. 'The Association between Blood Pressure, Age, and Dietary Sodium and Potassium: A Population Study'. *Circulation.* 77:53–61.

Khuraibet, A. M., and F. A. Attar. 2000. 'Preliminary Assessment of Indoor Industrial Noise Pollution in Kuwait'. *Environmentalist.* 20:319–24.

Kroemer, K. H. E., H. B. Kroemer, and K. E. Kroemer-Elbert. 1994. *Ergonomics: How to Design for Ease and Efficiency.* Englewood Cliffs, New Jersey: Prentice Hall.

Kroemer, K. H. E., and J. C. Robinette. 1968. *Ergonomics in the Design of Office Furniture: A Review of European Literature.* AMRL-TR-68–80.

Lang, T., F. Christiane, and Jacquinet S. Marie-C. 1992. 'Length of Occupational Noise Exposure and Blood Pressure'. *Int Arch Occup Environ Health.* 63:369–72.

Lang, T. et al. 1983. 'Blood Pressure, Coffee, Tea and Tobacco Consumption: An Pidemiological Study in Algiers'. *Eur Heart J.* 4:602–07.

Lee, R. A., and A. I. King. 1971. 'Visual Vibration Response'. *J Appl Physiol.* 30:281–86.

Levy, W. C., M. D. Cerqueira, G. D Harp, K. A. Johannessen, I. B. Abrass, R. S. Schwartz, and J. R. Stratton. 1998. 'Effect of Endurance Exercise Training on Heart Rate Variability at Rest in Healthy Young and Older Men'. *Am J Cardiol.* 82:1236–40.

Liao, D., R. W. Barnes, L. E. Chambless, R. J. Simpson, P. Sorlie, G. Heiss et al. 1995. 'Age, Race, and Sex Differences in Autonomic Cardiac Function Measured by Spectral Analysis of Heart Rate Variability: The ARIC Study'. *Am J Cardiol.* 76:906–12.

Loimaala, A., H. Huikuri, P. Oja, M. Pasanen, and I. Vuori. 2000. 'Controlled 5-mo Aerobic Training Improves Heart Rate But Not Heart Rate Variability or Baroreflex Sensitivity'. *J Appl Physiol.* 89:1825–29.

Lovesey, E. J. 1975. 'The Helicopter: Some Ergonomic Factors'. *Appl Ergon.* 6:139–46.

Magid, E. B., R. R. Coermann, R. D. Lowry, and W. J. Bosley. 1962. *Physiological and Mechanical Response of the Human to Longitudinal Whole-Body Vibration as Determined by Subjective Response.* Biomedical Research Laboratory's Technical Document. MRL-TDR-62–66.

Makoto, T., A. Hisashi, H. Yuji, F. Yoshihisa, and I. Tsutomu. 2000. 'Association between Alcohol Intake and Development of Hypertension in Japanese Normotensive Men: 12-year Follow-up Study'. *Am J Hypertens.* 13(5):482–87.

Malik, M., A. J. Camm, edited. 1995. *Heart Rate Variability.* New York: Futura Publishing Company.

Malik, M., edited. 1998. *Clinical Guide to Cardiac Autonomic Tests.* Netherlands: Kluwer Academic publishers.

Malliani, A., M. Pagani, F. Lombardi, and S. Cerutti. 1991. 'Cardiovascular Neural Regulation Explored in the Frequency Domain'. *Circulation.* 84:428–92.

Mantysalo, S., and J. Vouri. 1984. 'Effects of Impulse Noise and Continuous Steady State Noise on Hearing'. *Br J Ind Med.* 41:122–32.

Martin, F. N. 1986. *Introduction to Audiology (3rd edition).* Englewood Cliffs, New Jersey: Prentice-Hall, Inc.

McAtamney, L., and E. N. Corlett. 1993. 'RULA: A Survey Method for the Investigation of Work Related Upper Limb Disorders'. *Appl Ergon.* 24:91–99.

Melamed, S., K. Boneh, and P. Froom. 1999. 'Industrial Noise Exposure and Risk Factors for Cardiovascular Disease: Findings from the CORDIS Study'. *Noise Health.* 1(4):49–56.

Mikell P. Groover. *Work Systems and the Methods, Measurement, and Management of Work.* ISBN 0-13-140650-7.

Miller, M. R., J. Hankinson, V. Brusasco, F. Burgos, R. Casaburi, A. Coates, et al. 2005. 'ATS/ERS Task Force: Standardisation of Lung Functions Testing (Standardisation of spirometery)'. *Eur Respir J.* 26: 319–38.

Molgaard, H., K. Hermansen, and P. Bjerregaard. 1994. 'Spectral Components of Short-term RR Interval Variability in Healthy Subjects and Effects of Risk Factors'. *Eur Heart J.* 15:1174–83.

Monahan, K. D., F. A. Dinenno, H. Tanaka, C. M. Clevenger, C. A. DeSouza, and D. R. Seals. 2000. 'Regular Aerobic Exercise Modulates Age-associated Declines in Cardiovagal Baroflex Sensitivity in Healthy Men'. *J Physiol.* 529:263–71.

Moore, J. S., and A. Garg. 1998. 'The Effectiveness of Participatory Ergonomics in the Red Meat Packing Industry, Evaluation of a Corporation'. *Int J Ind Ergon.* 21:47–58.

Morris, N. M., D. B. Lucas, and M. S. Bressler. 1961. 'Role of the Trunk in Stability of the Spine'. *J Bone Joint Surg.* 43A:327–51.

Murata, K., S. Araki, and H. Aono. 1990. 'Central and Peripheral Nervous System Effects of Hand-arm Vibrating Tool Operation'. *Int Arch Occup Environ Health.* 62(3):183–88.

Murrell, K. F. H. 1971. *Ergonomics: Man in His Working Environment.* London: Chapman Hall.

Narlawar, U. W., B. G. Surjuse, and S. S. Thakre. 2006. 'Hypertension and Hearing Impairment in Workers of Iron and Steel Industry'. *Indian J Physiol Pharmacol.* 50(1):60–66.

National Institute of Occupational Safety and Health (NIOSH). 1981. *Work Practices Guide for the Design of Manual Handling Tasks.*

Nelson, D. I., R. Y. Nelson, B. M. Concha, and M. Fingeruht. 2005. 'The Global Burden of Occupational Noise Induced Hearing Loss'. *Am J Ind Med.* 48:446–458.

Ni, C. H., Z. Y. Chen, Y. Zhou, J. W. Zhou, J. J. Pan, N. Liu, et al. 2007. 'Associations of Blood Pressure and Arterial Compliance with Occupational Noise Exposure in Female Workers of Textile Mill'. *Chin Med J.* 120(15):1309–13.

Nicholas, C. L., E. Sullivan, A. Pfefferbaum, J. Trinder, and I. M. Colrain. 2002. 'The Effects of Alcoholism on Auditory Evoked Potentials During Sleep'. *J Sleep Res.* 11:247–53.

Olesen, B. W. and K. C. Parsons. 2002. 'Introduction to Thermal Comfort Standards and Proposed Version of EN ISO 7730'. *Energy Build.* 34:537–48.

Onder, M., S. Sarac, and S. Onder. 2005. 'A Study of Heat Stress Parameters at Kozlu Coalmine: Turkey'. *J Occup Health.* 47:343–45.

'OSHA's Noise Standard Defines Hazard Protection'. Accessed April 27, 2015. http://multimedia.3m.com/mws/media/918620/oshas-noise-standard-defines-hazard-protection.pdf

Pandve, H. T., and P. A. Bhuyar. 2008. 'Need to Focus on Occupational Health Issues (Letter to Editor)'. *Indian J Commun Med.* 33(2):132.

Parsons, K. C. 2008. (Editorial), 'Industrial Health for All: Appropriate Physical Environments, Inclusive Design, and Standards That Are Truly International'. *Ind Health.* 46:195–97.

Paul, L. E., S. L. Vaschillo, E. Dwain, V. Bronya, S. Anthony, and H. Robert. 2006. 'Effects of Age on Heart Rate Variability, Baroreflex Gain, and Asthma'. *CHEST.* 129(2):272–84.

Peter, L., H. Josef, and W. K. Walter. 1993. 'Work Noise Annoyance and Blood Pressure: Combined Effects with Stressful Working Conditions'. *Int Arch Occup Environ Health.* 65:23–28.

Peterson, H. R., M. Rothchild, C. R. Weinberg, R. D. Fell, K. R. McLeish, and M. A. Pfeifer. 1988. 'Body Fat and the Activity of the Autonomic Nervous System'. *N Engl J Med.* 318:1077–83.

Pingle, S. R. 2007. 'Do Occupational Health Services Really Exist in India?' Accessed April 27, 2015. http://www.occuphealth.fi/NR/rdonlyres/04399102-514B-4444-AC38-C90DCC3D9A3D /0/7 Do OH services really existinIndia.pdf.

Popelka, M. M., K. J. Cruickshanks, T. L. Wiley, T. S. Tweed, B. E. Klein, R. Klein, et al. 2000. 'Moderate Alcohol Consumption and Hearing Loss: A Protective Effect'. *J Am Geriatr Soc.* 48(10):1273–78.

Porte, R. et al. 1985. 'Coronary Heart Disease and Total Mortality. Recent Developments in Alcoholism'. 3: 157–63.

Pottier, M., A. Dubreuil, and H. Mond. 1969. 'The Effects of Sitting Posture on the Volume of the Foot'. *Ergonomics.* 12:753–58.

Rai, R. M., A. P. Singh, T. N. Upadhyay, S. K. Patil, and H. S. Nayar. 1981. 'Biochemical Effects of Chronic Exposure to Noise in Man'. *Int Arch Occup Environ Health.* 48:331–37.

Razay, G., K. WEeAton, C. Htolton, and Oeughes. 1992. 'Alcohol Consumption and Its Relation to Cardiovascular Risk Factors in British Women'. *Br Med J.* 304:80–83.

Rennie, K. L., H. Hemingway, M. Kumari, E. Brunner, M. Malik, and M. Marmot. 2003. 'Effects of Moderate and Vigorous Physical Activity on Heart Rate Variability in a British Study of Civil Servants'. *Am J Epidemiol.* 158:135–43.

Rodahl, K. 2003. 'Occupational Health Conditions in Extreme Environments'. *Ann Occup Hyg.* 47(3):241–52.

Rodger, A. and P. Cavanagh. 1962. 'Training Occupational Psychologists'. *Occup Psychol.* 36:82–88.

Rongo, L. M. B., F. Barten, G. I. Msamanga, D. Heederik, and W. M. V. Dolmans. 2004. 'Occupational Exposure and Health Problems in Small-scale Industry Workers in Dar es Salaam, Tanzania: A Situation Analysis'. *Occup Med.* 54:42–46.

Rosenhall, U., S. Eva, V. Sundh, and A. Svanborg. 1993. 'Correlations between Presbyacusis and Extrinsic Noxious Factors'. *TIJA.* 32(4):234–43.

Rossy, L. A., and J. F. Thayer. 1998. 'Fitness and Gender-related Differences in Heart Period Variability'. *Psychosom Med.* 60:773–81.

Saiyed, H. N., and R. R. Tiwari. 2004. 'Occupational Health Research in India'. *Ind Health.* 42:141–48.

Salonen, J. T., J. Tuomilehto, and A. Tanskanen. 1983. 'Relation of Blood Pressure to Reported Intake of Salt, Saturated Fats, and Alcohol in Healthy Middle-aged Population'. *J Epidemiol Community Health.* 37:32–37.

Savdie, E., G. M. Grosslight, and M. A. Adena. 1984. 'Relation of Alcohol and Cigarette Consumption to Blood Pressure and Serum Reatinine Levels'. *J Chron Dis.* 37:617–23.

Schroeder, K. L., and M. S. Chen. 1985. 'Smokeless Tobacco and Blood Pressure'. *N Engl J Med.* 312:919.

Schuit, A. J., L. G. V. Amelsvoort, T. C. Verheij, R. D. Rijneke, A. C. Mann, C. A. Swenne, et al. 1999. 'Exercise Training and Heart Rate Variability in Older People'. *Med Sci Sports Exerc.* 31:816–21.

Seals, D. R., and P. B. Chase. 1989. 'Influence of Physical Training on Heart Rate Variability and Baroreflex Circulatory Control'. *J Appl Physiol.* 66:1886–95.

Shannon, D. C., D. W. Carley, and H. Benson. 1987. 'Aging of Modulation of Heart Rate'. *Am J Physiol.* 22:874–77.

Shannon, R. M., H. Russ, S. Joel, L. W. Paige, J. S. Thomas, and C. C. David. 2001. 'Association of Heart Rate Variability with Occupational and Environmental Exposure to Particulate Air Pollution'. *J Am Heart Assoc Circ.* 104:986–91.

Shaper, A. G., S. J. Pocock, Mary Malker, N. M. Cohen, C. J. Wale, and A. G. Thomson. 1981. 'British Regional Heart Study: Cardiovascular Risk Factors in Middle-aged Men in 24 Towns'. *Br Med J.* 283:18.

Shin, K., H. Minamitani, S. Onishi, H. Yamazaki, and M. Lee. 1997. 'Autonomic Differences between Athletes and Non-athletes: Spectral Analysis Approach'. *Med Sci Sports Exerc.* 29:1482–90.

Singh, L. P., A. Bhardwaj, K. K. Deepak, and S. Sahu. 2008. 'Evaluation of Work Strain on Workers Working in Small Scale Forging Industry', *J Environ Physiol.* 1(20):83–92.

Singh, L. P., A. Bhardwaj, K. K. Deepak, and S. Sahu. 2010. 'Small & Medium Scale Casting and Forging Industry in India: An Ergonomic Study'. *Ergon SA.* 22(1):36-56.

Singh, L. P., A. Bhardwaj, K. K. Deepak, and S. Singh. 2009. 'Occupational Noise and Hearing Conservation of Industrial Workers in Casting, Forging Industry in Northern India'. *Int J Hum Geogr Environ Stud.* 1:1.

Singh, L. P., A. Bhardwaj, K. K. Deepak, S. Shahu. 2008. 'Evaluation of Work Strain on Workers Working in Small Scale Forging Industry'. *J Environ Physiol.* 1(20): 83–92.

Singh, Lakhwinder Pal, Arvind Bhardwaj, and Kishore Kumar Deepak. January–March 2010. 'Occupational Exposure in Small and Medium Scale Industry with Specific Reference to Heat and Noise'. *Noise Health.* 12:(46):37–48.

Smeatham, D. and P. D. Wheeler. 1998. 'On the Performance of Hearing Protectors in Impulsive Noise'. *Appl Acoust.* 54(2):165–81.

Sollers, J. J., T. A. Sanford, R. N. Oberg, C. A. Anderson, and J. F. Thayer. 2002. 'Examining Changes in HRV in Response to Varying Ambient Temperature'. *IEEE Eng Med Biol* July/August.

Starck, J., E. Topilla, and I. Pyykko. 2003. 'Impulse Noise and Risk Criteria'. *Noise Health.* 5(20):63–73.

Stephens, John-Paul, A. Vos Gordon, Edward M. Stevens Jr., and J. Steven Moore. 2006. Test Retest Repeatability of the Strain Index. *Appl Ergon.* 37:275–81.

Subroto, S. N., and V. D. Sarang. 2008. 'Occupational Noise Induced Hearing Loss in India'. *Indian J Occup Environ Med.* 12(2):53–57.

Sunkaria, R. K., V. Kumar, and S. Chandra. 2010. 'A Comparative Study on Spectral Parameters of HRV in Yogic and Non-yogic Practitioners'. *Int J Med Eng Inform.* 2(1):1–14.

Tabuchi, T., S. Kumagai, M. Hirata, H. Taninaka, J. Yoshidai, H. Oda, et al. 2005. 'Status of Noise in Small-scale Factories Having Press Machines and Hearing Loss in Workers'. *SanEiShi.* 47(5):224–31.

Tafti, M. F., and G. Ravindran. 1996. 'Effect of Age in Brainsem Auditory Evoked Potential Monitoring'. Accessed April 27, 2015. doi: 10.1109/SBEC.1996.493215.

Tambs, K., H. J. Hoffman, H. M. Borchgrevink, J. Holmen, and B. Engdahl. 2006. 'Hearing Loss Induced by Occupational and Impulse Noise: Results on Threshold Shifts by Frequencies, Age and Gender from the Nord-Trondelag Hearing Loss Study'. *Int J Audiol.* 45(5):309–17.

Task Force of the European Society of Cardiology and the North American Society of Pacing and Electrophysiology Heart Rate Variability, 1996. 'Standards of Measurement, Physiological Interpretation, and Clinical Use'. *Circulation.* 93:1043–65.

Taylor, F. V., and W. D. Garvey. 1966. 'The Limitations of a 'Procrustean' Approach to the Optimization of Man-Machine Systems'. *Ergonomics.* 9:187–94.

Taylor, W., B. Lempert, O. Cincinnati, P. Peter, and J. Kershaw. 1984. 'Noise Levels and Hearing Thresholds in the Drop Forging Industry'. *Acoust Soc Am.* 76(3):807–19.

Teo, K. K., S. Ounpuu, S. Hawken, M. R. Pandey, et al., on behalf of the INTERHEART Study Investigators. 2006. 'Tobacco Use and Risk of Myocardial Infarction in 52 Countries in the INTERHEART Study: A Case-Control Study'. *Lancet.* 368:647–58.

Thakur, L., J. P. Anand, and P. K. Banerjee. 2004. 'Auditory Evoked Function in Ground Crew Working in High Noise Environment of Mumbai Airport'. *Indian J Physiol Pharmacol.* 48(4):453–60.

Thatcher, A., J. James, and A. Todd. 2005. 'Ergonomics Analysis of the Thermal Environment: A Case Study for Companies in the Alpaca Textile Sector of Arequipa Peru'. Proceedings of CybErgh 2005, the Fourth International Cyberspace Conference on Ergonomics. University of the Witwatersrand.

Thomas, J. C. 1965. 'Use of Piezo–accelerometer in Studying Eye Dynamics'. *J Opt Soc Am* 55:534–37.

Togo, F. and M. Takahashi. 2009. 'Heart Rate Variability in Occupational Health: A Systematic Review'. *Ind Health.* 47:589–602.

Toppila, E., I. Pyykko, J. Strack, Kaksonnen, and H. Ishizaki. 2000 'Individual Risk Factors in the Development of Noise-induced Hearing Loss'. *Noise Health.* 8:59–70.

Tremolieres, C., and R. Hetu. 1980. 'A Multi-parametric Study of Impact Noise-induced TTS'. *J Acoust Soc Am.* 68(6):1652–59.

Universities Occupational Safety and Health Educational Resource Center. 2003. NIOSH Spirometry Training Guide. Accessed April 27, 2015. http://stacks.cdc.gov/view/cdc/11336.

Urstad, K., F. Jensen, Bouvier, B. Saltin, and M. J. Urstad. 1998. 'High Prevalence of Arrhythmias in Elderly Male Athletes with a Lifelong History of Regular Strenuous Exercise'. *Heart*. 79:161–64.

Waters, T. R., V. Putz-Anderson, A. Garg, and L. Fine. 1993. 'Revised NIOSH Equation for the Design and Evaluation of Manual Lifting Tasks'. *Ergonomics*. 36(7):749–76.

Webber, R. C. W., R. M. Franz, W. M. Marx, and P. C. Schulte. 2003. 'A Review of Local and International Heat Stress Indices, Standards and Limits with Reference to Ultra-deep Mining'. *J S Afr Inst Mining Metall*. 313–24.

Wu, T. N., F. S. Chou, and P. Y. Chang. 1987. 'A Study of Noise-induced Hearing Loss and Blood Pressure in Steel Mill Workers', *Int Arch Occup Environ Health*. 59(6):529–36.

Yasushi, O., M. Toshiaki, S. Yasushi, K. Etsuko, and N, Koji. 2001. 'Alcohol Consumption and Blood Pressure in Japanese Men'. *Alcohol*. 23(3):149–56.

Yataco, A. R., L. A. Fleisher, and L. I. Katzel. 1997. 'Heart Rate Variability and Cardiovascular Fitness in Senior Athletes'. *Am J Cardiol*. 80:1389–1291.

Yoshihisa, Fujino, Iso Hiroyasu, and Akiko Tamakoshi. 2007. 'A Prospective Cohort Study of Perceived Noise Exposure at Work and Cerebro Vascular Diseases among Male Workers in Japan'. *J Occup Health*. 49: 382–88.

Web References

http://www.indianforging.org/indian_forging_industry.aspx.
http://www.indiaprwire.com/pressrelease/mining-metals/200804038515.htm.